Automotive Audio Systems

by
Homer L. Davidson

PROMPT® PUBLICATIONS

DELMAR

THOMSON LEARNING

Automotive Audio Systems
Homer L. Davidson

Business Unit Director:
Alar Elken

Executive Editor:
Sandy Clark

Managing Editor:
Deborah F. Abshier

Executive Marketing Manager:
Maura Theriault

Channel Manager:
Fair Huntoon

Marketing Coordinator:
Brian McGrath

Executive Production Manager:
Mary Ellen Black

Production Coordinator:
Sharon Popson

For permission to use material from the text or product, contact us by
Tel.   (800) 730-2214
Fax   (800) 730-2215
          www.thomsonrights.com

Library of Congress Cataloging-in-Publication Data:
2001095789

ISBN: 0-7906-1235-6

NOTICE TO THE READER

Acknowledgment

A great deal of thanks goes to Sams Technical Publishing for the Auto Radio Series (AR), and to the many electronics technicians who have provided service to the car radio industry over the last 60 years. Last but not least, I dedicate this book to the arrival of a new great-granddaughter, named Drew.

Table of Contents

Introduction

Servicing the automobile's receiver, cassette, and CD player are quite easy when compared to TV and VCR units. The most difficult problems are with the front-end and high-powered amplifier systems. The RF, converter, and IF circuits are very easy to service within the early car radio, but today, the varactor diode and system control circuits might appear more difficult. All of the circuits in between are right down the same alley as the TV circuits, which the electronics technician has mastered. Troubleshooting the audio circuits is very easy, and sometimes you can listen to the music while troubleshooting the dead stereo channel.

The purpose of this book is to help those interested in electronics to be able to service and repair yesterday's and today's car radios and high-powered amplifier systems. Each chapter is filled with hints, practical servicing data, and actual case histories. Locating the defective components takes up most of the service time, while replacing the new or old part takes only a few minutes. Servicing the auto radio, cassette, and CD player can be a lot of fun, and can be accomplished by the novice, handyman, electronics student, and electronics technician.

Most of the servicing problems found in the car radio can be located with only a few tools, DMM, and a semiconductor test instrument. The digital multimeter can test diodes and transistors with a diode-junction test. The only additional required pieces of test equipment found on the electronics technician's bench are a 14.4-volt heavy-duty power supply and a signal generator. The audio signal generator might also help to locate and solve weak, distorted, and intermittent audio circuits. The specialized auto receiver and high-powered audio amplifier service firms should have a distortion meter to help locate and test distortion in the high-powered amplifier systems.

The basic car radio servicing data is provided in Chapter 1 for the early car radio, cassette, CD, and CD trunk player. Servicing the new SMD components and locating a defective miniature component are found in this chapter. How to connect and test the speakers to the various amps and speaker loads is also located in Chapter 1.

The car radio antenna system might consist of a side, top cowl, rear fender mount, and motor-powered antenna for the auto receiver, and this can be found in Chapter 2. How to find the early car radio service problems and required tests after removing several PCB is most difficult. Help with locating the noisy problems within the early car radio is given in Chapter 2.

How to check the early power supply and defective components within the early car radio are found in Chapter 3. Help with locating the bad electrolytic, choke coil, and transformers with voltage regulation circuits are located in these pages. It tells you how to troubleshoot the Mosfet power supply, the Mosfet high-powered amp supply, the cassette power, and the CD power supplies found in today's power supply systems.

Servicing the front-end circuits of the early and present-day auto receivers is found in Chapter 4. How the early capacitor tuning performed with permeability tuning, and how today's varactor tuning operates, is all found here. Electronic tuning and digital synthesizer tuning is explained within the PLL (VCO) tuning systems. The different AM and actual radio front-end problems are also found in this chapter.

Servicing the FM circuits of the auto radio within the early and present radio circuits is found in Chapter 5. It describes how the FM circuits function with separate RF, oscillator, and mixer stages. Many early and present-day actual FM radio front-end service problems are found throughout this chapter. It tells how troubleshooting the FM circuits with the signal generator can quicken car radio servicing.

Troubleshooting the intermediate Frequency (IF) transformer circuits and present-day IF stages is found in Chapter 6. You'll learn how to repair the intermittent and weak FM IF circuits, which are made easy with actual IF symptoms and problems. A typical FM/MPX alignment chart is also located in Chapter 6.

The different service problems related to the audio AF and preamp auto radio circuits are located within Chapter 7. Servicing the tape head to the audio driver or output stages are found in this chapter. You'll learn how signal-tracing the stereo AF and preamp circuits with the external amplifier can easily locate a defective AF stage.

Repairing the auto receiver's audio output stages is one of the most important chapters in the book. Chapter 8 shows how to repair the transistor and IC output stages within the car radio. You can easily spot a defective audio output circuit with the service methods found in this chapter. Solving the distorted, weak, and noisy speaker indications can lead you to a defective transistor or IC output stage.

The many service problems related to the auto cassette player are found in Chapter 9. Here you'll learn how to locate the defective component from the tape head to the speaker. Slow and fast speed problems of the tape-player operations with actual speed histories are located in these pages.

Troubleshooting the many problems found in the auto CD player and actual case problems are located in Chapter 10. How the laser and pickup circuits operate, along with the many tracking and focus error circuits, are found in this chapter. You'll learn how the disc, spindle, and loading motors operate and how actual motor problems are solved, which can help the electronics technician do a better and quicker repair job.

One of the most interesting and troublesome problems with the high-powered amplifier is found in Chapter 11. You'll discover how to repair the DC-DC converter power supply with Mosfet transistors. Finding and replacing those high-powered transistors are located within this chapter. Servicing high-powered amp circuits and high-wattage speakers is a whole different set of service problems within the electronic field.

How to repair and locate the different obsolete components are given in Chapter 12. This chapter deals with everything from the early tube chassis to the all-transistor auto car radio. The vibrator high-output voltage transformer is discussed and explained with antique car radio problems. In here you'll find out about the early capacitor and permeability tuning along with directions on stringing that dial cord. This chapter will fully explain how those hard-to-find choke and choke transformers can be repaired and placed back into service. There are many individual circuits of the early tube and transistor car radios found in this chapter.

The different speaker systems and hookup connections are found in the last chapter of the book. Chapter 13 deals with the different speaker sizes, wattages, and power-handling capabilities. The many problems that cause the speaker to become defective and to need replacement are found in this chapter. The defective speaker can be damaged by too much power applied, by weather conditions, and by a DC voltage placed on the voice coil and common ground.

Besides providing numerous car radio circuits of yesterday's and today's auto receivers and head units, this book is well illustrated with photos and diagrams of different circuits. You will find many actual case histories in the old and new auto receivers, cassette players, and CD players. A lot of the old car radio circuits were taken from the auto Howard Sams Photo-Fact series. Let's now have some fun in tackling and troubleshooting that car radio of yesterday and today.

CHAPTER 1

Car Radio Basics

Servicing the car radio is not as difficult as one might think. There are many small repairs that just about anyone can make with a few added suggestions, service tips, and case histories found throughout these pages of the book. Each chapter gives several electronics repairs upon each section of the auto receiver. This book was gathered and written for the handyman, hobbyist, electronics beginner, and student, and the electronics technician who might want to add another service line to his or her present electronics service business.

THE EARLY CAR RADIO

The early car radio was quite large and cumbersome with a variable-plate tuning capacitor to bring in those local radio stations within a superhet circuit. The early radio consisted of vacuum tubes that operated at a higher DC voltage and developed in a bulky power supply. A noisy vibrator was tied to a high-voltage transformer that took the car battery voltage and boosted 12 volts into 90 to 250 volts DC for tube operation. The famous OZ4 tube served as the high-voltage rectifier.

The early car radio had flexible cables that extended from the radio mounted upon the firewall to an off/on switch and tuning dial. The flexible metal cables were operated from a faceplate and knobs mounted under the cowl or front dash of the automobile.

Later on, like all electronic auto products, the car radio had a separate tuning unit that mounted in the dash, and the radio was bolted to the firewall. The early auto antenna had two metal standoff side mounts that cleared the side of the car and extended upward. Later, a top cowl antenna was designed to mount on top or with a top cowl flexible mount antenna that had an outside spring-mounted support. Often, the top cowl antenna was mounted on the side, close to the car radio.

As the years went by, the auto receiver grew from only an amplitude modulation (AM) receiver to a combination frequency modulated (AM/FM) receiver. Permeability coil tuning replaced the bulky variable capacitor for tuning in the AM/FM stations.

A few years later, the auto receiver became smaller with the advance of semiconductors and transistors. Now the auto receiver mounts directly into the front dash of the automobile. At first only the audio output stage was transistorized, and then the whole semiconductor audio circuits replaced the audio tubes. Because the transistors worked at low voltage, the entire radio operated directly off of 12 or 14 volts of battery power. Different designed tubes were also made to operate directly from the car battery. The noisy vibrator and large power supply circuits were no longer required.

Later, the cassette player was found in the portable radio and large console models. Likewise, the cassette and stereo 8-track players were designed to provide your favorite recorded music in the car radio. The new audio circuits provided two different sources of audio and ended up with stereo music sounding off with speakers mounted on each side of the automobile.

The auto stereo 8-track player and AM/FM radio were in play for many years until the cassette player offered separate tape recordings and playback operations. When the cassette player-radio was turned on, the auto antenna slowly extended from the fender to the sky with a motor-operated antenna system.

The audio circuits were changed again, when integrated circuits were found in the table model radio. At first the IC component was only found in the audio output circuits; two different ICs were found driving separate speakers in the audio stereo circuits. Like the transistor used in the audio circuits, the whole audio circuit might include only one IC component that included all of the audio stereo circuits. Today, one large IC component might include all of the audio circuits with a very large wattage output that drives several different size speakers.

Besides the audio circuits, you might find the RF, converter, intermediate frequency (IF), and detector circuits in one large IC component. Sometimes only two or three large IC components make up the entire car radio circuits. The AM/FM MPX stereo receivers might appear with a cassette player, a separate CD player, or a combination of all three different operations. Brand-new car radio circuits and radio designs have made the entire auto receiver much smaller in size, and they deliver great outputs of sound. Surface-mounted devices (SMD) have combined to make the radio smaller with several layers of PC boards.

With the AM/FM/MPX stereo circuits, the new auto receiver might have a cassette and CD player with a separate high-powered amplifier system. The in-dash car receiver might include a detachable faceplate that lets you remove the control panel and take it with you to prevent theft of the car radio. The auto stereo radio is useless to thieves without a faceplate, and so the possibility of theft is greatly reduced.

Today, the auto radio might have a separate auto theft circuit, so that when the faceplate is removed, you activate a built-in car alarm that trips when someone opens the car door. The speaker or horn produces a loud high-pitched alarm tone that will quickly drive a thief away.

Today's AM/FM stereo receiver might include a CD player that mounts directly in the auto's front dash. A multidisc CD changer can be mounted within the trunk and operated from the dash plate controls. Some of the latest auto receivers have a remote control system that selects radio programs, cassette, and CD operations away from the radio. The 12-disc changer mounted in the trunk has a full-floating shock absorption system that helps to eliminate skipping on rough roads.

High-powered amplifiers and different size speakers can be hooked to the auto receiver to produce up to 1000 watts in 2-ohm load speakers. The MAX Thunder 1000 kHz amplifier provides 500 watts at 1 channel with 12.5 volts, 850 watts of peak power X 1, and 1000 watts RMS X 1 at 2 ohm load. Likewise, a Kenwood MAC-728S amp provides 100 watts X 2 channels at 14.4 volts, 300 watts bridged power in watts at 300 X 1, a peak power of 200 watts at 2 channels or an RMS at 170 watts at 2 channels.

FOUR DIFFERENT AUTO RECEIVERS

Pioneer KEG-1900 Stereo Cassette Receiver

The Pioneer KEG-1900 car stereo cassette receiver is a low-priced radio under $100. The key features include a detachable faceplate, green display, super tuner III, 24 radio presets, travel presets, seek tuning, soft-load mechanism, loudness, and preamp outputs. The 17-watt output has an RMS 40 and a peak X 4 channels, a tape frequency response from 40-14 kHz, and a tape S/N ratio of 22 dB.

A soft-load, power-eject mechanism helps to protect your cassettes from damage. The radio has 24 presets for station storage, seek tuning to help you find stations quickly, and radio recall to keep the music playing while you're fast-forwarding or reversing a tape. For music that is loud and clear, add an external amplifier to the receiver's set of preamp circuits.

Clarion Pro-Audio DRX7675Z Car Stereo CD Series

The Clarion Pro-Audio DRX7675Z stereo CD player has a detachable fold-down face, dot-matrix readout, front and rear sets of 4-volt preamp outputs, 4-volt nonfading subwoofer output, 2-Enchancer with adjustable effect levels, CD changer controls, TV control, disc titling, and a wireless remote. The auto receiver has 14 watts RMS 145 peak X 4 channels, a CD frequency response of 5-20 kHz, CD S/N ratio of 100 dB, and an FM selectivity of 11 dBf.

The DRX7675Z can deliver great sound with eloquent appearance. Dial in just the right level of equalization with the 2-Enchancer's adjustable effect feature. Tone bypass disables the bass and treble circuitry to maximize the receiver's signal to noise ratio (S/N) and clarity. Three sets of 4-volt preamp outputs (one is nonfading) are provided for upgrading your system with component amplifiers and subwoofers. A wireless remote is included for your convenience.

Panasonic CQ-DVR909U DVD Audio/Video Receiver

The Panasonic CQ-DVR909U auto receiver has a detachable, motorized fold-down faceplate, multiformat display (compatible with DVD audio groups/tracks and DVD video title changer), day and night metal-finish faceplate, faceplate/center speaker angle adjustment, and AM/FM tuner with 24 preset stations. The receiver has a center channel speaker, three sets of outputs, plus mono center-channel output, video output, auxiliary input, CD changer controls, Dolby Pro Layer, Dolby Digital and DTS decoding, dynamic range compression, and a wireless remote.

The receiver specs are 18 watts of RMS/45 peak X 4 channels plus 5 watts RMS (center channel), DVD frequency response from 2-88 kHz, CD frequency response from 2-20 kHz, DVD S/N ratio of 100 dB, and FM selectivity of 12 dBf.

This Panasonic CQ-DVR909U receiver provides in-dash DVD audio and video in a compact, single-Din-size package. The revolutionary component even has Dolby Pro-Logic, Dolby Digital, and DTS decoding built in for awesome video soundtracks; the receiver will play favorite audio and video CDs as well. Use the speaker setup menu to balance the output level at the receiver's 5-watt center speaker.

Once you have added the video monitor of your choice, you can even bring your video DVDs along for home theater wherever you go. Dynamic Range Compression (Dolby Digital only) reduces the difference between the loudest and softest sounds, to enchance low-volume listening. You can also listen to the different sources on the front and rear speakers by pressing the Multi-Source Selector. Some of these functions are workable by the wireless remote control.

The Kenwood KMD-42 Minidisc Receiver

The Kenwood KMD-42 Minidisc radio has a detachable face, multicolor display, contrast adjustment, CR-2 tuner, 24 receiver presets, station naming, 10-second digital anti-skip, and CD/MD changer controls. The minidisc receiver has one set of rear preamp outputs, with 22 watts of RMS/45 peak X 4 channels, MD frequency response of 20-20,000 Hz, MD signal-to-noise ratio of 90 dB, and FM sensitivity of 9.30 dBf.

The digital sound quality, ease of rise, and editing flexibility are just a few of the reasons why some people are ditching their tapes and hopping on the minidisc bandwagon. The KMD-42's 10-second digital anti-skip, for worry-free performance on rough roads, is a plus. You can add an auxiliary input for your portable MPS, CD, or cassette player with an added adapter. You can also add on a multidisc CD changer to the minidisc receiver.

THE BASIC AUTO CASSETTE PLAYER

You might find the early auto cassette player in a separate dash mount or combined inside with the radio components. The cassette player mechanism is mounted below the receiver chassis (**Fig. 1-1**). By removing the bottom receiver's cover, the cassette flywheel, motor, and drive belts are easily inspected and tested. The capstan/flywheel is usually driven by a motor pulley and belt by the cassette motor.

The tape is rotated between the capstan and pressure roller. A supply spindle supplies the excess cassette tape from the cassette to the capstan and pressure roller assembly. The take-up spindle takes up the played tape and winds it back into the cassette.

In fast-forward operation, the take-up reel rotates at a higher rate of speed, disengages the pressure roller assembly, and rewinds the tape inside the cassette. Usually in reverse, the tape operates at a slower speed and the excess tape is wound around the supply spindle in reverse order. In most cassette players, the fast-forward and reverse operations are accomplished by an idler wheel or pulley that rotates the supply reel in reverse, switching over and engaging the take-up reel for fast-forward operation.

When the tape passes over and against the tape head in play mode, the music is picked up by the tape head and amplified by a transistor or IC preamplifier circuit (**Fig. 1-2**). The cassette preamp audio is mechanically switched into the audio amplifier circuits and out of the radio circuits when the cassette is inserted into the radio chassis. The cassette audio may be switched with fixed diodes placed into the audio AF amp circuits instead of a mechanical switch. Nowadays, the audio power output circuits amplify to the different speakers in each audio stereo channel.

FIGURE 1-1. The cassette player mechanism is found at the bottom area of the 1997 Delco auto receiver.

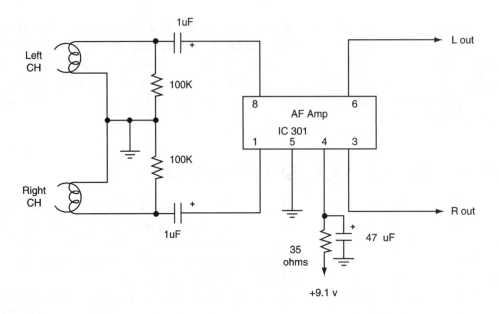

FIGURE 1-2. The tape head picks up the music from the tape and is fed to a transistor or IC preamp stage.

Most of the cassette tape problems are related to slow or uneven speeds, the pulling or eating of tape, and intermittent operations. A binding pressure roller, dry capstan bearing, defective motor, and loose or broken motor drive belt account for most of the speed problems. A worn pressure roller or a slow and intermittent take-up reel can cause the eating or pulling of excess tape. Usually the excess tape will continually spill out until the cassette motor is stopped or the mechanism is bound up with excess tape. Intermittent operations can result from a loose belt, oil on the belt area, a dry capstan bearing, or a defective cassette motor (**Fig. 1-3**).

BASIC CD AUTO RADIO PLAYER

Besides a regular AM/FM stereo receiver, the CD mechanism is added to play the compact disc through the audio stereo system. The optical pickup assembly picks up the music recorded upon the disc without any part touching the disc itself. When the disc is inserted into the CD slot, a loading motor is energized and it loads the disc upon a rotating platform. The CD may be loaded with a drive belt, gear mechanism, or both, in the loading process.

The optical pickup assembly picks up the slants and pits recorded at the under-sided area of the disc (**Fig. 1-4**). This rainbow side is placed downward and the label side is upward, when placing the disc in the slotted area. The CD platform is rotated

FIGURE 1-3. Intermittent operation of the tape can result from a loose or dirty motor drive belt, or from oil on the belt area, in the auto cassette player.

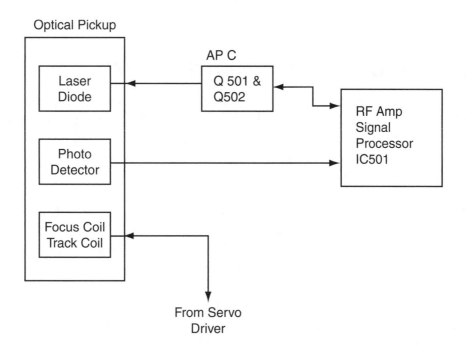

FIGURE 1-4. A block diagram of the digital pickup assembly picks up the pits of the CD disc and is amplified by the RF amp circuits.

by a spindle or disc motor that starts at a fast rate of speed of about 500 rpm and slows down as the laser pickup assembly moves outward toward the outer rim of the CD (approximately 200 rpm).

The laser optical pickup assembly moves from the center of the disc toward the outside rim by a slide or sled motor. The slide motor is usually gear-driven and moves the laser assembly down two sliding bars, keeping the object lens constantly in line with the center of the optical axis. The slide or feed motor might have a fast-forward and rewinding operation. The spindle and sled motor are controlled by the servo control IC circuits, while the loading motor is operated from a mechanism control IC.

The optical pickup assembly signal is fed into a transistor or IC RF amp circuit, where the EFM signal is fed to a digital signal processor and the focus error and tracking error signals are fed directly to the focus and tracking coils to keep the laser beam on focus and on the right track. The digital signal processing IC changes the digital signal into an audio or analog signal and is then switched to each stereo amplifier or the left and right output line jacks.

The basic minidisc receiver is built somewhat like the auto CD player with only a very small disc. The MD auto receiver player can be recorded in an MD recorder and then played within the MD player mounted in the dash of the automobile. The clarity, detail, and fullness of a quality-type MD recording sounds great within the auto MD player. The minidisc changer is small enough to fit into most auto glove compartments or the changer can be mounted in the trunk area.

THE BASIC CD TRUNK CHANGER PLAYER

The CD trunk changer has practically the same electronic components as the regular CD player, except that a larger mechanism is added. Usually the compact disc changer is mounted in the auto trunk area. The compact disc changer should not be installed where temperature may become high, where it is exposed to direct sunlight, or in places where vibrations are heavy and the unit cannot be installed securely.

The CD trunk changer may have 10 standard discs or a combination of 6/10 CD and MD discs. Some auto CD changers play a mixed load of CDs and minidiscs. A magazine assembly holds the CDs for loading.

The compact disc changer might have four or more different motor operations. The roulette or tray motor rotates the turntable or magazine and stops at the correct disc selection. The disc motor selects the correct disc in the tray or magazine. The disc or spindle motor rotates the CD at 500 to 200 rpm. An elevator and disc motor might be operated from the same driver IC, while the up/down and magazine motor are operated from the same dual-motor driver IC. The slide or sled motor moves the laser from the center to the outside of the CD on sliding rods.

The CD trunk changer can be controlled from the faceplate of the auto receiver. Some auto CD changers load directly, while others are contained in a disc magazine. The changer-controlling receiver can select your favorite disc or music with fast CD-to-CD changing. Most CD changers include disc and track scan, random, and repeat play operations.

SURFACE-MOUNTED DEVICES (SMD)

You will find many different types of ICs, microprocessors, and transistors throughout the auto receiver, cassette, and CD changer. A new kind of transistor with small resistors included within the base and emitter circuits are called digital transistors (**Fig. 1-5**).

The SMD transistor might also be a digital-type semiconductor. Some transistors, microprocessors, diodes, capacitors, and resistors are found in tiny SMD components. These SMD components are found mounted on the same side as the PC wiring. You might also find LEDs with a compact, thin, leadless component.

The surface-mounted transistor might appear in a chip form with flat contacts or terminals on one side, top, and bottom, or on both sides. The SMD transistor might

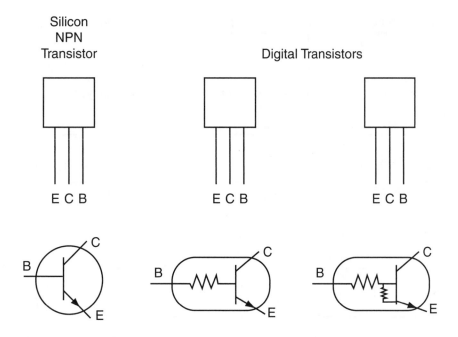

FIGURE 1-5. The regular digital transistor might contain resistors inside the transistor of the base and emitter terminals.

have gullwing terminal connections. The fixed silicon diode might appear in the same type of component as the transistor. Two or more silicon diodes might be found in one chip. Test each SMD transistor and diodes like any regular transistor or diode with the diode test of the DMM.

A surface-mounted resistor might appear as a flat or round, leadless device. The fixed resistor might appear in the same type of flat package as the fixed ceramic capacitor. You might find more than one resistor in the same SMD component. The SMD fixed resistor has a number for identification with lines at the ends, while ceramic capacitors have a line at the top with a letter of the alphabet and numbers (**Fig. 1-6**).

The SMD capacitor is a flat, solid device with the terminal connections at each end with tinned soldered ends. Sometimes these fixed capacitors have a shiny flat surface while others might appear black upon the PC wiring. Check each fixed capacitor with the ESR meter or a capacitor tester by pushing the test prods at each end of the SMD capacitor. Try to obtain these SMD parts from the manufacturer or from kits available at electronic wholesale and mail order catalog firms.

The SMD IC or microprocessor might have gullwing flat terminals that solder directly to the PC wiring. The SMD IC might have four or more gullwing terminals.

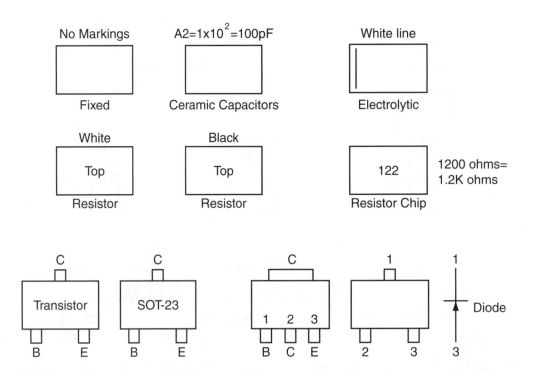

FIGURE 1-6. The identification of the SMD resistors and ceramic capacitors.

FIGURE 1-7. The microprocessor has many gullwing terminals, and it mounts directly on the PC wiring in the auto receiver.

A surface-mounted microprocessor might have 80 or more gullwing terminal connections (**Fig.** 1-7). Terminal 1 of most ICs or processors is identified by a white or black dot and indented circle. Make sure terminal 1 is marked on the PCB before removing the defective IC.

Testing SMD Components

SMD components can be tested like any regular electronic component. The test meter probes must be dug into the end of gullwing-type terminals. Test each SMD capacitor with a capacitor tester or ESR meter (**Fig.** 1-8). Check each fixed resistor with the ohmmeter scale of the DMM. The transistor can be tested with in-circuit tests of the diode test of the DMM or with a transistor tester.

Check for leakage and voltage tests on the suspected IC supply voltage terminal to common ground for leakage and the correct voltage. Make sure the test instrument probes are really sharp so that they will bite into the end terminals of the SMD resistor or capacitor. File or grind down each test probe to a fine point for easy testing.

The DMM is the ideal test instrument in making resistance, leakage, continuity, voltage, and diode tests. The diode test of the DMM can quickly determine if a diode or transistor is leaky. Check each transistor for leakage with the diode test to determine

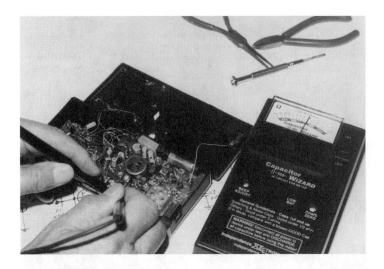

FIGURE 1-8.   The ESR meter can quickly test capacitors with in-circuit tests.

if the transistor is shorted, leaky, or open between the base and collector, base and emitter, and collector to emitter. Most defective transistors become leaky or shorted between the collector and emitter terminals.

Removing SMD Components

Small SMD components such as transistors, resistors, and capacitors can easily be removed with a soldering iron and solder wick material. The mesh-type solder wick helps to pick up the excess solder. Hold the solder wick against the end terminal and apply the solder iron tip to the mesh material, removing the excess solder and picking it up with the solder wick.

Insert a small screwdriver or knife blade under the component and lift up the soldered contact end (**Fig. 1-9**). Clean off all soldered ends of the PC wiring pads so the new SMD part can be mounted flat on the PC wiring. Be very careful not to lift the PC wiring pad or to trace and splash excess solder on the PC wiring.

Remove the gullwing terminals of the transistor, IC, or microprocessor with solder wick and soldering iron. Run the solder wick with the hot iron on the top and down each side of the row of pin terminals with solder wick and soldering iron. Go down each side of the microprocessor in the same manner. Remove all excess solder. Carefully inspect each terminal to see if the solder is free around each pin terminal, so as not to damage the trace or PC wiring.

You can remove each terminal, one at a time, by applying the hot soldering iron tip to the gullwing terminal. Pry up the loose terminal as the solder is melted with the

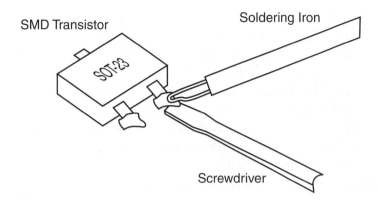

SMD Transistor

Soldering Iron

SOT-23

Screwdriver

FIGURE 1-9. Remove each SMD terminal with the soldering iron and lift the terminal up with a screwdriver or knife blade.

iron. Likewise, remove each terminal in the same manner. Make sure to mark terminal one upon the PCB before lifting the defective SMD component.

Another method is to cut each microprocessor terminal really close to the body of the component with a sharp blade or Xacto knife. Carefully remove the body of the microprocessor. Now, apply the soldering iron to each cut terminal to be removed with a pair of long-nose pliers or tweezers. Make sure that each gullwing tip is removed and thrown away. Do not leave any cut tips on the PC wiring to possibly short out other electronic components. Do not try to reuse a removed SMD part, even if it tests okay.

The defective SMD component might break down the same as any other electronic part. The body of the SMD resistor or ceramic capacitor can become cracked or broken if too much pressure is applied when replacing the SMD part. Intermittent or broken connections at each end of the SMD component can cause intermittent chassis operation. Sometimes the soldered end connection will break or separate from the PC wiring, causing an intermittent or a dead chassis. A touch-up of the soldered end connection can solve the intermittent or poorly soldered connection.

Installing the New Chip

The new chip component can be damaged when poorly soldered into the PC wiring. If too much heat is applied, the chip-end connection separates and the new part will not function. Soldering should be done rather rapidly. Do not apply heat to the chip itself, only to the end terminals. Too much solder can cause extra applied heat to the end connections and might slop over to another PC wiring or trace. Too much heat can also lift the PC wiring up from the PCB. Test each SMD component to make sure the correct part and the new component are okay (**Fig. 1-10**).

Do not apply too much pressure to the SMD chip. Do not use the soldering iron tip to push down one end of the chip after one connection is made; you can destroy the chip. Make sure all excess solder is sucked up from the end mounting pad so the chip can lie flat on the PCB pad. If the small ceramic capacitor, resistor, or diode chip wants to slide around, hold it in position with the eraser end of a wooden pencil. Always mount the SMD chip horizontally on the PC wiring.

Handle all IC and microprocessor chips with care. Do not remove these static-prone chips from the package until they are ready to be mounted. Make sure terminal 1 is at the right spot. Clean off all IC terminal pads with solder wick and iron. Tack in one IC terminal at opposite ends, and make sure each terminal is over the correct pad. Use the smallest diameter of solder for each transistor or IC gullwing terminal.

Double-check each terminal to common ground with the ohmmeter. Make sure any two terminals are not soldered together. Run a knife blade down between the IC terminals to clean out any sharp points or excess solder.

Double-check the schematic when two side-by-side terminals are soldered together. Check out the two terminals with the low ohm scale of the DMM. Make sure the sharp DMM test probes are not touching each other. If two terminals are accidentally soldered together, remove the excess solder with solder wick and iron upon both terminals. Apply the hot tip to pick up the excess solder. Sometimes, by holding the chassis at an angle, the excess solder will run down into the mesh material for easy removal.

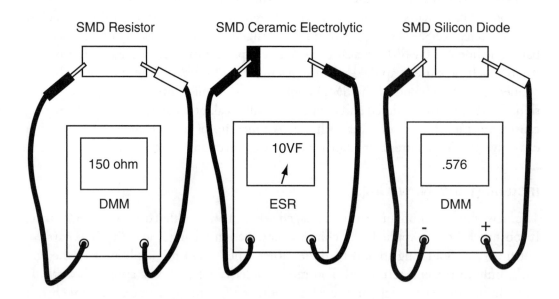

FIGURE 1-10. Test each SMD component before installing and soldering into the car receiver or changer.

CORRECT TEST EQUIPMENT

The handyman, electronic hobbyist, or student can make many auto receivers, cassette, and CD changer repairs with only a DMM and adequate power supply (**Fig. 1-11**). The different repairs upon the auto receiver are simple breakdowns that might occur and can be serviced with continuity and ohmmeter measurements. A digital multimeter (DMM) is an ideal test instrument in taking continuity, ohmmeter, voltage, and current tests. The DMM is a lot more accurate than the volt-ohm-meter (VOM). Digital multimeters can be purchased for as little as $25 to more than $200.

The DMM

Choose a DMM that will measure resistance, current, and voltage, and will make diode tests. You can purchase a DMM that can measure resistance, voltage, and current, and can check diodes, transistors, frequency, capacitors, and logic tests in one package. The resistance range starts at 200 and tests up to 2000 megohms. The AC-DC voltage range starts at 200 MV up to 1000 volts DC and 750 volts AC. A current range might start at 2 milliamperes and go up to 20 amps. The frequency range starts at 2 kHz to 200 kHz. Most combination DMMs will only test capacitors from 2 nF up to 20 uF.

   The diode test can check about any type of diode for shorted, leaky, or open conditions. The diode test can also be used to check leakage and open junctions in the NPN and PNP transistors. A diode test of a silicon diode will indicate a normal diode

FIGURE 1-11.   Only a DMM is required to make service repair problems found in the auto receiver.

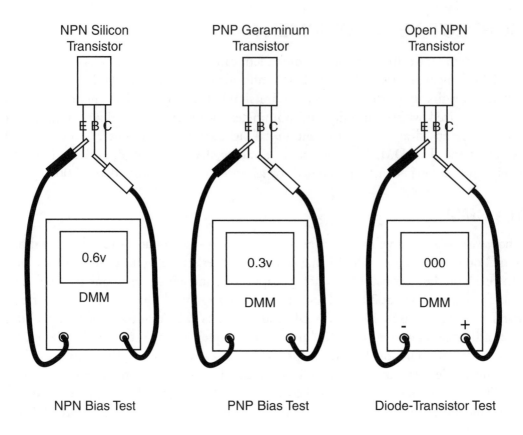

FIGURE 1-12. Critical forward-bias voltage tests between base and emitter terminals can indicate if the transistor is normal, leaky, or open.

with a low resistance measurement in one direction and no reading in the opposite polarity with a normal diode. A leaky diode will show leakage in both directions, when the test probes are reversed on the fixed diode. RF and detector diodes will have a much higher normal measurement than silicon diodes.

The 2-volt range of the DMM is ideal when checking forward bias in a transistor (**Fig. 1-12**). A normal NPN silicon transistor has a 0.6-volt measurement between emitter and base terminals with in-circuit tests. The normal PNP germanium transistor will have a 0.3-volt measurement between base and emitter terminals.

Simply measuring the forward-bias voltage on a transistor with in-circuit tests can indicate a normal or defective transistor. No or lower bias voltage indicates a leaky or shorted transistor. A rather high voltage measurement between emitter and base terminals indicates an open transistor. The multipurpose DMM can make most of the required tests when servicing auto receivers.

The Bench Power Supply

Those on a limited budget can purchase a 4-6 amp regulated 13.8-volt bench power supply for less than $30. (See **Fig. 1-13**.) The power supply should have high-surge current capacity, short circuit, and overload protection. This power supply can provide the correct voltage to the auto receiver, cassette, and CD players. The low-voltage bench power supply operates from the 120 V AC power line and provides regulated 13.8 V DC for servicing the auto receiver.

For technicians who work in an electronics service business, choose a 30- or 40-amp heavy-duty power supply. This power supply is ideal for servicing and powering high-current auto sound products. The output voltage is variable from 1-15 V DC. The output current is stable at 30 amps of operation. High-current binding posts are required with a cigarette lighter receptacle output.

The power supply has two sets of low-current outputs with an overload indicator and over-current protection. A lighted voltage and current meter indicates the operating voltage and current drawn by the attached auto receiver. By varying the output voltage, intermittent radios and auto sound products can and will act up. Excessive current drain on the current meter can indicate a leaky audio output transistor or IC component (**Fig. 1-14**).

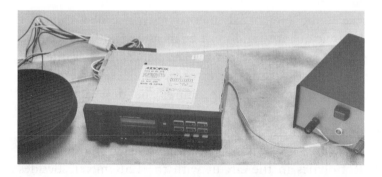

FIGURE 1-13. A low-priced power supply can be used to power the car radio while taking voltage measurements.

FIGURE 1-14. A variable heavy-duty power supply is a requirement on the electronics technician's service bench.

The Transistor Semiconductor Tester

Although the diode or the transistor test is included in the DMM and can accurately test transistors, FETs, SCRs, and diodes, a low-priced semiconductor tester for less than $100 can make outside semiconductor tests out of the circuit. Also, the semiconductor tester identifies NPN/PNP transistors, N-channel FETs, germanium diodes, makes continuity tests, has an audio tone, and can operate from a 9-volt battery. The audio tone indicates a short under 20 ohms. Only a few test instruments are required in servicing the auto radio, when knowing how the different circuits perform and how to make correct tests with any electronic test instrument.

ADDITIONAL TEST EQUIPMENT

The ESR Meter

The in-circuit capacitor tester is ideal for testing electrolytic capacitors that have an Equivalent Series Resistance (ESR) condition. The ESR meter can find the defective capacitor; a regular capacitor tester cannot test for ESR problems. The ESR meter can accurately measure ESR of capacitors of 1 uF and up. Capacitors from 0.1 uF to 1 uF can also be tested with a given chart.

Equivalent Series Resistance is the sum of all internal resistance of a capacitor measured in ohms. ESR is a dynamic quantity and must be measured with an accurate AC test signal. When there is no significant capacitive reactance, ESR is the AC resistance of a capacitor. The ESR of a capacitor is affected by conduction techniques, defective properties, frequency of operation, and temperature. The ideal capacitor has zero ohms of ESR.

The ESR meter is ideal for testing SMD capacitors for the correct capacity (**Fig. 1-15**). This meter can check capacitors for good or bad while soldered into the PCB. You can quickly check all electrolytics in the circuit with the ESR meter. Besides capacitors, the ESR meter can check and measure small inductors, transistors, and diodes, measuring small ohm resistors, and checking for breaks in the PC wiring. The ESR meter can find shorted PC traces, cracked or broken PC wiring, and poor SMD component connections.

A beeper tone inside the ESR meter is a quick method of identifying good capacitors without looking at the meter. If the ESR meter beeps, then the capacitor is good. The beeper can be used to find intermittent capacitors and connections. A raspy sound indicates an intermittent connection or a poorly soldered joint. Like using the ohmmeter in any circuit, the voltage or capacitor must be discharged before performing ESR tests. Pull the DC power from the auto radio or CD changer before taking in-circuit ESR tests. If not, you might damage the ESR meter.

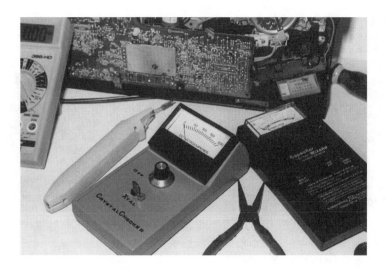

FIGURE 1-15. The ESR meter can check large and small electrolytic capacitors found in the auto receiver and CD changer.

The Oscilloscope

Although the following test equipment is not necessarily required for servicing the auto radio, changer, or cassette player, it can speed up the process, make accurate tests, and quickly help to locate a defective stage or component. The dual-trace oscilloscope can be used to locate a weak and distorted stage in the audio circuits. The scope can be used to take EFM and critical waveforms throughout the CD changer circuits, to name a few (**Fig. 1-16**).

The sine/square wave generator and oscilloscope can locate weak and distorted audio electronics circuits in the CD changer, audio receiver, and cassette player. The scope is ideal when making alignment and adjustment procedures. A dual-trace scope can be used to check out two stereo audio circuits at the same time and compare the gain in each stage. By injecting an audio signal from the generator or test cassette into both stereo channels, the weak or distorted circuit can be located with a scope waveform. The dual-trace scope can be used in AM/FM/MPX, compact-disc signal tracing, and other critical adjustments.

The dual-trace scope is just right when taking waveforms or alignment in the audio stereo channels. The scope can check the stability of the amplifier and locate hum in the power-supply circuits. Select an oscilloscope with at least a 40- to 100-MHz bandwidth for audio stereo measurements. Some of the features might be dual channel, dual trace, dual time base, delayed sweep, auto triggering, X-Y operation, scale illumination, and intensity control.

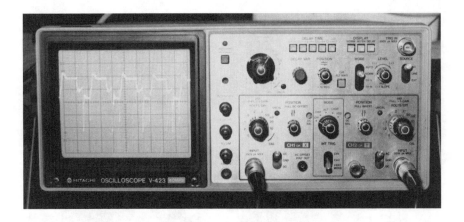

FIGURE 1-16. The dual-trace oscilloscope comes in handy to take critical waveforms in the stereo amp and CD changer circuits.

AM/FM Generator

The AM/FM generator is used in alignment of the RF, oscillator, and IF frequencies of the AM/FM car radio. Although today's auto radio IF stages have fixed ceramic filter components, no alignment is needed as the frequencies do not change. Of course, the AM/FM signal generator can be used to align those old antique radios found in the antique automobiles. The AM/FM signal generator can be used to troubleshoot the various front-end stages by injecting an RF signal into the RF, mixer, and IF circuits.

The AM/FM signal generator should have a wide frequency range of 100 kHz to 450 MHz. A high-stability feature can be obtained with internal crystal-controlled oscillator circuits. The modulated frequency is usually around 1 kHz. The AM/FM signal generator can be purchased for around $150.

Frequency Counter

The frequency counter might be included in a handheld, battery-operated test instrument or within the digital multimeter. A frequency counter might be included within a function generator or audio oscillator test instrument. The frequency counter counts signal cycles or pulses in a frequency measurement. The frequency counter can check frequency, speed head alignment, and overall frequency response of cassette players and amplifiers. With a cassette, the frequency counter can tell if the speed of the cassette is running slow or fast (**Fig.** 1-17).

The audio oscillator/frequency counter has a low distortion of 20 Hz to 200 kHz audio oscillator with sine and square-wave (TTL) output. A variable amplitude and output attenuator (-10 dB, -20 dB, -40 dB) is found here. The output impedance is

FIGURE 1-17. Besides frequency response tests, the frequency counter can check to see if the speed is running slow in the cassette player.

at 600 ohms with external sync. A built-in 1-MHz frequency counter has a frequency counter range of 10 Hz to 1 MHz.

Test Speakers

The test speakers should have the correct output impedance connected to the amplifier for a perfect match. Because the common speaker impedance is 8 ohms, just about any size of speaker will do. Several speakers can be connected in parallel to match a 2- or 4-ohm amplifier output impedance. A pair of compact-shelf speakers is ideal where they can be moved and clipped to the audio chassis. A 10- or 12-inch pair of woofer speakers that have an average power of 50/100 watts might be required when servicing high-powered stereo amps.

Most speaker enclosures have RCA-type plugs into the amplifier. Some audio chassis have lug-to-lug, pin connectors, or tinned wire ends. A pair of alligator clips on each speaker might make quick and solid connections. Make sure the connections are good and tight. Check the volume control setting before firing up the amplifier. Do not accidentally apply too much power to the test speakers so that the voice coils are not damaged at once.

Speaker Loads

Attach a speaker to the radio or amplifier before attempting to service any audio product. The amp must have a load attached before the volume control is rotated, or else the amplifier output circuits can be damaged. High-powered dummy loads are required for servicing high-powered auto or car amp systems. Sometimes the audio output circuits become defective and place a DC voltage on the speaker voice coil,

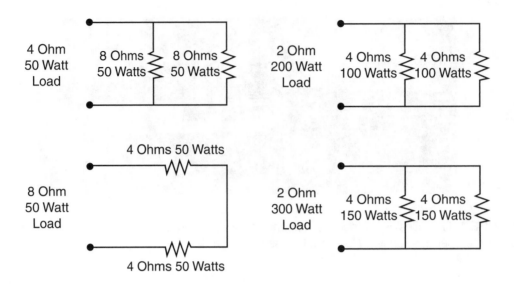

FIGURE 1-18. Connect large 50- to 100-watt resistors together to acquire the correct speaker impedance and wattage output.

destroying the speaker. Always check the speaker terminals or jacks for a DC voltage before attaching any PM speaker.

Because many of the latest audio amplifiers have high wattage output systems, a test-speaker dummy load must be able to stand up to 1000 watts or more. Also, by using a dummy load, you do not destroy an expensive speaker if a DC voltage is found on the speaker terminals. Connect several 50- or 100-watt resistors in series and parallel to acquire the correct resistance and wattage. For instance, two 50-watt 8-ohm resistors can be paralleled to achieve a 100-watt resistance to match a 4-ohm impedance.

Simply connect two 150-watt-at-4-ohm resistors in series to match an 8-ohm, 150-watt dummy load (**Fig. 1-18**). A variable load resistor is designed for testing noninductive loads for bench testing amplifiers from 0-8 ohms at 90 watts and 0-18 ohms at 225 watts. Adjust the wiper ring for the correct resistance upon the wattage load resistor.

The full wattage of a high-powered amplifier is never turned on full while being serviced. A bank of 20- to 25-watt resistors can be added for a 100-watt output. Often the electronics technician keeps the volume turned down as low as possible while trying to locate the weak or distorted circuit. A 100- to 250-watt 8-ohm dummy-load resistor on each stereo channel can provide adequate loading of most powerful amplifiers. Just remember to keep the volume control turned down for adequate audio from the amplifier.

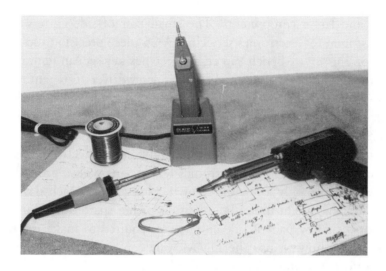

FIGURE 1-19. A temperature-controlled soldering iron and a battery-operated iron are ideal for soldering in microprocessors and SMD components.

Soldering Equipment

You can do a lot of service work with a 30-watt handheld soldering iron. For the electronics technician who requires many different soldering contacts and unsoldering large components, the temperature-controlled electric soldering station works best. The new soldering stations have digital display for temperature control and adjustment range from 320 degrees F to 900 degrees F (160 degrees C to 480 degrees C). Just dial up the precise heat needed for the work to be done. A built-in tray with cleaning sponge helps to keep the soldering iron tip clean and bright. A grounded tip is required when soldering static-sensitive devices and SMD components (**Fig. 1-19**).

A cordless quick-charge soldering iron works great when soldering up transistors, ICs, and microprocessor terminals. A long-lasting nickel cadmium battery withstands high-rate charging without damage. A partially discharged battery returns to full capacity in an hour or two. A fully charged unit can make up to 125 solder jobs per charge. A fine replacement tip is required to solder up those tiny SMD components. The battery-operated soldering iron can be taken anywhere without cords dangling from the cordless soldering iron.

Bench Tools

The required bench and hand tools are usually found on the service bench for making auto radio and changer repairs. A set of nut drivers quickly helps to remove those outside covers and larger components. The small pair of side cutters and long-nose pliers are a must item for electronics servicing. A 25-piece precision tool set is ideal for fieldwork and on the service bench. An eight-piece hex key set can remove those special sunken set screws. The large and small Phillips screwdrivers can remove and replace those hard-to-get metal screws. The Torx and Star screwdriver is needed to remove those difficult bolts and screws. Special tools and test cassettes or CDs can be found in the various chapters.

The test instrument is no better than the handyman or electronics technician who takes critical measurements. Some electronics technicians with a great deal of knowledge and experience use only a few different test instruments each day. A handyman or electronics hobbyist might use only one in his or her spare time. Sometimes only a Phillips screwdriver and a pair of side-cutters are required to make the most difficult repairs. By knowing how to use the DMM, one can service many repair problems found in the auto radio, amplifier, and CD player.

CHAPTER 2

Auto Antenna and Hookup Problems

The early antenna system for the car radio was a side-mount bolted to the car body with a flexible shielded lead-in cable. Later on, the top cowl or top/side universal mount fit most automobiles. Today you can purchase a different top cowl-mounted antenna that will fit most GM cars, OE style, with mounting heads for fender angles of 3 to 30 degrees and a 64-inch cable.

A custom antenna is made to fit the VW Bug, Super Beetle (1978 and up) with a three-section mast and a 114-inch black cable. You can find a custom antenna to fit the Honda Civic (1988-91) in two sections with an 80-inch black cable. Several different antenna adapters are made to fit the standard radio to a GM antenna, standard antenna to a Nissan radio, and with a Motorola "F" connector to a 36-inch cable.

When the antenna cable is not long enough, antenna extensions for the auto radio with a Motorola plug and jack are available. The in-line AM/FM antenna filter helps to block out static noise received by the car antenna that can be generated by the vehicle electrical system. The AM/FM antenna filter might also help to improve the radio reception. You can maximize your AM/FM reception with an AM/FM signal booster as well as with the standard car antenna. The signal can be increased by 20 dB. Of course, the signal amplifier requires a 12-volt DC power source. The antenna booster works well with long antenna cable leads, and it also works well when the antenna is mounted on the rear fender.

Besides the top cowl and side-mounted antennas, a fully automatic power antenna fits under the rear fender, and the antenna motor turns on when the in-dash auto receiver is turned on. These fully automatic power antennas have a 4-section mast that extends upward to 31 inches with silent operation. When the auto radio is first turned on, a lot of static and weak reception is noticed until the antenna is fully extended upward (**Fig. 2-1**).

Suspect a bad motor, a blown antenna fuse, or a broken cable wire where it enters into the antenna base. Check the DC voltage at the antenna motor assembly with a normal fuse and with partial or no operation of the motorized antenna mount.

Several years ago, some of the new automobiles came out with the auto antenna molded right inside the front windshield. If the antenna cable wires broke off at the bottom of the windshield, the whole windshield had to be replaced, or another top cowl antenna was installed alongside of the automobile. Most of these molded glass car antennas caused a lot of noise in the radio reception, especially on the AM band.

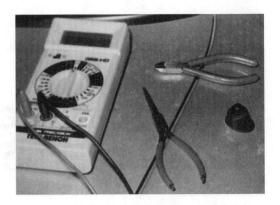

FIGURE 2-1. A power-driven rear 1998 Oldsmobile auto stereo antenna on the rear fender.

AUTO RADIO PROBLEMS

Most problems with the auto antenna were antenna rod breakage and dead or noisy reception. When the top cowl or side-mount antenna had a poor ground where it was mounted, extreme motor noise was picked up from the engine. A broken shielded cable at the body mount, where the antenna plugged into the radio, picked up motor noise from the vehicle ignition system. Both the noisy antenna and the motor noise can only enter the auto receiver from the antenna or the power "A" lead from the battery or ignition system. Always check the antenna mount that might loosen up while traveling.

By pulling out the antenna plug from the auto radio when the motor noise is still present, the motor noise is great enough to come through the "A" power wire (**Fig. 2-2**). A car antenna dummy load can be used to troubleshoot noisy car stereo installations. Plug the dummy load into the car stereo's antenna jack, and if the noise is eliminated, then the antenna is the noisy source. These dummy loads have a standard 50-ohm Motorola antenna plug at one end. Troubleshoot the auto ignition system when the noise is not picked up by the antenna system.

You can make your own dummy load from a 1-watt 50-ohm carbon resistor and a Motorola antenna plug. Insert one wire terminal lead of the resistor into the plug end and solder the male end to the fixed resistor. Now solder the other end of the resistor to the outside metal ground area of the male plug. The dummy 50-ohm antenna load can now be used to check for a noisy antenna system.

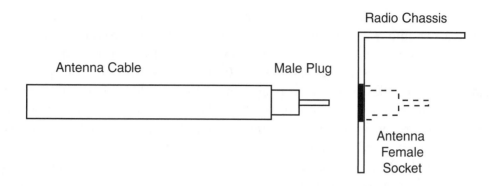

FIGURE 2-2. Remove the antenna plug to determine if the motor noise is coming through the antenna or "A" lead.

Make sure the top cowl antenna metal base bites into the car metal body to make a very good ground. If not, the antenna can pick up the noisy car motor noise. Check for a break in the metal shield of the lead in at the antenna mount or at the male plug. If the shielded cable is broken at the antenna plug, cut off the plug end and install a new Motorola-type antenna male plug. Insert the center insulated conductor wire to the center of the antenna plug and then feed molded solder into the male end. Solder up the shielded cable to the outside of the antenna plug. Push the insulation up into the plug so that it will not short out. Wrap plastic tape over the soldered metal part of the plug and cable.

Dead—No Reception

Determine if the dead radio reception is caused by a defective antenna system or by a bad component in the car radio. Simply remove the antenna plug from the radio and turn up the volume on the radio. If a rushing sound is heard with no local stations, suspect a bad antenna system. The outside antenna cable might be broken into, the inside shield wire might be torn from the antenna mast, or you might have a bad metal plug and possibly a grounded outside antenna. Do not overlook excess water down inside the antenna cable.

If another top cowl antenna is handy, plug the antenna into the radio, and hold the antenna by the plastic mount outside the car window. Now, try to tune in your local radio stations at each end of the dial. If the radio comes to life, suspect a bad outside antenna system. When the car radio remains dead with another antenna plugged into the radio, remove the car radio for proper repairs.

Check out the mounted top cowl antenna with the low-ohm scale of the DMM. Set the ohm scale to 200 ohms and touch one meter probe to the outside antenna rod and the other probe to the antenna male plug in the auto (**Fig. 2-3**). The center conductor is okay if a low-ohm measurement is less than 5 ohms. Either no resistance measurement or a very high resistance measurement indicates an internal broken lead-in.

Now take a high resistance measurement (in megohms) from the male plug to the shielded antenna cable or metal male plug. Replace the outside antenna if the resistance shows any type of reading under 100K ohms. Sometimes excess water or moisture collects inside the shielded cable and shorts out the picked-up stations. Often the inside conductor will break at the antenna mount or at the male plug, right where it fits inside the antenna jack. Replace the antenna male plug with a metal Motorola antenna plug when the break is at the male plug end.

Replace the whole top cowl antenna when the internal wire or conductor is broken at the bottom of the antenna. Select the correct top cowl or motor-driven antenna to fit a particular automobile. Several universal top cowl antennas can mount correctly on to most cars. If the wholesale or local dealer does not have the correct antenna, go to your local car dealer that handles the same make of automobile for the exact model antenna.

A shorted outside car antenna will have a resistance of only a few ohms. Most leaky or shorted lead-in cables occur at the bottom mount or where it fits into the metal male

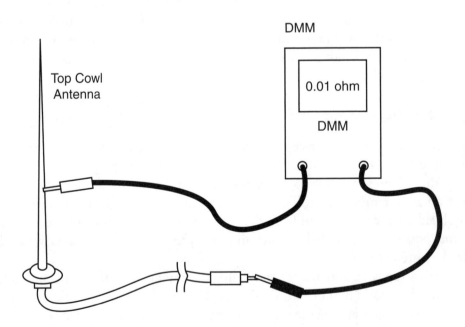

FIGURE 2-3. Check the car antenna for open inner conductor with the low-ohm scale of DMM.

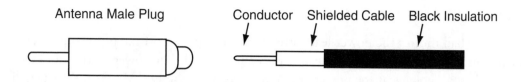

FIGURE 2-4.   Cut off the old antenna plug and install a new Motorola auto antenna plug on the antenna cable.

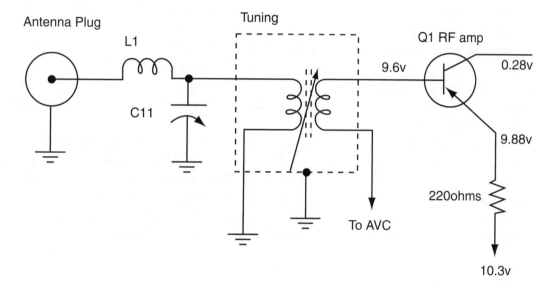

FIGURE 2-5.   Check for broken connections on coil L1 inside the auto for only a rushing sound in the auto radio.

plug. Replace the whole auto antenna when the lead-in cable is shorted at the base mounting area. Cut off the male plug and install a new one if the conductor is pulled out and broken at the male plug (**Fig. 2-4**).

Suspect a broken wire or a small choke coil broken off inside the car radio when you hear only a rushing noise and no stations can be tuned in. If only a local radio station can be tuned in, suspect a broken wire or input connection and a defective RF amp transistor (**Fig. 2-5**). Check for a broken antenna wire where the input circuits connect to the female antenna jack. Sometimes a small choke coil or capacitor is found rolling around in the bottom of the auto radio. Check for broken wire connections on L1 connected to the antenna jack and connecting the picked-up radio station to the RF amp circuits of the car radio.

Power-Up Antenna Problems

Suspect a bad motor, the power lead either broken or pulled off the power-up antenna, or the antenna rods binding, when the auto antenna will not power or rise up. Check the car battery voltage at the power antenna motor with the DMM. Turn the car radio on; the power-up antenna should start to rise upward. If not, check for no battery voltage at the antenna. Sometimes the gears inside the motorized antenna become stripped and will not raise the antenna rod. Try spraying silicon oil on the antenna rod assembly if the rod comes up only partway.

Most of the power-up antennas are mounted at the rear of the automobile and will extend upward when the car radio is turned on. Replace the power-up antenna when the antenna rod is broken off or bent out of shape. Replace the power antenna when the motor will not raise the antenna with normal battery voltage (13.8 to 14.4 volts) at the motorized antenna. Sometimes a universal motorized antenna can be fitted under most auto fenders. It's wise to pick up the exact motorized antenna at your local car dealer; they fit every time.

If only a local radio station or a rushing noise is heard in the auto receiver, with the antenna built right into the windshield, check the lead-in cable for possible breakage. First check the male plug for conductor breakage. If the shielded conductor is broken right where it enters the windshield and there is not enough wire to reconnect to the shielded cable, then it's best to mount another top cowl antenna outside of the car. Choose the correct top cowl antenna to fit the same model of automobile.

Blown Fuses

Suspect a blown fuse when no stations can be tuned in on the auto radio. Check for an open fuse if the dial lights do not light up or if you hear no noise from the radio. Rotate the volume control up and down quite rapidly and listen for a rushing sound, indicating that voltage is applied to the radio. Make sure both front doors are closed on the new cars, as in some of the latest GM models the radio shuts off when either door is opened. In some new autos, the radio will play for a few minutes after the ignition key is turned off; this prevents running the battery down with the car locked up and the radio is still playing.

Locate the car's fuse-block assembly, usually under the front dash on the driver's side of the auto. The fuse block might be mounted on the firewall or on a separate block for easy fuse replacement. Sometimes you might have to stand on your head to locate and remove a suspected fuse. Locate the fuse marked "radio" on the fuse compartment. If the fuse has turned black or if you can see the fuse link is blown, replace it with the exact amperage (**Fig. 2-6**).

Check the fuse with the low-ohm scale (200 ohms) of the DMM, if in doubt; when you cannot determine by sight, the fuse is open. No reading on the DMM indicates

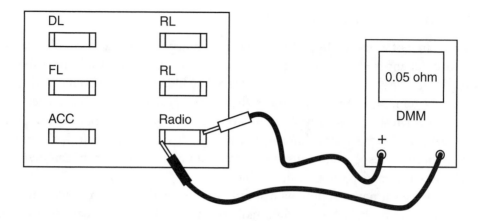

FIGURE 2-6. Check for an open or blown fuse with the low-ohm range of the ohmmeter.

that the fuse is bad. Replace the blown fuse with the same amperage. Some of these blown fuses are special Littelfuse types and cannot be fitted with a larger or incorrect fuse. Check the markings on the end of the fuse for correct amperage. Double-check the fuse rating stamped on the fuse-holder assembly.

If the fuse is too small in amperage, the fuse will blow right away. Sometimes if the fuse is too large, the fuse will not open if an output transistor or IC component becomes leaky or shorted in the radio. Often the radio overheats, smoke begins to roll out of the radio, and the "A" power-lead wire becomes burned, all the way from inside the radio to the power switch. Always replace the car radio fuse with the correct amperage to prevent radio fires and the burning of connecting cables.

When you do not know what size of fuse that should be installed for the car radio or changer, check inside the automobile operator's booklet. If no operation booklet is handy or is lost, most fuse holders will only hold a certain type of fuse. The cassette stereo player operates with a 3-amp fuse, a car radio cassette player has a 3- to 5-amp fuse for protection, and the early cartridge 8-track player contains a 10-amp fuse. The early antique radio with a vibrator might have a 15- to 20-amp fuse. The 2X50-watt amplifier might operate with a 15-amp fuse, while the 170-watt amp must have a 30-amp protection fuse.

Special cartridge fuses that fit into a specially sized holder will only contain a certain size fuse. These fuse holders might range in size from 5, 7.5, 10, 15, to 20 amps. You cannot insert a 15-amp fuse into a holder that holds only a 7.5-amp fuse. Popular mini-auto fuses used in the latest autos have a 2, 4, 5, 10, 15, and 30-amp rating.

Suspect a leaky or shorted output transistor or power output IC when the fuse will not hold and when it opens up at once. The auto radio must be removed from the auto and repaired if the auto fuse keeps blowing as soon as the radio is turned on.

Burned or Charred Cable Wires

When too large of a fuse is clipped in or partially plugged into the fuse holder and the car radio or changer shorts out internally, the connecting "A" lead and cables will become burned or charred. You might find that the blown fuse is covered with tin foil to serve as a temporary fuse until the right one is located. Often the power-output transistors, the Mosfet power-supply transistors, and the power-output IC components become leaky or shorted and will normally cause the fuse to blow open. If the fuse does not open up, the excess current caused by a shorted internal component can destroy the "A" or power lead and corresponding cables.

Often the bench power supply with shutdown circuits that has overloaded protection with excessive current, pulled by the auto radio or changer, begins to groan and then shuts down. With an overloaded car radio, the current meter hand hits the peg and the voltage meter lowers at once. The auto radio should be repaired before turning on the power supply again. Go directly to the audio output circuits and power supply for leaky components.

The charred wires must be repaired or replaced after the internal repairs are made. Sometimes a separate "A" lead can be spliced into the existing cable with a new fuse holder. Simply replace all speaker and wire cables that might be melted together with separate flexible hookup wire. Try to replace the color-coded wires, when possible. The quickest method is to repair the cable harness instead of trying to locate a whole new harness replacement.

Speaker Wiring and Connections

The speaker harness and "A" lead might consist of separate cable wires, speaker male and female plugs, and a sliding-type PC socket. A separate "A" lead might contain only an in-line fuse holder. American International wiring harnesses virtually eliminate any unnecessary splicing of the factory wiring in your car. These harnesses also come with an exact wiring diagram for easy installation of a certain marked radio.

American International wiring harnesses are available for the 1984-1996 Chrysler, 1983-1996 Ford, General Motors 1978-1996, Honda 1983-1996, Nissan 1985-1994, and Toyota 1978-1996 automobiles. These wiring harnesses and plugs come in many different sizes and shapes, and can be obtained from MCM Electronics in Centerville, Ohio.

The simple speaker wiring might plug into the auto radio socket with separate wires to each speaker (**Fig. 2-7**). Sometimes a two-way or four-way flat connector separates the speaker harness when the speaker cables come directly out of the radio receiver. Solder up all speaker terminals and "A" lead power wires when spliced into a cable network. Do not just twist the wires together and then tape them up; they can easily work loose and short out against the dash or with other cable wires.

The speakers, the battery in-line fuse holder, and the ground and power antenna wires might be soldered directly into the auto receiver. A separate speaker plug can disconnect the left and right speaker wires that remain connected inside the auto (**Fig. 2-8**). In some

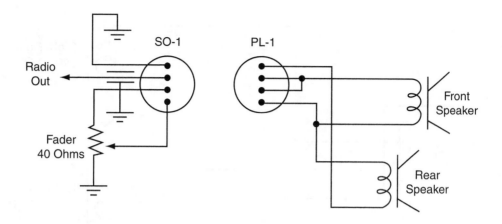

FIGURE 2-7. A single front- and rear-speaker hookup with a removable plug.

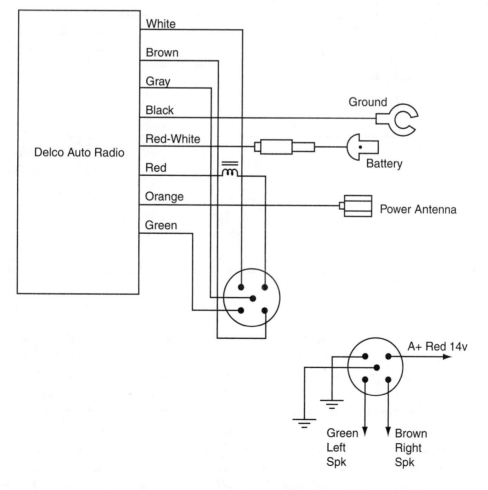

FIGURE 2-8. Disconnect the speaker plug when removing the auto radio.

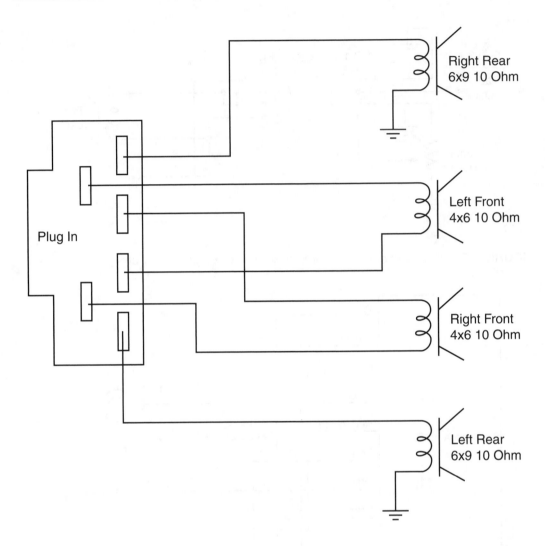

FIGURE 2-9.   Remove the plug-in strip of a Delco car radio to disconnect speakers from the auto receiver.

Delco car radios, a flat plug on a PC board can be pulled out that connects the right rear, left rear (6X9), left front, and right (4X6) front speakers in the automobile (**Fig. 2-9**). While in the auto, a multipurpose plug connects the fuse "A," lead, ground wire, left and right tweeters, and right and left woofer speakers inside the car (**Fig. 2-10**).

When removing the AM/FM compact disc (CD) player from the auto, there are many different cables to disconnect besides the antenna cable. Often the CD player might appear in two different sections. You might find three sets of cables going to the rear power amp while another three sets of cables go to the front amplifier. A separate spade-type wire-connected cable goes to the CD changer in the trunk area. The light

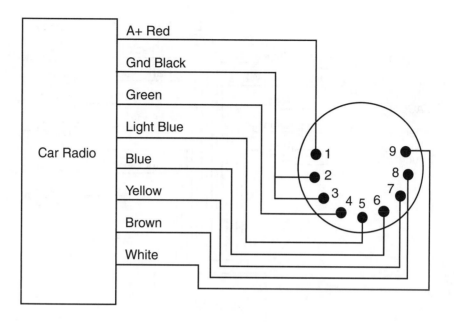

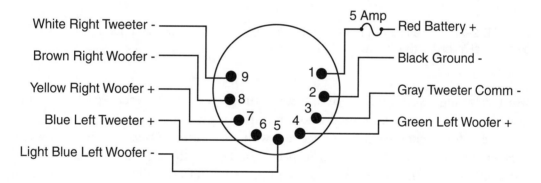

FIGURE 2-10. Remove the nine-prong plug to the woofer- and tweeter-speaker wiring.

switch and ACC wires have spade type connections. A 5-amp fuse holder is connected to the battery or ignition system. A black ground wire with a spade or ring terminal is provided for properly grounding the auto radio circuits (**Fig. 2-11**).

Poor Speaker Wiring Connections

A dead speaker might result from a bad speaker connection or a broken speaker wire. Sometimes the speaker wires can get caught on objects being loaded and can cause a set of speaker wires to go dead. A speaker wire lying under the seats and carpet can get

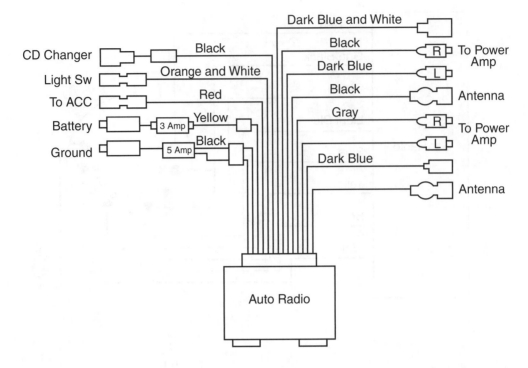

CD Changer — Black

Light Sw — Orange and White

To ACC — Red

Battery — 3 Amp — Yellow

Ground — 5 Amp — Black

Dark Blue and White

Black — R — To Power Amp

Dark Blue — L

Black — Antenna

Gray — R — To Power Amp

— L

Dark Blue —

— Antenna

Auto Radio

FIGURE 2-11. The CD AM/FM compact disc player has many wires and cables to disconnect in order to retrieve the auto receiver.

worn and might get broken into. Always run two wires to each speaker, as just grounding one speaker wire might not end up being a good ground.

Check for a poor speaker connection with intermittent music from one or two speakers on each side of the auto. Suspect a broken or loose ground wire when two different speakers on one side become intermittent. A defective PM speaker with a torn voice coil or broken flexible speaker wire from the voice coil can cause intermittent sound. Clip another PM speaker of any size to check for a dead or intermittent speaker. If both speakers are dead or intermittent, suspect a broken speaker wire or a bad connection. Replace the defective speaker when the original speaker is found open or intermittent.

Do not just wrap the wire around the speaker voice coil terminals. Carefully solder in each speaker wire with the cordless soldering iron. Poor speaker connections can become corroded when the auto is parked outdoors or in an unheated garage.

Noisy Motor Noises

After determining that the motor noise is in the radio and not picked up by the antenna, and that the ignition noise is coming into the radio through the "A" lead, try to isolate the noisy problem. Make sure that the car radio is properly grounded. If the auto radio does not have a noise choke in series with the "A" lead wire, then install a 3-amp noise choke transformer to filter out engine noise coming in the power-lead wire. Use the power-line choke with car stereos, power amps, CBs, marine radios, and tape players (**Fig. 2-12**).

Large shielded bypass noise capacitors (0.1 uF to 5 uF) can be shunted across the power lead to a good body ground to help eliminate motor noise that comes in the power lead. Automobile noise filters for use with radios, equalizers, tape players, and compact disc players to filter out engine noise come in 10-amp/150-watt, 15-amp/250-watt, and 25-amp/350-watt filters. These types of noise filters are perfect for automobiles, boats, and motorcycles with heavy double-wound coils.

There are many different power-line noise filters on the market. To cure pickup noise that might exist between separate amp and auto radio, a line-level filter isolates the ground between amp and radio. This unit eliminates noise due to signal path ground loops. If the noise filters do not eliminate the engine noise, no doubt a garage mechanic should check the noisy ignition system in order to adjust spark plug points and to make distributor adjustments on older autos.

In the older cars with motor ignition noises, a noise suppressor was placed in the center of the high-voltage distributor to help eliminate motor noise. Sometimes with excessive motor noise, each spark plug wire had a noise suppressor inserted in each high-voltage tension lead. Make sure the motor block is properly grounded to the car frame.

You can cure a whining noise caused by a noisy generator or an alternator by shunting a large bypass capacitor on the power lead to the generator and common ground. Most motor noises are caused by a poor ground system. Simply tie all ground wires of the car radio, changer, and amplifier to the car metal framework. Make a good ground by grinding a spot on the frame with the electric drill and grinding pad.

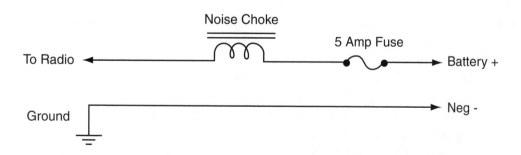

FIGURE 2-12. Install a noise choke to prevent engine noise from entering the car radio.

REMOVING AND REPLACEMENT

Sometimes you almost have to stand on your head to get at the bolts and brackets that hold or mount the auto receiver to the front dash. Removing the faceplate is rather easy, and in some radios, the radio can be removed from the front of the dash. Be very careful when removing the radio from the back of the dash to prevent shorting out wires and cables against the metal radio cabinet.

In most front-dash mountings, remove the large nuts holding the radio to the dashboard. Remove the bolts on each side and at the rear of the radio. Sometimes a large strap is bolted to the rear of the radio to keep it from bouncing or moving up and down. Either remove the mounting strap or pull it out of the way after removing from the car radio so that the radio can be removed from the rear area of the auto's dashboard.

Unplug the main fuse lead-in, power wire, and antenna plug from the radio. Remove the speaker plugs and changer connection that connects the radio to the changer and various speakers. Usually all wires and cables that are connected to the auto receiver are disconnected by plug-in cables or interconnector speaker plugs. Since the plug connector will only fit one way, you do not have to worry where each plug or wire is connected.

If the removal of the radio seems to be rather complicated or if one feels that he or she cannot successfully remove it, make an appointment with your local car dealer to have the radio removed. Most garages have a removable chart or method for removing and replacing the auto radio. Besides, they may have the correct tools for removing and replacing the auto receiver.

Component and Speaker Hookups

The most common auto receiver hookup connections with a cassette player are the power-supply lead (13.8 Volts), spade backup, power antenna clip, and a common black ground wire (**Fig. 2-13**). A separate ground wire with the left and right speaker wires are connected to each speaker. The power "A" or B+ lead connects to the ignition battery terminal. A clip-type lead connects to the antenna motor from a DC source.

The more deluxe AM/FM auto stereo receiver and cassette player may have several different cables and wires connected to a plug and pigtail leads attached to a male plug. Two separate line outputs (left and right) plugs are provided for a connection to the high-powered amplifier. The left speaker output has a white speaker output wire to the positive (+) left 2CH and left/front 4CH speaker. A green speaker output wire connects the left minus (-) 2CH and left/rear 4CH speakers. The green/black and white/black is the left common wire through an electrolytic (1000 uF) to chassis ground (**Fig. 2-14**).

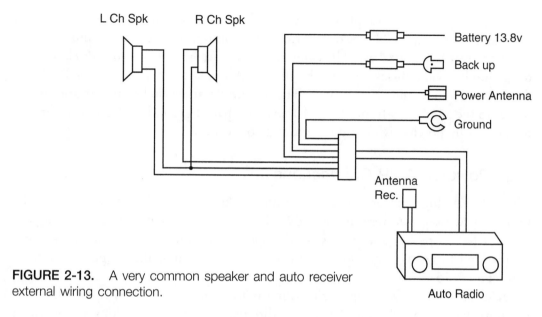

FIGURE 2-13. A very common speaker and auto receiver external wiring connection.

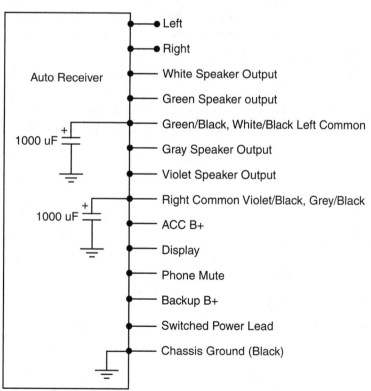

FIGURE 2-14. The deluxe AM/FM/MPX auto receiver with a cassette player and speaker connections.

The right power IC supplies a gray speaker output wire to the right positive (+) 2CH and right/front 4CH speakers. A violet right output wire goes to a minus (-) 2CH and right/rear 4CH speakers. The right common wire goes to a 1000 uF electrolytic to ground in a violet/black and gray/black wire. A separate wire connects the B+ ACC, display, and phone-mute connections. The backup B+ and switched power lead supply power to the auto radio from the car battery. A separate ground lead connects directly to the auto radio chassis for a good ground connection.

High-Powered AMP Connections

The 170-watt high-power amplifier output speaker and "A" lead terminals may be found on two separate terminal boards. The CP4 speaker screw-type terminal goes to the left (+) speaker and minus (-) left speaker terminals. The positive (+) right speaker and minus (-) right speaker terminals are marked at the output connection strip (**Fig. 2-15**).

A separate CP3 screw type strip-connects to a remote circuit with two B+ connections that tie to a 30-amp fuse. Two different ground connections are provided to the Mosfet transistors step-up power supply. The Mosfet transistor power supply has a 14-4 DC volt input with a positive (+) and minus (-) 34 volts applied to the high-powered audio amp transistors.

A deluxe high-powered auto receiver and CD player are shown in **Figure 2-16**. Large 12-inch and 8-inch PM speakers on the left and right are powered by the high-powered amps. A medium-sized power amp powers equalization to a set of tweeter speakers with a separate amplifier. Sometimes a six-disc changer can be installed in the

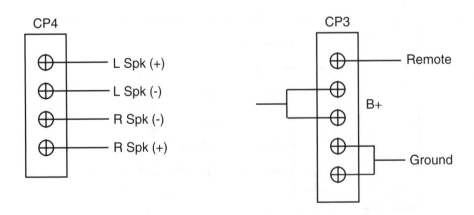

FIGURE 2-15. The high-powered amplifier speaker and power connections at one end of the amplifier.

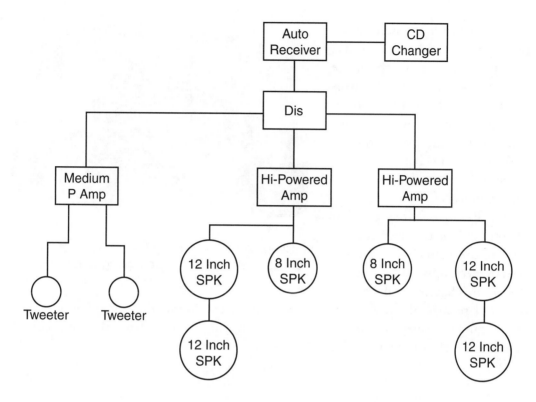

FIGURE 2-16. A high-powered auto receiver and CD player hookup with several different sizes of speakers.

glove compartment or in the front seat center console, while a 10-disc CD changer must be installed in the trunk area.

Removing the Covers

To remove the top cover, check for two or more metal screws holding the edge or top metal cover. Many of the top cover screws are Phillips-type metal screws. Pull back the cover and lift off the top cover. After viewing the radio-cassette player components, try to locate the defective section (**Fig. 2-17**).

Turn the radio upside down or over to get at the bottom cover. Sometimes the cover might be clipped on with no mounting screws; the metal lid just snaps into position over the bottom area.

Remove the detachable control panel from the front part of the radio to get at the pilot lights. Pull off the receiver control knobs. Carefully remove three to four point hooks from the main chassis. You might have to remove the detachable panel before the whole front panel can be removed. Pull the front panel away from the metal chassis.

Removing the cassette chassis might be more difficult and might take a little more service time. Some cassette mechanisms are removed from the bottom and others from the top area. Disconnect all connections and wires tied to the mechanism. Mark down each wire color and where they connect for easy replacement. Remove three to four screws holding the deck mechanism to the main chassis. Lift the cassette deck upward and out of the main chassis to make the

FIGURE 2-17. Removing the covers from an auto radio cassette player before making repairs

required repairs of the tape deck. In many cases, the moving idlers, pulleys, and motor can be cleaned and replaced without removing the deck from the auto receiver or the main chassis (**Fig. 2-18**).

FIGURE 2-18. The inside view of the cassette player and AM/FM stereo receiver after removing the top cover.

Light Bulb Replacement

Replacing the light bulb in the early auto radio was fairly easy compared to the latest receivers. You simply unscrewed the light bulb or gave it a half-twist and replaced it with a new one. Then you checked the bulb voltage and current rating. Today, you remove the top cover or faceplate assembly to get at the more-than-one light bulb or LED. The pilot light might have a natural color with several other lamps in amber, green, and red colors. You might see four different pilot lamps with different colors in the latest AM/FM stereo car cassette player.

The pilot lamps might be small LEDs with or without long connecting wires. Some of the latest colored lamps operate at 14 volts (40 MA) with a different colored cap. Usually the pilot lamps operate directly from the 12- to 14.4-volt battery source. Several different-colored LEDs might serve as indicators or an LED display might indicate the tuned-in station's letters and channel numbers. The time of day might also be included in the front display of the auto receiver.

The LED lamps might light up when in MPX, FM, Dolby, and in normal or reverse operations. Sometimes a pilot light operates at a lower voltage and has a small resistor in series to drop the required lamp operating voltage (**Fig. 2-19**). If in doubt

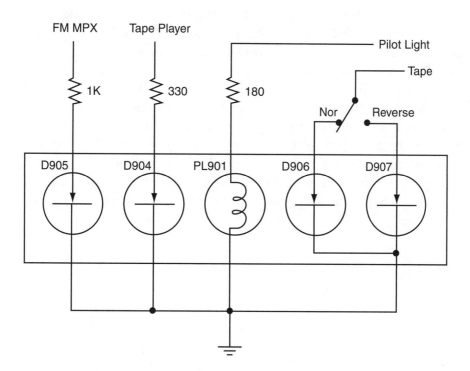

FIGURE 2-19. The different LED indicators and pilot lamps might have a voltage-dropping resistor in series with each indicator.

about a pilot lamp replacement, check the service literature for the correct part number and the type of lamp.

BEFORE ATTEMPTING TO REPAIR THE COMPACT DISC PLAYER

Extreme care should be exercised before attempting to service or repair the compact disc player. The compact disc player incorporates many ICs as well as the optical pickup laser assembly. These critical components are sensitive to — and easily affected by — static electricity. If the static electricity is high voltage, then these critical components can easily be damaged.

The optical pickup is made up of many different components and very high-precision components. Extreme care should be taken to avoid damage in storage and while handling and to avoid temperatures where there is high humidity, where strong magnetism is present, or where there is excess dirt and dust.

Disconnect the power-supply lead wire from the optical assembly, replacing any components of the laser assembly. Ground all test equipment, measuring instruments and tools. Make sure that the workbench is covered with a conductive sheet and is grounded. Anti-static field service kits that include a 24-inch square of electricity conductive workmat provide electrostatic protection for sensitive boards and optical laser assemblies during transport or installation. A wrist strap and ground cord are grounded through a 1-megohms resistor.

Do not remove the laser pickup from its protective bag until it is ready to be installed. Also, do not place the laser pickup on the protective bag when removing the bag, as there is a possibility of damage by static electricity.

Make sure the metal part of the soldering iron is grounded. Wear a grounded arm band or wrist strap to drain off static electricity from the body area. Keep the laser pickup away from clothing to prevent static charges in clothing, such as a wool sweater. Keep the laser pickup beam away from directly facing the eyes or bare skin.

CHAPTER 3

Troubleshooting the Power Supply

The defective output transistors, power output ICs, and shorted components in the power supply can cause the fuse to open in the main power supply. The heart of the auto radio is the voltage generated from the power supply. The power or input "A" lead supplies a 12- to 14.4-volt DC battery source to the power supply circuits. At first, most electronic technicians check the output voltage of the power supply to see if the required voltage is applied to the auto radio circuits. This DC voltage can easily be tested at the large filter capacitor in the power supply (**Fig. 3-1**).

Defective electrolytic capacitors in the power supply can cause many different service problems. A shorted electrolytic in the power supply can blow the main car fuse. Excessive hum and low output voltage can be caused by a defective electrolytic capacitor. Electrolytic capacitors that have black or white substance oozing out around the positive terminals can indicate power supply problems and should be replaced. Motorboat sounds in the speaker and low output voltage can result from electrolytic capacitors of changing capacity and with ESR problems. Check all electrolytics in the power supply with the ESR meter.

A defective transistor or IC voltage regulator in the power

FIGURE 3-1. Take a quick voltage test across the main filter capacitor to determine if a component in the power supply is defective.

supply can cause a dead radio with no display, no AM/FM functions, and no audio in the car radio. Suspect a shorted or leaky voltage regulator for no play, FF, or rewind operations within the auto radio-cassette player. A leaky voltage regulator in the tape deck can cause distorted audio. Intermittent loading problems can result from a defective voltage regulator transistor or IC in the auto radio and CD players. A defective voltage regulator in the power voltage source can cause improper focusing and oscillations in the focus error circuits of the optical laser assembly.

An open or leaky decoupling electrolytic can cause noise, a loss of power, or a dead audio player that will not play CDs in the auto CD player. Open or leaky decoupling capacitors within the voltage regulator circuits can cause a loss of display functions, loss of AM/FM functions, and result in its being stuck in record-mode of the auto cassette player. An open or leaky decoupling capacitor can also cause intermittent audio, loss of audio, and the car stereo player light going out.

Besides no radio or cassette operations, a defective electrolytic in the power supply can produce many different noisy operations. A no-volume symptom with a weak station reception and a motorboat sound can be caused by a 5000 uF electrolytic capacitor in the power supply. A squeaking noise when stations are tuned in can be caused by a 1000 uF electrolytic. The really loud whistling noise can result from a bad main filter capacitor. A slight or really loud hum can be caused by large 2200-6800 uF filter capacitors within the power supply.

THE EARLY POWER SUPPLY

The early auto radio power supply was quite simple with a filter choke, electrolytic filter capacitor, and voltage dropping resistors. The input voltage from the auto battery was around 12.6 volts with a 10-amp fuse in the power "A" wire lead. The 10-amp fuse provides fuseable protection to the auto radio circuit components. A filter choke helps to filter out any DC ripple in the power supply within the choke-input circuit. A bank of two or more electrolytic capacitors prevented oscillations and filtering action of the 12-volt source (**Fig. 3-2**). Voltage-dropping resistors provided several different voltage sources. A 12-volt pilot light was connected after the line filter choke.

The highest 12.34-volt source (A) supplied voltage to the AF amps, second AF amp, and audio output transistors. A 9-volt (B) source went to the RF amp, converter, and IF amp circuits. Voltage source (C) supplied voltage to the AF driver and detector circuits. C1 (500-300-100 uF) provided filter action to the three different voltage sources.

You might find two different power supplies within the early stereo cassette player with two separate power output IC components. The red battery source wire is protected by a 5-amp fuse to pin 1 of a 9 plug in the left and right speaker and power lead socket. A separate choke input with a 1000 uF electrolytic provides filtering action to voltage supply pin IC702. IC802 supply pin is fed from another filter choke and a 1000

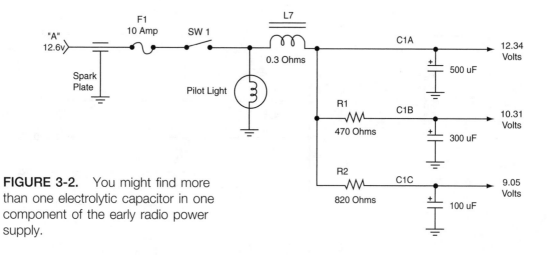

FIGURE 3-2. You might find more than one electrolytic capacitor in one component of the early radio power supply.

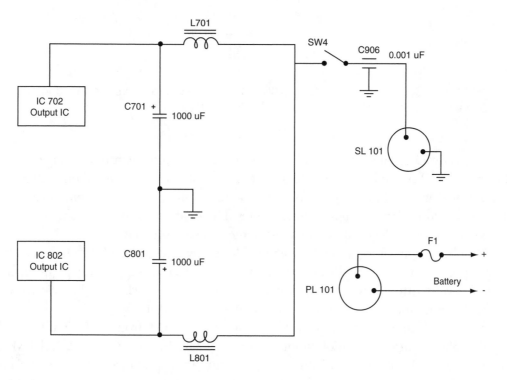

FIGURE 3-3. Check for separate electrolytic mounted capacitors in the latest auto-cassette receivers.

uF capacitor from the same battery source. Sometimes both electrolytic capacitors are found in one capacitor component and other times separate electrolytics are found mounted upon the PCB (**Fig. 3-3**).

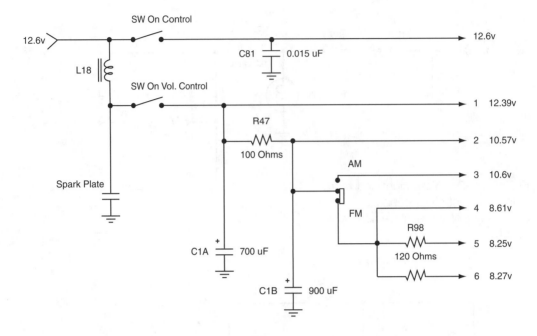

FIGURE 3-4. C1A (700 uF) and C1B (900 uF) electrolytic capacitors are found in one component of an early Delco auto radio and 8-track player.

The early Delco auto radio and 8-track player has more than one voltage source. The positive 12.6-volt battery source is fed to an on/off switch mounted on the volume control, and it switches the DC voltage to a 12.6-volt source to the antenna motor through a 500-millampere power filter choke. The dual switch on the volume control turns on the DC voltage source to six different voltage sources.

Number 1 voltage source (12.3 volts) is filtered by L18 and C1A (700 uF) electrolytic capacitor to a push-pull transistor power output stage. Supply voltage number 2 feeds 8.5 volts through R97 (100 ohms) and is filtered by C1B (900 uF) electrolytic to the AF amp and driver transistor audio circuits (**Fig. 3-4**).

Number 3 voltage source provides voltage to the AM front-end circuits, while supply voltage number 4 supplies voltage to the FM RF, oscillator, and AM/FM IF circuits. An 8.61-volt source of power supply number 5 provides voltage to the FM limiter and discriminator stages. The 8.27-volt source (number 6) supplies voltage to the 10.7 MHz FM IF circuits.

POWER SUPPLY PROBLEMS

Most auto radio receivers supply problems are caused by a blown fuse, resulting in a dead receiver, shorted output transistors, and power output ICs that can overload the power supply. If the protection fuse is replaced by a much higher amp fuse, usually the "A" lead power wire becomes overheated and can be burned or charred. Besides burned wires, the overheated power supply begins to smoke and the volume switch is quickly turned off.

A shorted electrolytic capacitor can break down and cause the main fuse to blow. If a filter or decoupling electrolytic becomes leaky or shorted and the fuse does not open, the filter choke can overheat and become burned. Likewise, small voltage-dropping resistors within the filter network can become overheated and end up with burned and cracked body components.

A shorted electrolytic capacitor can break down and cause the main fuse to blow. If a filter or decoupling electrolytic becomes leaky or shorted and the fuse does not open, the filter choke can be overheated and contain burned windings. Likewise, small voltage-dropping resistors within the filter network can be overheated and end up with burned or cracked resistors.

Sometimes electrolytic capacitor terminals will break internally and cause an open capacitor. An open filter capacitor can cause a lower voltage source and extreme hum from the low-voltage power supply. An open decoupling capacitor can cause a lower voltage source resulting in intermittent and weak radio reception. A dried-up or open decoupling electrolytic can cause a screeching noise when a station is tuned in, and a slight hum can be heard in the audio.

Check the noisy or constant hum in the speaker with the volume control turned down. Suspect a dried-up or open electrolytic with excessive hum in the speakers. A really loud whistling noise and excessive hum can be caused by a defective electrolytic capacitor. Check for a bad filter capacitor when a loud popping noise is heard in the sound, with the volume turned up. An excessive motorboat sound in the audio is a direct result from a bad electrolytic within the low-voltage power supply.

One of the old methods of checking suspected filter capacitors is shunting a new electrolytic across the suspected capacitor to eliminate hum in the speakers. In the early radio tube circuits, you can shunt these capacitors while listening for a reduction of hum in the speakers without having to turn off the auto radio; no damage will result.

Today, you can damage other semiconductors in the radio circuits by shunting filter capacitors with the auto radio turned on. Shut the radio off and clip a known electrolytic across the suspected capacitor with test clips; then see if the hum has disappeared by turning the radio on. Shunt a 1000 uF 35-volt electrolytic with the same or greater capacity and voltage across each filter capacitor and notice if the hum is still in the speaker.

Checking the Electrolytic with the ESR Meter

The ESR meter is the ideal instrument to check for correct capacity and ESR problems within the low-voltage power supply. You can quickly check all electrolytic capacitors within circuit tests inside the power supply. The ESR meter will note the correct capacity and whether any ESR problems exist in the green, yellow, or red scale of the meter. Make sure each electrolytic is discharged before touching any capacitor with the ESR meter (**Fig. 3-5**).

Simply place the two ESR meter probes across the electrolytic terminals for correct ESR tests. You

FIGURE 3-5. Checking electrolytic capacitors in circuit with the ESR meter

do not have to observe the correct capacitor polarity because now you know that each capacitor is good. Always replace any capacitor that the ESR meter shows has a problem. You do not have to look at the meter's hand if an okay tone or beep can be heard from the ESR meter. Check for a shorted electrolytic capacitor with a regular capacitor checker or low-ohm range of the ohmmeter. You can check all electrolytics in the power supply within seconds with the ESR meter.

You can check the electrolytic capacitor with the DMM or VOM meter. A good electrolytic will charge up when the meter probes are applied to each capacitor terminal. Reverse the test probes and the numbers on the DMM will roll downward. The electrolytic capacitor will charge up and then discharge through the meter. Of course, the meter hand of a VOM meter can easily be followed when checking an electrolytic within the low-voltage power supply (**Fig. 3-6**).

Excessive Hum

Excessive hum in the auto radio speakers can be caused by a change in capacity of an electrolytic capacitor in the low-voltage power supply. Hum problems can also be caused by a leaky or shorted output transistor or IC component. Replace all electrolytic capacitors that have bulging sides or substance oozing out around the capacitor terminals. An open and broken internal connection or an electrolytic capacitor lowering in capacity can cause hum in the speakers.

Besides a slight or loud hum in the sound, a defective electrolytic capacitor can cause motorboat sounds, popping, whistling, and a high-pitched sound in the speakers. A slight hum might be found with any of the above sound problems. You might find more than

FIGURE 3-6. The charging or discharging of an electrolytic capacitor is easily detected as normal on the VOM meter.

one bad electrolytic within the same metal or cardboard container. Excessive hum can be detected by placing your fingers on the speaker cone area with the volume turned clear down.

Excessive hum in the power supply can be located with a waveform taken across the main filter capacitor. Clip the ground lead of the scope to the common ground or metal chassis and the other probe to the positive terminal of the filter capacitor. Turn the auto radio on and notice the increase in spikes on the scope with an open or dried-up filter capacitor. Besides looking at the waveform, you can hear excessive hum in the radio speakers.

Filter Capacitor Replacement

Replace the entire metal can or all of the electrolytic capacitors in one capacitor component. Do not just clip or solder in the new capacitor across the defective one. Often, when one electrolytic capacitor loses its capacity or it dries up, it will not be long until the remaining capacitors in the same can will go bad. Dual or triple electrolytics within one package are hard to locate, unless they are replaced with factory original replacements (**Fig. 3-7**).

Single radial-type aluminum electrolytics are easy to find, but units with more than one capacitor in a container are more difficult to locate. For instance, replace a defective 900 uF 25-volt electrolytic with a 1000 uF 25-volt replacement if a direct replacement is not available. Likewise, replace a 1000 uF 16-volt single electrolytic with a 2200 uF at 25 volts if there is enough room on the PCB.

When a direct filter replacement is not available, replace any electrolytic with the same or higher capacity and working voltage. Do not replace an electrolytic capacitor with a lower capacity or lower working voltage than the original. Sometimes, if a lower voltage capacitor is replaced, the capacitor will become really warm and then blow apart. You might locate a defective electrolytic with a blown top and cracked outside plastic container. Sometimes gas will build up from the components inside the capacitor and finally explode.

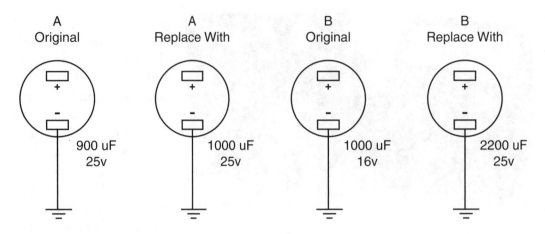

FIGURE 3-7. Replace the original defective capacitor with ones of the same voltage and capacity or with a higher capacity and voltage.

Excessive Current

Most overloaded circuits within the auto receiver or cassette player are caused by leaky or shorted filter capacitors. Leaky or shorted output transistors and IC components can also cause overload problems. A directly shorted output transistor, leaky MOSFET transistor, and power output IC components can cause the main fuse to blow or open. If the replacement fuse is too large, the power lead and wiring harness become burned and charred with a smoking chassis.

When the defective radio and cassette player are connected to the bench power supply, excessive current can show on the current meter. Often, an overloaded chassis will pull down the bench power supply and the voltage will lower on the voltage meter of the power supply. In Sams Technical Publishing's Photofact auto service manuals, the input voltage and operating current are found on the schematic.

In the Automatic CPV-7154 auto receiver, with an input of 12.6 volts, the radio has a 450 millampere operating current. The Lincoln Delco auto receiver and cartridge player has a 12.6-volt input and a normal operating current of 500 milliamperes. The auto radio should never pull more current than the main fuse rating. Go directly to the power supply and take critical voltage measurements and then check the output audio circuits, when the auto radio keeps blowing the main fuse.

Burned Choke Coil

A shorted output IC or transistor can cause damage to the filter choke coil or transformer if the fuse does not open and if the receiver remains switched on. Carefully inspect the outside area of the filter choke for burned or scorched areas. Feel the outside of the filter choke after the radio has operated for several hours. Notice if the filter capacitor feels really warm; they should run cool. Notice the current rating on the power supply bench supply if the choke coil is running really hot.

If extreme hum is still found in the speakers with the volume turned completely down, and if all electrolytics have been shunted with no results, suspect internal burned windings of the choke coil or transformer. Sometimes the choke coil winding runs quite hot and the enamel is burned off of the wiring and the windings short against one another. Most choke coils or transformers have a really low ohm measurement because they are wound with large-diameter enameled wire. Replace the suspected choke coil that measures less resistance with a burned cover and that produces hum in the speakers.

Diode Protection

The latest auto receivers and cassette players have a diode protection circuit so that a reversed battery polarity will not damage semiconductors in the radio circuits. A silicon diode is either inserted in series with the fused power lead or paralleled across the input wiring. Sometimes the car battery might accidentally be charged up backward, and the radio might operate in the early tube chassis. Critical transistors and IC components can easily be damaged if the car battery polarity is switched or charged up backward (**Fig. 3-8**).

In the cassette or CD auto receiver, if the tape motor runs backward without a protection diode, suspect that the battery is charged up backward. When the car radio is connected to a bench power supply and the power lead polarity is reversed, the radio can be damaged without a protection diode. Replace the shorted or leaky protection diode when the fuse keeps blowing.

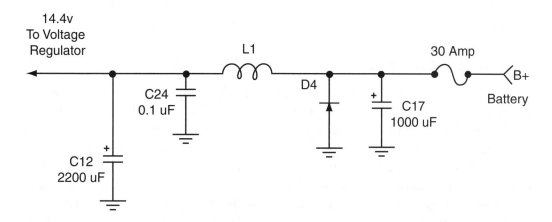

FIGURE 3-8.   Diode D4 prevents the incorrect battery polarity from being applied to the auto radio-cassette player.

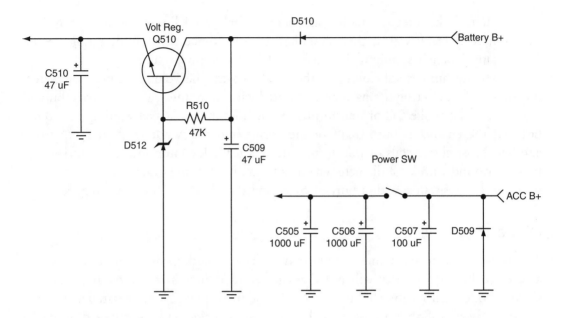

FIGURE 3-9. D510 provides diode battery protection to the low-voltage regulator transistor and other semiconductor circuits.

In a deluxe AM/FM stereo cassette car radio, diode protection is provided by D509 in the ACC B+ power lead and D510 in the voltage regulator circuit of the backup B+ power supply. D509 provides a direct short to chassis ground if the positive battery polarity is reversed. The radio is protected in the automobile, not only in this circuit, but also when the radio is connected to the bench power supply for required service. D510 silicon diode protects the voltage regulator circuit of Q510 (**Fig. 3-9**). Inspect the protection diode for cracked or burned marks in case the car battery is charged up backward.

Correct Voltage Tests

Take critical voltage measurements on each electrolytic capacitor located in the low-voltage power supply. Connect the auto radio to the 13.6-volt bench power supply and clip the negative probe of DMM to the negative or common ground area of the electrolytic capacitors. Place the positive probe to the positive terminal of the filter capacitor. You can take critical voltage measurements with the DMM and not have to worry about the correct polarity across the capacitor; the meter will indicate if it is a positive or negative measurement (**Fig. 3-10**). Of course, you cannot do this with the VOM meter.

Likewise, take a quick voltage measurement on each electrolytic, which might be located in several different voltage sources. Compare the voltage reading with those found on the schematic. If the voltage is really low or is lower than normal, suspect an over-loaded or leaky component in that particular voltage source. Isolate the lower voltage source and see where this voltage is applied in the radio or audio circuits. Usually, decoupling electrolytic capacitors are found in lower voltage and separate critical front-end circuits.

FIGURE 3-10. You do not have to worry about voltage polarity when taken with the DMM.

Decoupling Circuits

A decoupling capacitor provides a low impedance path to ground to prevent common coupling between the stages of the radio or audio circuits. The resistor-capacitor (RC) decoupling filter circuit separates the lower power supply voltage in the car radio to the radio input, audio preamp, and output power circuits. A decoupling capacitor might be found in the same aluminum container as the main filter capacitor or mounted on the PC board as a separate capacitor.

Suspect a slight hum in the speaker and oscillations in the RF and IF circuits with a defective decoupling capacitor. Decoupling capacitors usually have lower capacity and working voltage. In **Figure 3-11**, an RC decoupling network of the R701 (330 ohms) resistor and C701 (220 uF) electrolytic provide a separate decoupling network that isolates the supply voltage from the preamp IC tape head circuit of the car radio and cassette player.

If C701 becomes leaky or shorted, R701 might overheat and burn with a really low voltage on pin 4 of IC701. Suspect pickup hum with a low level of audio at the output of IC701 if C701 changes capacity or becomes open. Tape head oscillations might occur with an open C701. Check for a dead or weak audio when R701 increases in resistance and with a lower voltage applied to pin 44 of IC701. A shorted or leaky decoupling capacitor can pull down or lower the voltage of the main power supply.

Voltage Regular Circuits

In large and deluxe car receivers, cassette, and CD players, voltage regulator circuits are found in the low-voltage power supply circuits. The voltage regulator circuits might consist of transistors, diodes, and IC components. The regulated voltage circuit supplies critical voltage to VDD circuits of the microprocessor control circuits in the radio and

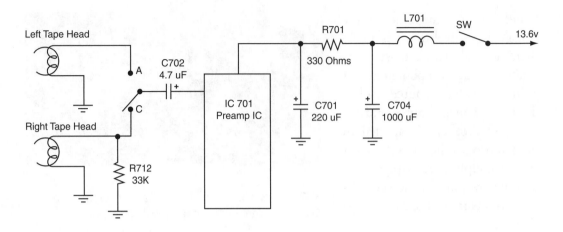

FIGURE 3-11. The RC filter network consists of R701 and C701 with the help of choke coil L701 to filter out hum in IC701 preamp circuits.

cassette circuits. Voltage regulator circuits might be found in the main B+ voltage, ACC B+, and backup power circuits (**Fig. 3-12**). Critical regulated circuits are found in the microcomputer, servo, and signal processing circuits of the auto CD player.

Voltage regulator circuits provide adequate and regulated voltage to many circuits of the auto cassette player. A loss of play, fast-forward (FF), and record circuits can be caused by a defective IC regulator in the power supply. The dead cassette player can result from a defective regulator transistor in the power source. A defective voltage transistor regulator can cause a loss of functions of the tape deck.

A loss of audio in playback can be caused by a shorted electrolytic in the B+ line of a leaky regulator transistor. Replace the voltage regulator with no sound from the radio and when only a slight noise is heard. Distorted audio in the tape deck can be caused by a leaky diode in the voltage regulator source. Extreme distortion in the cassette player can result from a leaky diode in the voltage-regulated circuits. After several hours of operation, the audio becomes distorted with a bad voltage regulator transistor.

A shorted regulator transistor can cause a loss of display function or operation. Poorly soldered transistor lead terminals of the voltage-regulated transistor might cause a no-display symptom. Replace the shorted motor regulated transistor when the capstan motor runs with the power turned off. Check the voltage regulator transistor in the motor circuits when the motor stops and slows down. Replace the 12-volt regulator transistor when the unit shuts off after operating.

Check the front-end voltage regulator when the scan-and-search shuts off after operating. Loss of AM or FM reception can result from a defective voltage regulator in the low-voltage source.

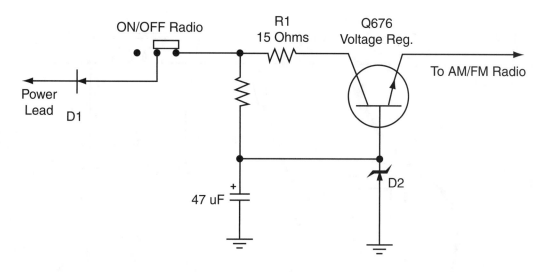

FIGURE 3-12. A simple voltage regulator transistor circuit might be found in the main power supply circuits.

Replace the voltage regulator IC if the loading process becomes erratic after warm-up in the CD player. Suspect the IC regulator when the unit will not accept the disc. Erratic loading and intermittent tracking can result from a defective IC regulator in the CD player. Check for a defective voltage regulator when the focus coil oscillates and operates erratically. Replace the 5-volt regulator if the CD player stops in the middle of a program after the auto CD player has operated for several hours. Suspect a shorted 5-volt regulator when the disc will not rotate and the lens assembly moves up and down.

A Bad Power Switch

The bad on/off switch can cause a dead auto radio chassis with an arcing noise in the speaker. A defective power switch will cause extreme audio distortion in the auto stereo circuits after several hours of operation. Replace the bad power switch when the switch will not stay on (**Fig. 3-13**). Check the on/off switch when sometimes the radio can be turned on and at other times it has no switch action. Suspect a defective power switch with no audio in the speakers. When the power switch will not turn off, check for a defective transistor in the low-voltage power supply.

Check for a worn power switch when the car hits a bump and the radio sound comes and goes. A badly soldered connection on the power switch can cause intermittent operation. A buzzing sound within the speakers can result from dirty or worn switch contacts in the power switch. Replace the power switch if the radio is really noisy

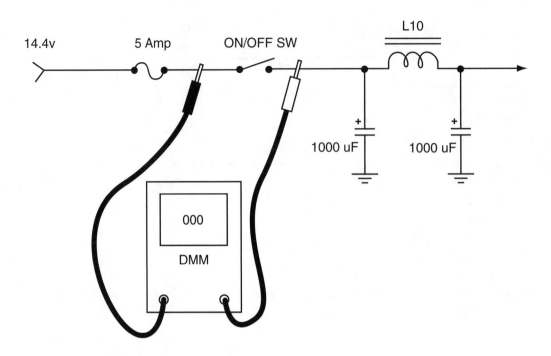

FIGURE 3-13. Check the power switch for dirty or worn contacts with no ohmmeter measurement.

when it's first turned on. Extreme noise all of the time in the radio cassette player can be caused by a bad function switch. Rotate the power switch off and on, and notice if it does not click on sometimes. Replace the radio/tape switch if there is no sound in either tape or radio operation.

The Noisy Power Supply

Replace the on/off switch if there is a buzzing noise in the speaker when the auto receiver is first turned on. Check the main filter capacitor with a buzzing and extreme humming noise in the speakers. Suspect a bad decoupling capacitor when the cassette player seems really noisy. Replace the noisy volume control when rotated. Check the lower decoupling capacitors with the ESR meter when a scratching noise is heard as the radio is turned on. A loose terminal lead of the filter capacitor can cause a popping sound, even with the audio control turned down.

Motorboating with a putt-putt noise can be caused by a defective filter capacitor. An open 1000 uF electrolytic in the power supply can cause a screeching noise as the radio is turned on. Excessive motorboating with a chirping sound can result from a large main filter capacitor. Replace the main filter electrolytic with a really loud whistling noise in the speakers. Check both the main filter and decoupling capacitors when a high-pitched whistle is heard as the volume is turned up. An extremely loud hum in all speakers can be caused by dried-up or open filter capacitors within the low-voltage power supply.

A Fuse That Keeps Blowing

Most blown fuses are caused by shorted or leaky power output transistors and IC components. Check for a shorted electrolytic in the main filter circuits for blowing the fuse when the radio is turned on. A shorted polarity diode that is connected from the B+ 13.6-volt lead to chassis ground can keep blowing the fuse. Check all electrolytic capacitors in the power supply for no power and a dead auto radio chassis. Do not overlook a bad spark plate that keeps shorting out and blowing the main fuse. A shorted diode or voltage regulator transistor in the low-voltage power supply can blow the main auto radio fuse.

Try to locate the component that is drawing heavy current and blowing the main fuse. Discharge all electrolytic capacitors in the low-voltage power supply. Take a resistance measurement across the main filter capacitor for a low ohm measurement. If the "A" lead is burned or charred, and the PC wiring might be burned or raised up going to the output transistors or power ICs, cut out a section of the PC wiring.

Now, take another low-ohm measurement at the filter capacitor. Suspect a leaky output or power transistors when the DMM begins to charge up. If the reading goes above 500 ohms, you might assume that the power source components are normal. When the DMM measurement is lower than 500 ohms, suspect a defective component in the low-voltage power supply. Recheck the resistance measurement on the power lead of the output IC or transistors, indicating a leaky output component. Do not forget to splice the cut PC wiring or trace with a piece of hookup wire after all repairs are made. Always replace the blown fuse with the correct amperage.

Burned Traces

Carefully look over the entire PCB very slowly when the chassis is dead and has a smell of burned resistors and raised foil traces. Sometimes a silicon diode will short out and burn a voltage-dropping resistor on the PC wiring to where the component was tied. Notice if a brown spot is found around the large resistor or transistor terminals on the PC wiring. A leaky or shorted transistor can cause the terminal connection to overheat. Replace the suspected transistor and form a loop to a piece of hookup wire and join it to both transistor and PC wiring.

Clean off the burned area. If the PC trace has lifted and been burned into, and the board shows signs of overheating, cut out the burned areas of the board. Use a pocketknife or Xacto blade to chip away the burned area. After replacing all burned and defective components, join the PC wiring with a piece of hookup wire. Likewise, if the burned area is around a large power output IC or transistor, clean up the burned area with flux remover and heavy-duty defluxer to clean up the PCB, relays, switches, and other electronic assemblies.

Blown PC Wiring

Pulled off or burned traces of PC wiring are usually caused by directly shorted components, such as power output transistors, ICs, and MOSFET transistors. A shorted protection diode, a power on/off switch, and large wattage resistors all might cause burned traces of the PC wiring. A shorted diode and transistor within the voltage regulator circuit can produce burned traces (**Fig. 3-14**).

Try to repair the bad traces of PC wiring with regular hookup wire. If the power "A" fuseable lead is burned where it is attached to the radio and on to the on/off switch, replace the bad wiring with a piece of flexible hookup cable. Short burned traces of wiring can be repaired with solid hookup wire. Remove the burned area, clean and scrape off the normal PC wiring cut ends, place a dab of solder paste on the ends, and solder in a short piece of hookup wire.

Make sure the piece of hookup wire is at least number 22 or larger to be able to carry the required current. Double-check the burned area with the low-ohm scale of the DMM. Select the correct fuse holder and fuse to replace the burned power cable. Replace the power and fuse holder with a new one and attach to the on/off switch terminal.

Broken PC Boards

Sometimes you might find a broken section of the PCB that might occur during transportation and does not act up until several months later. Cracked boards or sections can occur if the auto radio has been dropped, especially around heavy components mounted on the board or tied to the PC board. A cracked board section might be found at the

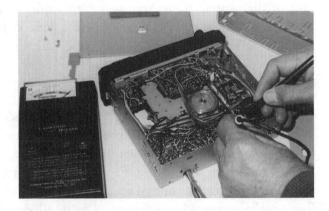

FIGURE 3-14. Check for broken or intermittent PC traces with the ESR meter.

corners of the board where the ground terminal is soldered to the metal chassis. Check for cracked areas where a high-powered transistor or IC terminals are soldered to the PC wiring and the component is mounted on the metal heat sink.

Determine what PC wiring or traces are cracked into on the board and repair with solid hookup wire. Clean off the traces or wiring with a pocketknife blade to remove any residue, PC wiring spray, or accumulated dirt and dust. Place on a dab of rosin paste to make a good soldered joint. You can replace the broken area by connecting hookup wire from the closest component soldered to the broken trace. Join the other piece of hookup wire to another component tied to the other end of the PC wiring.

Double-check all soldered wiring connections with the low-ohm range of the DMM. The ESR meter is ideal in checking for broken traces and PC wiring on the PCB. When the ESR meter shows no sign of a high reading or when the tone cannot be heard, suspect a cracked wiring at that point.

The ESR meter will sound off with a beeping tone with a good soldered connection. Do not just clean up the cracked wiring and apply solder over the joint; often, this does not make a good contact and can later cause the radio to act up. Always mend the broken trace with a piece of solid hookup wire. Now check out the joint with the ESR meter or low-ohm range of the DMM.

Display Not Lighting Up

When a lamp or LED does not light up, check the continuity of the lamp and the voltage source at the lamp terminals. Also, check the DC voltage applied to each LED terminal. If the LED is suspected of being open or dead, check out the LED with the diode test of the DMM. Compare these measurements with one or two other LEDs in the auto display assembly. The LED will have a higher ohm reading than a regular diode. Double-check all voltage-dropping resistors in series with each lamp or LED for proper resistance.

A complicated display is controlled by a controller (IC1002) in an auto AM/FM cassette player (**Fig. 3-15**). The segment divider (IC1003) and digit divider (IC1004) control the numbers within the display. A Black Key Matrix board does the switching with Q1002 providing DC voltage to the black key matrix assembly. Voltage regulator (Q1000) provides voltage to IC1003 and the black key assembly. The controller supply voltage is found on pin 13 from the 14.4-volt battery source.

The liquid display (LCD01) in an AM/FM stereo CD player is controlled by a microcomputer IC1001. The microcomputer has S1 through S30 wiring contacts and three display contacts to control the segments within the liquid display. A positive (+) volt source (VDD) is applied to pin 56 and common ground to pin 55 (VDD) for the supply voltage of IC1001. Simply measure the 5-volt source at pin 56 when no numbers light up on the liquid display (**Fig. 3-16**).

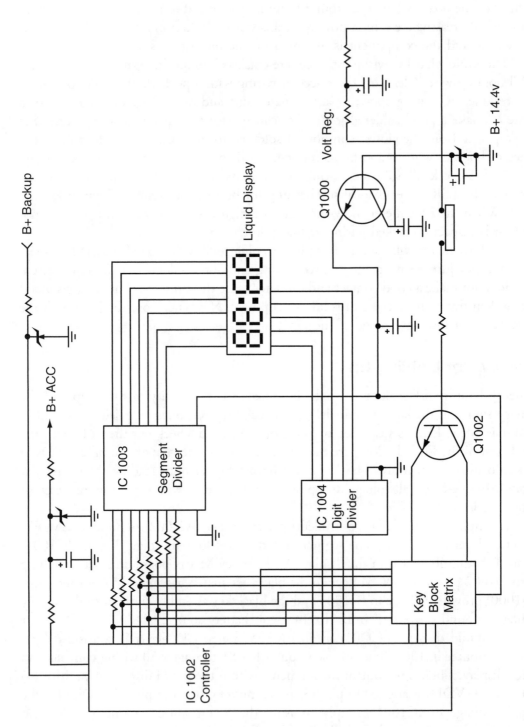

FIGURE 3-15. The IC 1002 controller provides signal to the liquid display panel.

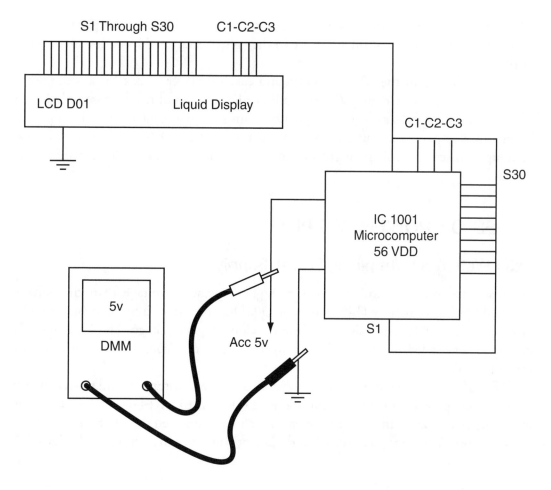

FIGURE 3-16. Check the voltage of the microcomputer IC1001 for correct VDD 5 volts with the DMM.

Most liquid display problems are caused by improper or no voltage applied to the microcomputer or Micro IC. Check the voltage regulator transistor with no light of the liquid display. A poorly soldered joint on the power regulator transistor can cause an intermittent or no liquid display. Suspect a shorted electrolytic capacitor in the voltage source to the microcomputer with no display function. An open diode in the DTS regulator board can cause lack of radio or display light-up.

A defective focus assembly can cause a blanking display and a laser pickup that does not focus in the auto CD player. Suspect a defective microcontroller IC with no display function or with a display that will not light up. A shorted or leaky zener diode in the supply voltage can cause a low level of illumination of the display assembly. Replace the defective liquid display when there is no display but all other functions are normal.

Check the main power voltage source with no display in the auto CD player. Check for an intermittent zener or silicon diode in the voltage source with no display or if the display fades out.

For no display in the AM/FM radio and cassette player, check for a leaky zener diode in the regulator circuit. With no audio or lighting up of display assembly in the car stereo, suspect a leaky voltage regulator transistor in the main power supply. A broken display panel can cause the display to not light up. When one digit does not light up, replace the voltage regulator and check the microcontroller for the panel display.

TYPES OF POWER SUPPLY

2x50 Watt Trunk Amplifier Power Supply

High-powered output transistors require higher operating voltage than the voltage supplied by the car battery (14.4 volts). A DC-DC converter circuit consists of high-powered and MOSFET transistors to help develop a higher voltage. The 14.4-volt DC battery voltage is fed to the DC-DC converter circuit that connects to four different transistors (**Fig. 3-17**).

The MOSFET transistors feed the signal to a power transformer, and the secondary winding voltage is rectified by silicon diodes D11 to D13. A 14.4-volt DC supply is fed to the center tap of the transformer with each leg connected to Q32 and Q33. D11 provides correct battery polarity. L1 and C22 provide filtering action to the DC-DC

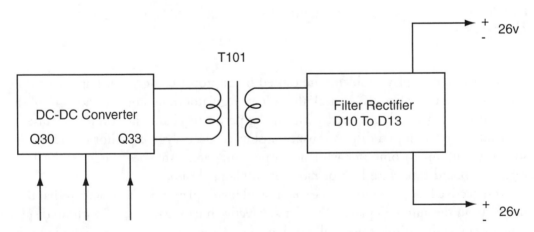

FIGURE 3-17. The battery 14.4-volt DC source is fed to four different DC-to-DC converter transistors.

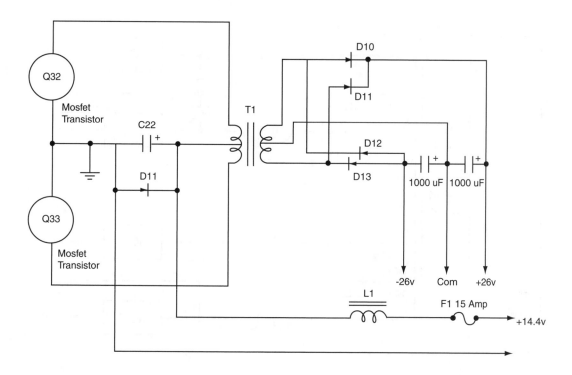

FIGURE 3-18. A positive and negative 26-volt source is fed to the output transistors in a 2X50 watt amplifier.

converter circuits. D10 through D13 rectifies the output voltage (26 volts) applied to the power output transistors in the 2X50 watt amplifier (**Fig. 3-18**).

Like most power supplies, a breakdown of the MOSFET and power transistors occurs within the DC-DC power supply. Check all zener and silicon diodes with the diode test of the DMM or with a semiconductor tester. Suspect C22 for excess hum and oscillations within the DC-DC power supply. A blown fuse can result from a leaky polarity diode (D11) and MOSFET transistors Q32 and Q33. Check for leaky diodes within the base to collector terminals of the output transistors and also within the remote section.

DC-to-DC Power Supply

The DC-to-DC converter is fed from the 14.4-volt battery source of the automobile, and the transistor or IC oscillators provide a pulsating voltage to the transformer T101. The secondary winding might have a half-wave silicon diode or bridge rectifier circuit (**Fig. 3-19**). You might find transistors, zener diode, and IC components as voltage regulators. Several electrolytic capacitors provide filtering action in the DC circuits.

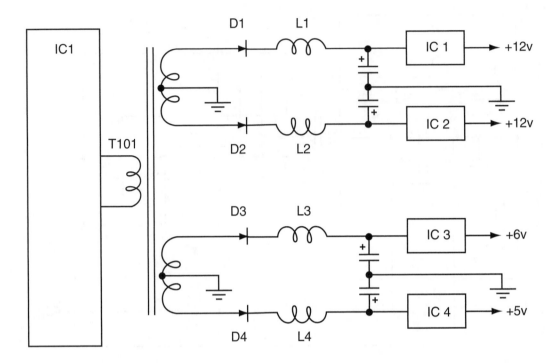

FIGURE 3-19. Four different supply rectifiers and IC regulators are found in some auto CD players.

Check the voltage supply source feeding the audio and headphone amplifiers within the CD player. Suspect a defective transistor or IC regulator with no voltage output. Measure the voltage into the regulator component. Check the DC voltage at the silicon diodes in the secondary winding of the power transformer. Accurate voltage measurements and scope waveforms of the IC or transistor-operated DC-to-DC converter solves most CD and high-powered supply sources.

MOSFET Power Supply

Higher operating voltages are required to produce higher wattage within the 100- to 250-watt stereo amplifiers. The metal-oxide-silicon field-effect transistor (MOSFET) is a field-effect transistor in which the gate electrode is not a PN junction, but a thin metal film insulated from the semiconductor by a thin oxide film. The Gate control element is electrostatic, while the Drain and Source electrodes are PN junctions. Several of these MOSFET transistors are paralleled to provide a higher current voltage found in the auto stereo high-powered amplifier.

Because the battery source in the automobile is only 14.4 volts DC, a DC-to-DC MOSFET power converter circuit was designed to provide higher voltage to the output transistors within the 100- to 250-watt amplifier. A DC voltage of 30 to 35 volts can be generated to operate the high-wattage amp by adding MOSFET transistors. The 14.4-volt battery voltage is fused by a 30-amp fuse and fed into the primary winding of the power transformer T101. The positive and negative 34-volt source might provide voltage to 10 to 14 transistors of only one stereo channel in the high-wattage amplifier.

The PWM signal is fed from U3 to Q401 and Q405, with MOSFET transistors Q409 to Q411 connected in parallel with power transformer T401. Likewise MOSFET transistors Q406 to Q408 are connected in parallel with the other split T401 winding. The DC-DC converter transistors, U3, and T401 provide a positive and negative (30 to 35) volt source. D405 and D406 rectify the output voltage of the secondary winding of T401 (**Fig. 3-20**). A 2200 uF electrolytic capacitor filters the DC voltage source.

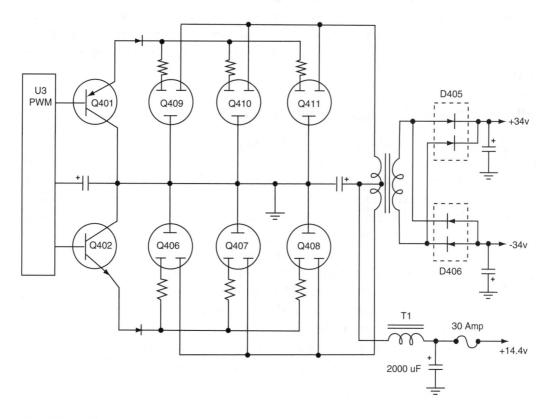

FIGURE 3-20. Q406 to Q411 consist of six Mosfet transistors in a DC-to-DC 170-watt amplifier power supply.

Suspect a blown fuse when the pilot lamp is out and there is no sound from the high-wattage amp. If the fuse keeps blowing, check for a shorted or leaky MOSFET transistor (Q406 to Q411). You might find more than one MOSFET transistor leaky or shorted. Check diodes D405 and D406 with the diode test of the DMM. Notice that two diodes are found in one component for a bridge rectifier circuit. If the dual diodes are not available for replacement, use two single silicon diodes to complete the circuit.

Do not overlook an overloaded power output transistor in the high-powered amplifier channels. Inspect transformer T401 and C406 when a hum is heard in the speakers. Replace C402 and C403 with a 35-volt working voltage and C405 and C406 at 50 volts.

Check for an open circuit or a cold-soldered joint of L1 and T1 for no output voltage. A defective PWM (U3) component can shut down the whole DC-DC converter circuits. Test all fixed silicon and zener diodes in the MOSFET power supply with a diode tester. Double-check each electrolytic capacitor in the power supply circuits for ESR problems with the ESR tester.

Remove the positive output voltage from the power output transistors with overloading of possible shorted power output transistors within the high-powered amplifier circuits. Now take another voltage test of the power source to determine if the power supply is okay.

Auto Radio and Cassette Player Power Supply

The auto radio and cassette player is quite simple to repair with filter capacitors, transistors, or IC regulators. The unregulated power supply source (14.4 volts) is applied to the power output transistors and ICs, AM, FM, and tape circuits. The positive 5-volt regulated voltage source is fed to a PLL control IC circuits (**Fig. 3-21**). The 14.4-volt source is fitted with two separate 1000 uF electrolytics to common ground. D1 provides correct battery polarity applied to the power source. SW1 switches the battery voltage to the main auto radio and cassette player.

The backup power lead is fed through a polarity silicon diode (D2) to a transistor voltage regulator (Q411). Zener diode D3 with Q411 provides a regulated 5-voltage output (VDD). An open voltage regulator (Q411) can cause a really low or no voltage source applied to the PLL control circuits. The leaky or shorted voltage regulator transistor might produce a low output voltage with overheated diodes D2 and D3. Check all electrolytic capacitors in the low-voltage power supply sources with the ESR meter.

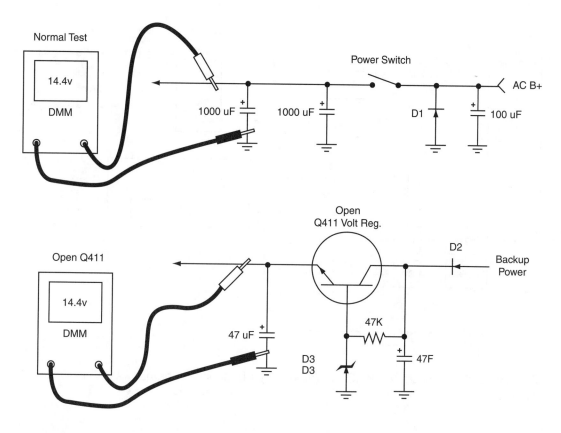

FIGURE 3-21. An open transistor regulator might have really low or no voltage upon the emitter terminal of Q411.

Auto Radio and CD Player Power Supply

Like the high-powered amplifier, the auto AM/FM receiver might contain a DC-DC converter fed from a 14.4-volt battery source. The converter circuits might be generated by an IC oscillator feeding into an output transformer with a rectified secondary voltage source. Full-wave and half-wave silicon diode rectification can be found in the transformer secondary sources. The different voltage sources might have IC, transistor, and transistor-zener diode regulation circuits.

The different voltage sources feeding the many different circuits within the AM/FM CD auto receiver might be a positive and negative 12-, 8-, and 5-volt regulated sources. Besides the power-supply regulated sources, the Disc Sensors might have a separate 5-volt transistor regulator, and the servo IC can have another separate 5-volt regulated source. The positive + 8-volt regulator circuits furnish working voltage to the D/A and audio output circuits. The + 5-volt regulated supply connects to the signal processor, servo,

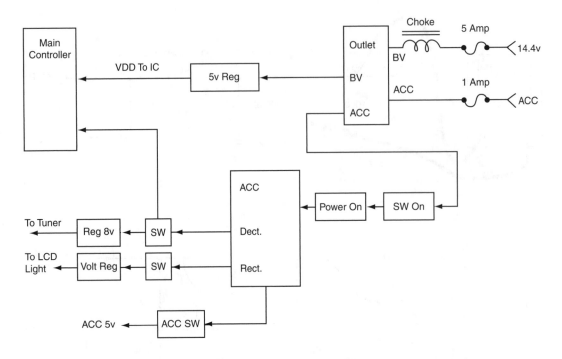

FIGURE 3-22. Transistor and IC regulators might be found in the auto AM/FM/CD stereo player.

and optical circuits. A positive + 10-volt regulated source might provide voltage to the motor driver circuits. You might find a combination of regulated transistors and IC components in one voltage output source (**Fig. 3-22**).

Inspect the different fuse for open or burned conditions. Check each regulated voltage source with a voltage measurement across each filter capacitor within the regulated source. Suspect a defective regulated transistor or IC if the voltage is quite low or has no output voltage. Check the suspected regulated transistor with the diode test of the DMM. Test each electrolytic capacitor within the power supply with the ESR meter.

Disconnect the voltage regulated transistors output terminal from the regulated source to determine if the circuit is overloaded with a leaky or shorted component. Check the output circuits for an overload if the regulated voltage source is fairly normal after removing the suspected loaded terminal. Do not overlook poor connections at the CN terminal connections. Use the low ohm scale of the ohmmeter or ESR meter to check terminal connections.

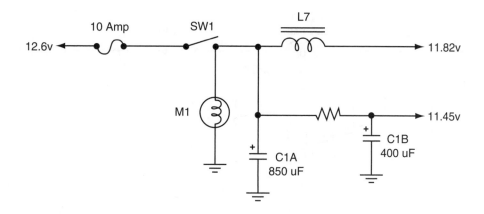

FIGURE 3-23. C1A and C1B were replaced with a dual capacitor 1000 uF and 500 uF.

TROUBLESHOOTING THE AUTO POWER SUPPLY

Inspect and check all power line fuses. Quickly measure the voltage across each filter capacitor terminals in the main power supply. Check the output voltage of the voltage regulator or IC if only a given circuit is defective. Test each regulator transistor and zener diode with the diode test of DMM. Check each electrolytic with the ESR meter. You should be able to quickly locate the defective component within circuit voltage and component tests within the low-voltage power supply.

Oldsmobile Model 93BPB1 Hum Problem

Excessive hum was noticed in an Oldsmobile car radio with the volume control turned down. When electrolytic capacitor C1A was shunted, the hum disappeared. A 1000 uF 35-volt electrolytic was clipped across C1A and common ground with the volume control switched off. The hum noise was no longer heard in the speakers when the switch was turned on. C1A and C1B were replaced with a dual 1000 uF and 500 uF at 25 volts (**Fig. 3-23**).

Bad Alpine 7212 Protection Diode

The 7-amp fuse in an Alpine 7212 auto receiver kept blowing the line fuse with the power switch shut off. In this model, the main tape and radio switch were not supplied with voltage until a relay was energized. A 0.17-ohm measurement was found from the fuse terminal to the metal chassis.

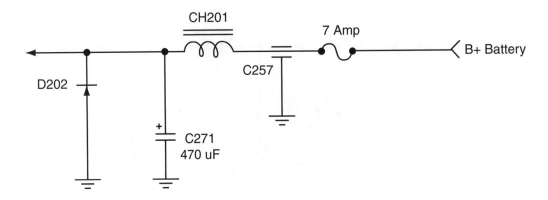

FIGURE 3-24. D202 kept blowing the fuse in an Alpine 7212 auto radio.

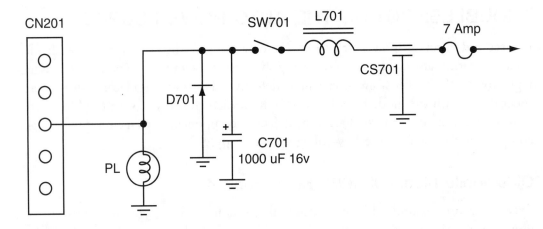

FIGURE 3-25. L201 was replaced in a Panasonic car radio for a low humming noise.

Upon checking the schematic, filter capacitor (C271) or the spark plate (C257) were suspected of possibly shorting out the power lead. Instead, the protection diode D202 was found shorted across the battery source. No doubt, the car battery was charged up backward and shorted out D202 (**Fig. 3-24**).

Panasonic CR7787EU Hum Problem

The Panasonic car radio came in with no sound and a loud humming noise. C701 was suspected of being defective (**Fig. 3-25**). Power output IC251 was replaced and that cured the audio problem. The humming noise was still heard in the speakers after C701 was clipped across with a new 1000 uf electrolytic. At first the power output IC251 replacement was suspected of being defective. After checking all electrolytics in the AF amp IC circuits, perhaps the hum was developed within the power circuits. The wrapping on choke coil L701 had turned brown and a few black spots were noticed. Removing a section of the wrapping showed the coil windings were burned. L701 was replaced with a used choke coil found in a discarded auto receiver, and one chassis hole was redrilled for mounting. Replacing L701 cured the low hum problem in the speakers.

No AM/FM Reception

A loss of AM or FM reception was noticed in a Radio Shack digital synthesized AM/FM stereo cassette player. The line fuse was normal. The display seemed to light up and the cassette player worked okay. A close check revealed that very little voltage was measured at the AM/FM switch. Another voltage measurement at the voltage regulator (Q476) was found upon the collector terminal and very little voltage at the emitter terminal. Replacing the open voltage regulator (Q476) solved the problem of having no AM/FM reception.

No Audio Output — Pioneer CDX-FM45

A voltage measurement on the preamp power audio ICs from the Pioneer stereo changer was found to be really low. Tracing the voltage back to the power-supply circuits revealed the + 8-volt regulator transistor was leaky. Replacing Q709 solved the problem of having no stereo audio output.

CHAPTER 4

Servicing the Front-End Circuits

The early auto front-end tube circuits consisted of an RF amplifier, oscillator, or converter stage. When semiconductors were first introduced, transistors were found only in the audio output circuits of the car radio. Later on, transistors were found in the audio AF, driver, and output stages. Finally, the RF, converter, and IF circuits consisted of separate transistors in each circuit. Today, the integrated circuits might contain the RF, oscillator, mixer, detector, and matrix circuits in one large IC component.

The early front-end transistor circuits contained permeability tuning coils, where an iron slug was pushed in and out of each coil to tune in a broadcast station. A ganged permeability coil assembly tuned the RF, oscillator, or converter stage to an intermediate frequency (IF) of 262.5 kHz. Several IF stages amplified the radio signal, which was rectified or detected by the detector circuit (**Fig. 4-1**).

The last IF amp tube circuit consisted of the diode element inside the same vacuum-tube envelope. Later on, the detector circuit of the AM band separated the audio and passed it on to the first audio frequency (AF) stage. A fixed germanium diode, which detected the audio signal from the final IF circuit, replaced the diode tube.

When the radio dial was rotated to a local radio station at 1400 kHz, the oscillator frequency was tuned to a frequency less than the 262.5 kHz IF stage. Later on, the IF frequency was changed to 455 kHz. The AM band of any radio is tuned to an amplitude modulation (AM) frequency from 550 to 1700 kHz. Amplitude modulation is a method of modulation in which the amplitude of the carrier voltage is varied in proportion to the changing frequency value of an applied voltage, while the carrier frequency remaining is not altered in the process.

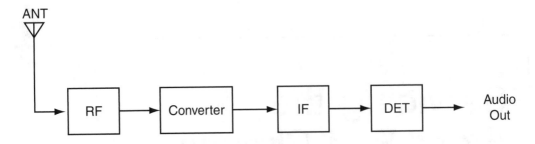

FIGURE 4-1. A block diagram of the AM RF, converter, IF, and detector stages in the auto radio.

The AM tuner that includes the RF, oscillator, and IF circuits in the auto receiver handles the amplitude modulation signals and supplies its low-amplitude audio output to a high-fidelity amplifier. The tuner circuits of an RF amplifier amplifies the AM band and is coupled to the oscillator or converter and IF circuits. The oscillator circuit selects the frequency of the broadcast station and supplies the different RF and converter signal to the intermediate frequency circuits.

A mixer or converter stage mixes the oscillator signal with the incoming RF picked-up signal from the auto antenna and supplies an IF frequency to the intermediate frequency transformer circuits. The IF circuits might consist of two or three IF stages that amplify the IF 455 kHz frequency. In the early auto radios, the IF circuits were tuned to 262.5 kHz frequency. Today, the PLL circuitry and IF frequency is 450 to 455 kHz.

THE TRANSISTORIZED RF STAGE

The AM radio signal is picked up by the outside auto antenna dipole and fed to a permeability-tuning coil through L1. Trimmer capacitor C1 is adjusted to tune the antenna to the car radio input circuit. C1 is adjusted for maximum reception with a weak radio signal tuned in around 1400 kHz. Permeability tuning coil L2 tunes in the RF radio station and is ganged together with the converter and oscillator permeability coils. The RF signal is transferred to the secondary winding of L2 and applied to the base of a PNP RF amp transistor (**Fig. 4-2**).

RF coil (L2) is tuned by a trimmer capacitor (C2) at the high end of the band, around 1600 kHz. Most permeability coils can be touched up by twisting the iron slug on the tuning bar for maximum reception. Usually this type of adjustment is made when one of the iron slugs is broken and needs replacement. The amplified AM RF signal by Q1 is capacity-coupled through coil L2 to the base terminal of the converter transistor (Q2).

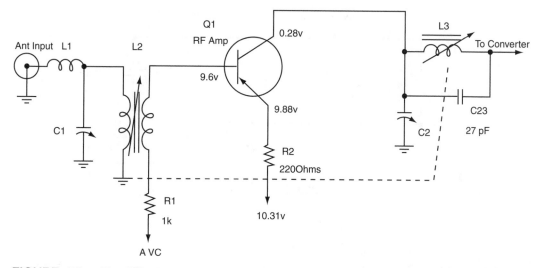

FIGURE 4-2. The RF amplifier stage connected to the antenna receptacle and to the AM converter circuits.

Most service problems found in the RF circuits include the loss of station reception, a weak signal, and only picking up a local station. The broken wire off of the RF antenna coil (L1) or from the antenna plug can cause a loud rushing noise with loss of AM reception. A broken wire on the AM tuning coil can cause loss of AM reception. A broken RF coil wire or a broken metal tuning slug can cause a radio to tune in only a really weak local radio station. Loss of AM reception can result from a leaky or open RF transistor. Loss of AM reception is found with an open RF transistor between base and emitter elements. Clean up the radio/tape switch for loss of AM reception.

Suspect a bad RF transistor with a weak reception or with only one local station tuned in. When making voltage tests on the base of the RF transistor and a local station pops in, suspect a defective RF transistor. A bad RF transistor can cause a loss of audio or of stations tuned in. A weak AM station can be caused by a poor antenna connection and bad IF stage. Suspect a leaky decoupling electrolytic for a weak local AM station. A defective voltage regulator transistor can cause weak AM and FM reception.

A leaky or intermittent junction within the RF transistor can cause intermittent sound. The local AM radio station that comes and goes can be caused by a badly soldered connection on the emitter terminal of the RF transistor. Spray coolant on the RF transistor if the AM band cuts in and out after operating for two or more hours. Sometimes, pushing up and down on the PCB close to the RF transistor can make the badly soldered joint act up.

Intermittent AM reception can result from a low DC supply voltage caused by a defective voltage regulator transistor or zener diode (**Fig. 4-3**). Suspect a dirty AM/FM switch for intermittent AM reception. Intermittent AM reception can be caused by a

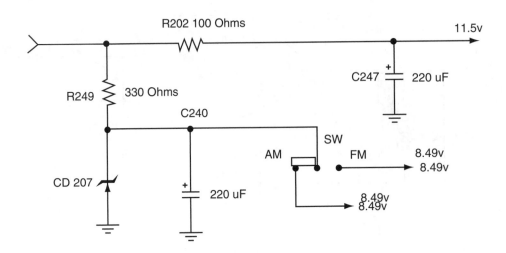

FIGURE 4-3. Zener diode CD207 caused intermittent reception in the Audiovox C-977 AM voltage source.

defective bypass capacitor off of the base of the first RF transistor. Suspect a 0.047-uF capacitor off of the emitter terminal of the RF transistor for weak volume and intermittent AM reception.

No AM Reception—FM Normal

Resolder the AM antenna input connector for no AM reception. Suspect poor contacts on the AM antenna coil for no AM and normal FM. A leaky RF transistor between base and emitter terminals can cause no AM and normal FM. Suspect no AM and normal FM reception when the AM RF transistor breaks down under load. A dirty AM/FM radio switch can produce intermittent AM with normal FM reception. A dirty band selector switch can cause no AM with normal FM reception. Loss of AM reception can result from a bad board connection on the AM detector diode. A microphonic RF amp transistor can cause microphonic AM reception.

Kraco KID575C—No AM Reception

Loss of AM with normal FM reception was noticed in a Kraco car radio. Only a local AM station could barely be heard with the volume control wide open. When taking voltage measurements on the base of the RF transistor, the local station came in much louder. The DMM diode test indicated leakage between base and emitter terminals with in-circuit tests. Because there was no schematic handy, the voltage measurements indicated a higher voltage on the collector terminal, and the emitter terminal was

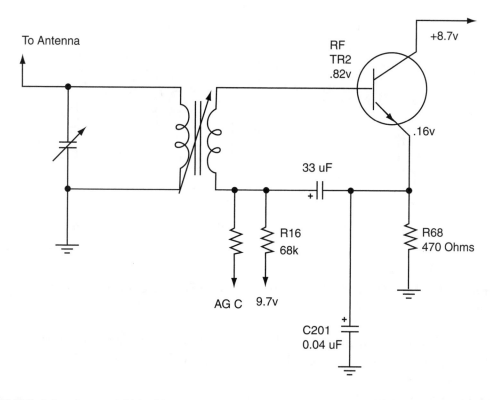

FIGURE 4-4. Loss of AM with normal FM was found in a Kraco KID575C with a open TR2 RF transistor.

connected to ground through a 470 ohm resistor. The 2SC710 RF NPN transistor can be replaced with a universal GE-39 or an SK3018 transistor replacement (**Fig. 4-4**).

TODAY'S AM RF CIRCUITS

Today the AM RF transistor might contain an FET RF transistor in both the AM and FM circuits. FET is an abbreviation for *field-effect transistor*. The FET has high-input impedance, somewhat like the vacuum tube. The FET has a Gate terminal (G) that has high impedance to the incoming RF signal. A Drain (D) terminal is comparable to the transistor's collector terminal with a higher positive voltage. The Source (S) terminal is at ground potential and is comparable to the transistor emitter terminal.

The AM RF signal is coupled by a small coil from the antenna input terminal to a tuned input coil L200. L200 is a capacity coupled to the gate terminal of the RF FET transistor Q201. The RF station signal is amplified by Q201 and has a tuned inductance coil in the drain circuit of the RF FET. The secondary winding of L202 inductively couples the

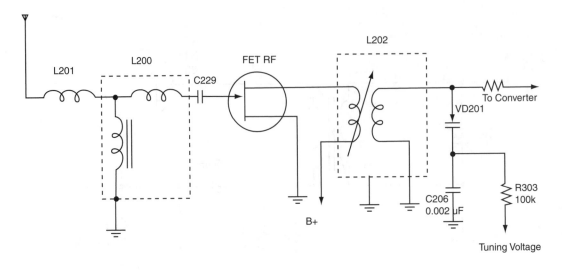

FIGURE 4-5. Varactor diode VD201 tunes in the RF FET transistor in the RF stage.

amplified RF signal to the gate circuit of another FET converter transistor. Notice that the secondary winding of L202 is tuned by a varactor diode VD201 (**Fig. 4-5**).

AM RF IC Circuits

The conventional RF IC circuits are incorporated inside one large front-end IC3 with the AM RF/OSC/IF circuits. The broadcast signal picked up by the antenna is coupled through coil L5 to the antenna input receptacle. CT-51 trimmer capacitor tunes the antenna to the input circuit of the car radio circuits. Simply tune in a weak station to around 1400 kHz and adjust CT-51 until the signal is the loudest in the speaker.

The picked-up signal is then capacity coupled with C52 to pin 1 of the AM RF IC3 circuits. The RF signal is tuned by the RF coil, attached to pin terminals 15 and 16 of IC3. Adjust CT52 for maximum with a weak station tuned in around 1400 kHz. The RF AM signal is amplified inside IC3 and mixed with the oscillator frequency to produce a 455 kHz output frequency to the IF circuits (**Fig. 4-6**).

AM RF Stereo-Cassette Circuits

The AM RF circuits in the auto AM/FM stereo cassette player might have a combination of RF silicon transistor and FET RF transistor fed to the input of an AM/IF/DET IC701. Both the AM and FM RF signals are picked up by an outside antenna dipole and fed to the FM tuner front-end TU101 and through an antenna loading coil to the AM RF stage (AM pack TU102). The amplified AM RF transistor amplifies the

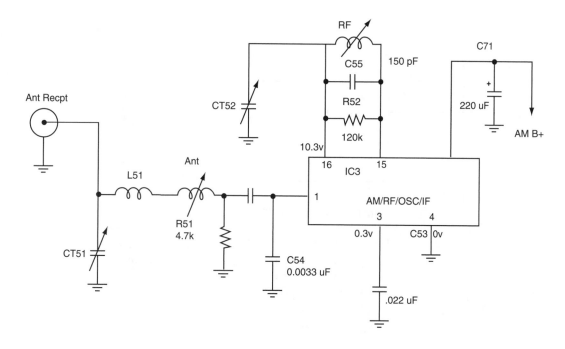

FIGURE 4-6. The tuned picked-up radio signal is mixed in IC3 to produce a 455 kHz IF output signal.

RF signal and applies this AM signal to pin 3 of the AM IC that includes a mixer, VCO, IF detector, and to the selector circuits (**Fig. 4-7**).

The RF amplifier transistor signal is tuned with varactor diodes and is powered by the PLL control circuits. The varactor diode is a special type of diode with a PN junction that has a certain internal capacitance. By varying the reverse-bias voltage, the diode acts as a voltage variable capacitor. The varactor diode takes the place of a variable capacitor in the very early auto radio chassis and the permeability tuner in the later auto receiver.

Typical AM RF Circuits in the Compact Disc Player

The RF AM and FM signal is picked up by the antenna and coupled to pins 1 and 2 of the tuning unit (TUN011). The tuning unit amplifies the AM and FM signal, contains the AM and FM oscillator circuits, mixer, IF, and the audio is coupled to the output circuits. A PLL IC provides signal to the LPF network and to pin 6 (VT) of TUN011. The AM/FM tuning control circuits go to the band switch and on to pins 8 and 9 of the tuning unit.

A B+ voltage from the PLL circuits connects to a tuner switch circuit and the supply voltage terminal (Vcc) of TUN011. The audio output is detected inside TUN011 and is coupled out of terminals 18 and 19 (**Fig. 4-8**). You might find digital transistors and IC components within the RF AM and FM front-end circuits.

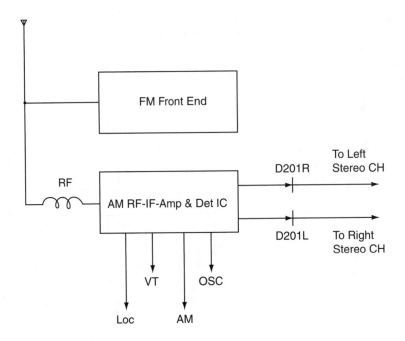

FIGURE 4-7. A block diagram of an AM RF/IF/DET IC701 with a right and left D201R and D201L audio stereo signal.

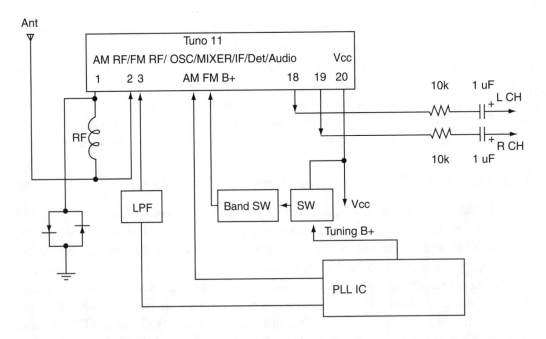

FIGURE 4-8. Block diagram of the RF AM/FM oscillator, IF, detector, and audio output of a tuning front-end from a radio-CD player.

The Converter or Oscillator Circuits

A converter circuit accepts the RF signal and mixes the incoming signal with the local oscillator stage. In the early transistor converter circuits, the tuned RF signal is coupled through a capacitor to the base terminal of the converter transistor. The local oscillator circuit occurs in the emitter circuit of the converter transistor. Both the RF and oscillator signals are mixed or connected to an early IF frequency of 262.5 kHz.

The converter stage is a heterodyne mixer in which two input signals of different frequencies are mixed to yield a third output signal of a different IF frequency. The intermediate frequency (IF) in the early transistor auto radio was 262.5 kHz and in the later transistor and IC IF circuits at 455 kHz. The local oscillator signal is at a lower frequency than the incoming RF signal. RF permeability coil (L3) can be adjusted for the 1000 kHz band, while IF coil (L5) is adjusted to a 262.5 kHz IF frequency (**Fig. 4-9**).

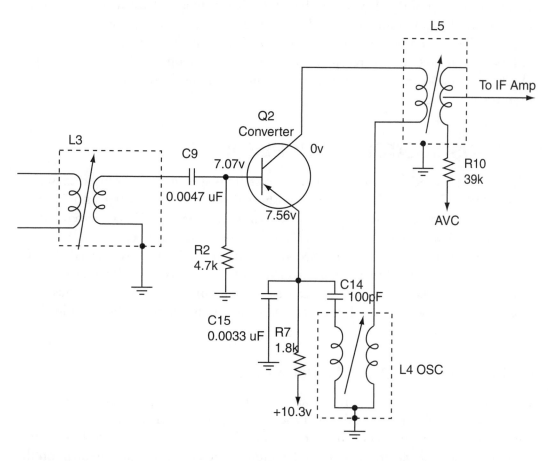

FIGURE 4-9. The IF frequency at the collector terminal of Q2 provides an IF 262 kHz IF signal to transformer L5.

The no AM symptom with normal FM reception can be caused by a badly soldered connection on the oscillator coil. No AM reception can result from a bad converter transistor with low bias voltage (0.20 volts) and should be 0.6 volts between the emitter and base transistor terminals. An intermittent converter transistor can cause intermittent AM reception. Replace both the RF and converter transistors for no AM reception. A dead AM auto radio can be caused by a defective AM converter transistor and IF front-end assembly.

Suspect a converter transistor when the AM stations drift off frequency. Check for water or moisture inside the oscillator trimmer capacitor for a shift in frequency. Adjust the oscillator frequency at the same point on the dial with a signal generator or with a known local radio station tuned in. A bad oscillator padder capacitor can cause intermittent AM reception. Check for a broken metal slug in the permeability oscillator coil with weak stations and with some stations shifting on the dial assembly.

Check for an open intermittent bypass capacitor in the converter circuits when the stations at the top of the dial are not heard. Only a couple of local radio stations can be heard with an intermittent RF or IF transistor. Suspect a broken or dirty AM/FM radio switch for intermittent AM reception. Spray cleaning fluid down inside the switch area. No or intermittent AM reception can be caused by an open IST IF AM transistor. Check for a bad contact or broken wire connection inside the IF transformer for intermittent AM reception.

IC RF-OSCILLATOR IF PROBLEMS

A loss of AM reception with normal audio cassette player operation can be caused by a bad AM/OSC IC. Check for an open or shorted voltage regulator transistor with no AM reception. Measure the supply voltage terminal of the AM RF, converter, and IF IC for no AM reception. With the volume control turned wide open and a rushing noise heard in the speaker with no AM reception, suspect a defective AM/RF/IF IC. Replace the AM RF/IF/Det IC with no AM control of tuned-in stations. A defective voltage regulator transistor or IC can cause a hissing noise in the AM band. Suspect a leaky IC regulator when the volume is turned up and only a hum can be heard in either the AM or FM band.

No Music From a Delco Model 70HPB1

The General Motors Delco car radio came in with no radio reception. A loud hum was heard when the volume control was turned wide open and when a screwdriver blade touched the center terminal of the volume control. No RF signal could be heard when the signal generator was attached to the RF input. A quick voltage measurement on the supply pin of the RF AM oscillator IC (DM32) was very low, indicating a possible leaky IC or improper voltage source. When the supply pin of the DM32 IC was removed from the PC board, the normal supply voltage returned. The leaky DM32 IC component was replaced with an ECG744 universal replacement IC.

Types of Tuning Permeability Tuning

The permeability tuning assembly consisted of the RF, oscillator, and converter tuning coils, with powered iron core slugs or metal cores. All three coils were tuned at the same time with the metal plungers entering each coil and tuning to a separate frequency. Some of the coils were enclosed in a metal cylinder, while others were mounted openly on a ganged tuning assembly (**Fig. 4-10**). The RF coil picked up the radio signal and tuned to the station that was tuned in. A second permeability coil served as a load on the collector terminal of the RF transistor and was tuned to the incoming signal, while the third coil was tuned to the oscillator frequency.

Later on, permeability tuning took the place of a three-ganged tuning capacitor. A variation of the frequency of an LC circuit can be changed with the position of a magnetic core within the inductor or coil. A permeability transistor or IC oscillator can vary the frequency by permeability tuning (moving a magnetic core in or out of the coil or LC circuits). When an RF metal core is broken inside the coil, the tuning can become weak or can change the inductance of the coil. The metal core inside an oscillator or converter coil might change the frequency of the radio and can drift off the station when the car hits a bump in the road.

The permeability assembly can be tuned in with a rotating gear assembly or with a dial cord-pulley arrangement. A dial pointer indicates the station location on the front dial and is fastened to the dial cord that slides along the dial assembly. Most problems with a mechanical gear assembly are improper meshing of gears and a dry movement of the gang assembly. A drop of 3-in-1 Oil or light grease can prevent the dial from sticking or sliding on the dial assembly. Broken dial cords can easily be replaced with a new dial cord or a piece of nylon fishing string.

FIGURE 4-10. Notice the three permeability tuned stages rotated by the tuning knob of an auto-cassette player.

Varactor Diode Tuning

Varactor tuning is found in many of the new auto receivers and CD players. A varactor diode is a semiconductor component voltage-variable capacitor; it is a special diode with a PN junction that has an internal capacitance. The varactor diode might be called a tuning diode. The varactor diode tuning is a method of changing the frequency of the RF and oscillator circuits in the auto radio. Varactor diodes are found in both AM and FM tuning methods.

Varactor diodes are used in radio circuits to tune a series or paralleled inductance of the RF and oscillator coils. The varactor diode looks like a small general-purpose diode, or in some cases, like a two-legged transistor. These diodes have cathode (K) and anode (A) terminals. Often, the anode terminal is at ground or negative potential and the cathode operates with a controlled voltage (**Fig. 4-11**).

Check the suspected varactor diode with the diode test of the DMM. Place the red probe of the DMM to the anode terminal and the black probe to the cathode terminal of the varactor diode. Like any fixed diode, the varactor diode will have a low-ohm resistance (.599 ohms) in one direction and infinite measurement with reversed test probes.

The varactor diode resistance measurement is a little higher in total ohms than the silicon diode with the DMM diode test. Replace the varactor diode when a low-ohm resistance measurement is noted in both directions. Suspect that the diode is defective when the voltage on the diode changes and no tuning is noted in the radio.

In **Figure 4-11**, varactor diode D702 has a variable tuning voltage applied from the PLL control IC, and tunes the RF coil circuits of T702. C702 prevents the DC voltage

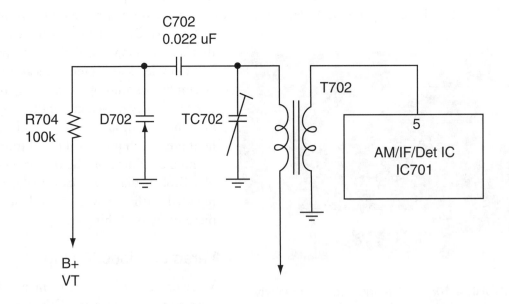

FIGURE 4-11. Varactor diode D702 tunes coil TC702 in the AM/IF/DET IC701 circuits.

from being grounded out through the primary winding of T702 and couples the varying capacity of the varactor diode in series with the primary inductance. When the voltage is varied at the collector terminal of the varactor diode, the change of capacity changes the frequency on T702.

PLL Synthesize Tuning

The digital phase-locked loop (PLL) circuitry is used to synthesize the AM and FM local oscillator frequencies in the auto receiver front-end circuits. In the AM operation mode, the PLL circuitry consists of IC610, reference oscillator, a crystal (4.5 MHz), and transistors Q613 and Q614 as an LPF (low-pass filter) circuit. The 4.5 MHz crystal is used as a frequency reference and is connected between pins 24 and 25 of microprocessor IC611 (**Fig. 4-12**).

The received broadcast signal is picked up by the auto antenna dipole, amplified by the AM RF amp, and fed to pin 3 of the VCO, mixer, IF, and detector IC610. The incoming signal and VCO oscillator signal are mixed by a mixer circuit and coupled

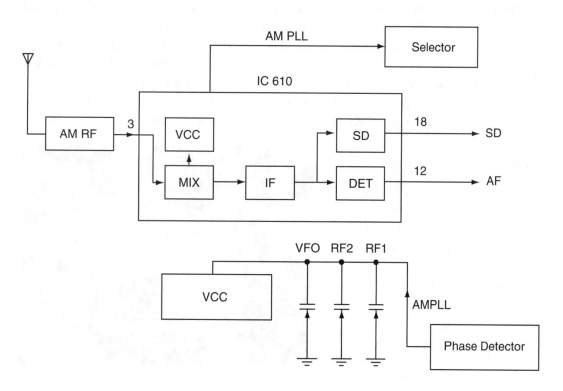

FIGURE 4-12. A block diagram of the AM PLL circuitry in the front-end circuits of a PLL Synthesize tuning microprocessor.

to the IF circuits. Here the intermediate frequency signal is rectified or detected by the detector stage with the analog or audio signal (AF) at pin 12 of IC610.

A 4.5-MHz crystal is used as a reference frequency and is connected between pins 24 and 25 of the microprocessor IC611. The code determines the "N" factor, the driver that produces the required frequency for each AM station, 10 kHz apart for the AM stations.

For example, when receiving the AM station at 1000 kHz, the VCO generates a 1450 kHz signal (1000+450 IF kHz), while the 1000 kHz is the received broadcast station and the 450 kHz is the intermediate frequency (IF). This 1450 kHz signal is applied to pin 5 of the PLL and microprocessor IC611 and divided by N=145 of IC611. The resulting output should be 10 kHz. The reference oscillator (4.5 kHz) frequency is divided by 450, resulting in another 10 kHz frequency.

The two 10 kHz signals are fed to the phase detector inside IC611. The phase detector generates an error voltage that is in proportion to the phase difference between the two 10 kHz signals. The error voltage is at pin 2 of IC611 and passes through the low-pass filter (LPF) network, where the error voltage is integrated, and the low-pass filter network filters out the harmonics and noise. The resulting DC voltage is applied to the varicap (varactor) diode (part of VCO), whose capacity varies with the supplied DC voltage. Now the output frequency of the VCO is corrected.

The VCO circuits are very accurate and precise in tuning the AM band. When the system is "locked in," the main phase detector senses no phase difference, and it generates a frequency that is accurate as the reference crystal oscillator. The LPF circuitry applies the DC voltage to the VCO circuits, which is converted from the output signal (10 kHz for AM) of the phase detector that is located inside IC611.

Push-Button Tuning

The early push-button tuning was a mechanical type button tuning assembly in which the tuning stopped on a selected station when the large push button was engaged. A normal setting of a set screw engaged the correct setting on the tuning dial (**Fig. 4-13**). Each station that was selected started adjustment at one end of the dial and proceeded to the other end of the dial. Most mechanical push buttons were difficult to push in and

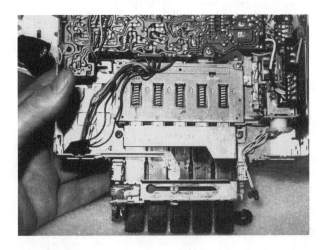

FIGURE 4-13. A manual push-button tuning assembly found in the early push-button car radio.

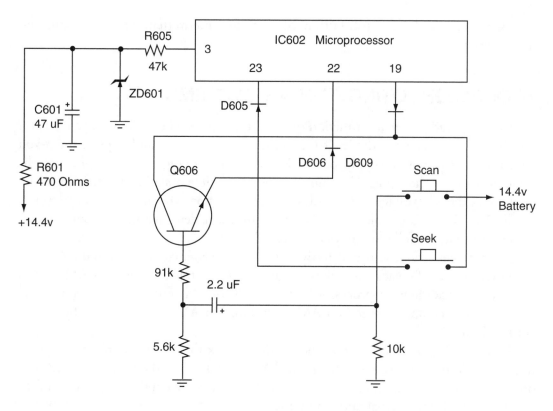

FIGURE 4-14. Suspect a bad zener diode ZD601, IC602, and Q606 when the scan/search system does not operate.

would easily change the manual setting of the station. Placing light grease on the sliding parts seemed to help when the lever became dry and sometimes would stick in one position.

Most of the electronic tuning is done with a microprocessor circuit and a matrix key operation. The key matrix system might consist of tactile switches. The different memory key, AM/FM band, manual up-and-down key, auto memory prescan key, local seek, and radio stereo/mono key work with transistors and diodes to select the various functions. The diodes are used for isolating certain signals from other signals and the different transistors are used as transistor switches. The various key push buttons select programs or stations and trigger the different selections with the large microprocessor.

Very few problems are found within the electronic process tuning system. Check for poor contact in the memory set key for a defective memory setting. Suspect a bad LED when the memory light does not light up. In **Figure 4-14,** when the scan/search does not operate in either AM or FM, suspect zener diode ZD601, IC602, and Q606.

Check the scan-and-seek push-button contacts for intermittent or erratic scan and select operations.

TROUBLESHOOTING TUNING SYSTEMS

When the car radio does not function or tune in a station, check the voltage regulator transistor or IC and zener diode. If the tuning presets do not hold, replace the 3-volt regulator on the power supply board. Measure the varying voltage applied to the varactor diode. If the voltage does not change as a different station is being tuned in, then check the VT voltage at the PLL or microprocessor. Measure the voltage supply terminal of the PLL or microprocessor IC. Suspect a defective power source or IC component with a really low power source (Vcc).

For no tuner action, suspect defective DC-DC converter circuits within the AM/FM stereo CD player. Replace the voltage regulator IC or transistor for no AM or FM reception. Check the microprocessor IC for no AM or FM control. Suspect a defective controller and components for no AM reception. No AM can be caused by a poor contact in the band selector.

Scope the VCO crystal for no tuner action. Check the voltage source going to the front-end tuning unit. Replace the jammed M1 Switch when the tuner is locked only on one AM station. Determine if the audio is present at either audio output pins of the tuner pack with an external audio amplifier.

Automatic Stop for Auto Tuning

A simple auto-stop tuning circuit is shown in **Figure 4-15**. When a signal is detected while in the auto-stop mode, a high-level signal is output from pin 17 of IC601 in the AM band. At the same time, a low-level signal is output at pin 5 from IC501 in the FM mode. Thus, the signal passes through R302, Q302, D302, and R611 to the microprocessor at pin 63 of IC611, where it becomes high. Now, seek-tuning is stopped and a radio station is received. The preset radio station keys are mounted on the front panel of the radio.

Suspect a dirty search switch contact when the auto seek does not operate. Check for a dirty switch contact on the auto seek/manual switch. A loss of scan or search can result from a defective AM tuner, shorted, or open front-end voltage regulator transistor and a 5-volt zener diode regulator. A defective front-end regulator can cause loss of scan search operation. A defective D302 or Q302 transistor can cause loss of auto seek signal to pass through to the microprocessor IC.

Low or no VT voltage on the tuner board can cause a loss of seek and auto stop. Check for correct data on the different pins 62, 63, and 64 of microprocessor IC611. For loss of auto stop and auto reverse functions, do not overlook a defective microprocessor

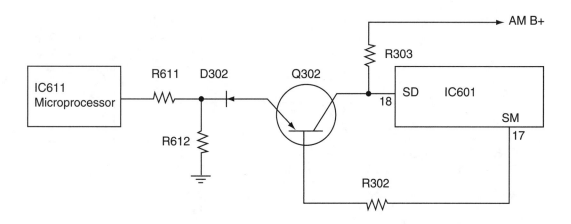

FIGURE 4-15. The AM automatic-stop circuit functions when the AM stations are tuned in and fed to the microprocessor IC601.

(IC611) or the IC in the main unit. Check for a poor contact in the auto seek search switch when the auto seek does not function.

Suspect a bad switch contact on the manual switch when the manual tuner does not work up or down. Clean up the dirty contacts on the manual up/down key for loss of manual tuner action. A poor contact in the auto memory set key can cause loss of memory setting operation.

Intermittent AM — Oldsmobile 93EPB1

The AM reception would come and go in an Oldsmobile auto radio. Sometimes the radio would operate for several hours before the local 1400 station would cut out. When the AM band cut out, the full 11.45 volts were found on the converter transistor Q2. Other times, the station would start up when the voltage probe was touched on the base terminal. The suspected Q2 was removed from the PCB and an SK3006 transistor was installed in its place. Replacing Q2 solved the intermittent AM radio reception (**Fig. 4-16**).

TYPES OF SWITCHING

Mechanical Switching

In the early auto radio-cassette player, a large mechanical switch was required to switch the radio and cassette audio functions in and out. Sometimes a large sliding switch was used, or a rotating switching arrangement occurred with several different contacts on

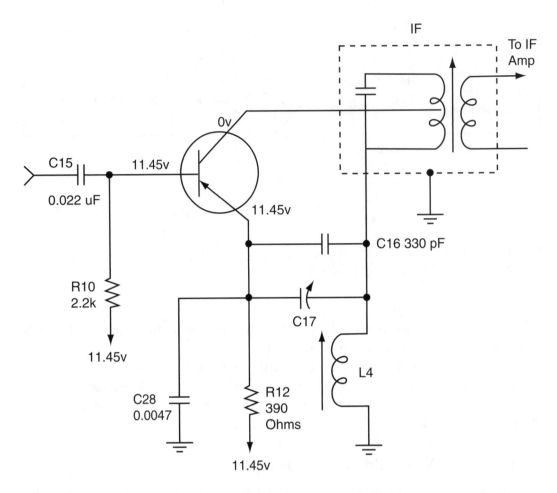

FIGURE 4-16. Converter transistor Q2 caused intermittent reception in an Oldsmobile 93EPB1 car radio.

wafer-type board terminals. In later car radio models, when the audio cassette was inserted into the cassette player, the supply voltage to the radio was cut out and then switched to the tape head and motor circuits. At the same time, a sliding switch moved the audio from the AM/FM stereo audio signal to the cassette amp circuits.

The tape head and preamp audio circuits were mechanically switched into the AF amp audio circuits, and the AM/FM stereo radio signal was switched out of the audio circuits within the transistor and IC audio circuits. When in radio operation, the stereo audio circuits are switched into the AF stereo circuits and the tape-head preamp circuits are switched out, as the cassette is removed from the radio-cassette player (**Fig.** 4-17).

The left and right tape head signal is capacity coupled into pins 1 and 8, while the amplified tape head music is fed out of pins 3 and 6 of the preamp IC. The amplified

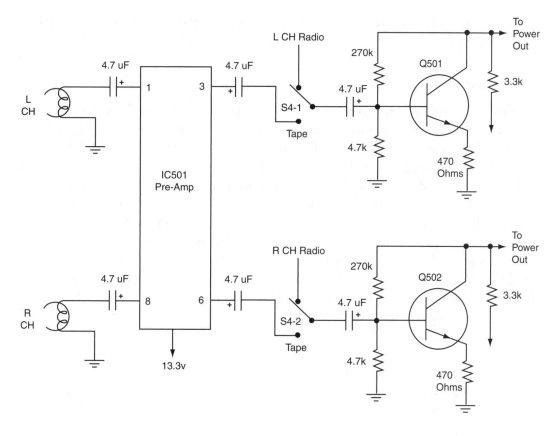

FIGURE 4-17. S4-1 and S4-2 provide mechanical switching of the tape or radio circuits to the AF amp transistors within the audio circuits.

music is then fed through a 4.7 uF electrolytic to a tape/radio switch S4-1 and S4-2. Again, the radio or tape signal is capacity coupled to the base circuits of the AF amp transistors (Q501 and Q502). S4-1 and S4-2 switching is engaged when the cassette is inserted into the radio-cassette player. Likewise, the radio circuits are switched in when the cassette is out of the auto radio player.

Because most switching contacts contain a silver-type switching connection, the contacts become tarnished and dirty. A dirty tape/radio switch can result in loss of tape or radio reception. The dirty or poor contacts within the switching of many contacts might produce a rushing or erratic operation. Sometimes a weak reception and hum can be heard in the speakers. A bad switch might cause a dead/no sound symptom. Garbled and distorted audio might result from a dirty radio-cassette switch.

Clean up the radio-cassette switch contacts with a silicone head cleaning fluid. Place the plastic tube from the spray can down inside the switching area. Work the switch back and forth to help clean up the contacts. Check for poorly soldered contacts if the switch is still erratic. Resolder all switching contacts on the PCB. The defective radio/tape switch with worn or broken contacts should be replaced.

Diode Switching

Four different fixed silicon diodes are found in switching the audio from the radio and the tape heads into the audio input circuits. When the cassette is out of the tape player, a tape/radio switch activates the AM/FM stereo radio. The stereo music is fed from the matrix IC to the input of the volume control in both left and right channels by D403 and D404. The radio signal is blocked out from going into the tape preamp IC401 circuits by D401 and D402. Thus, the radio signal is directed to both 4.7 uF electrolytic coupling capacitors (C401 and C402) to the left and right volume controls.

When the cassette is inserted into the auto-cassette player, the tape/radio switch is engaged and power is applied to the tape motor circuits. The stereo tape music is amplified by IC401 and fed to D401 and D402 (**Fig. 4-18**). Here the music is fed directly to C401 and C402 to the audio volume control circuits. D403 and D404 will not let the audio signal feed back into the radio circuits due to the polarity of the fixed

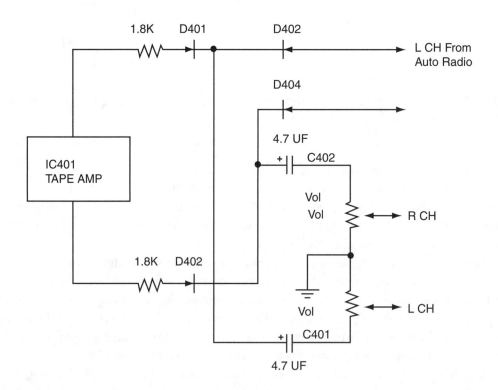

FIGURE 4-18. Silent fixed silicon diodes provide tape and radio switching without any mechanical switch involved.

diodes. Without any type of mechanical switching or with dirty contacts, the tape and radio signals are silently switched into the audio input circuits.

Radio-Tape-CD Switching

A separate AM radio and FM radio push-button switch is engaged to turn on either the AM or FM reception within the stereo CD player. Again, silent diode switching is accomplished with separate AM and FM radio stages. When battery power is applied to the FM radio circuits, the FM stereo signal is fed out of pin 5 of IC102 to the right channel fixed diodes D103R and D105L. The FM radio signal is fed directly from these diodes to C501 and C502, and to the audio preamp IC501. D301R and D305L block the FM radio stereo signal from entering the tape circuits. D201R and D201L block the FM radio signal from entering the AM radio circuits, routing the FM radio stereo signal directly to the audio preamp IC501.

The AM radio signal is fed from pin 6 of IC701 to the fixed diodes D201L and D201R. Both of these diodes are connected directly to C501 and C502, feeding the AM signal into the audio preamp IC501. D103R and D105L block the AM radio signal from entering the FM matrix circuits of IC102. Also, D301R and D501L block the AM radio signal from entering the tape head circuits (**Fig. 4-19**).

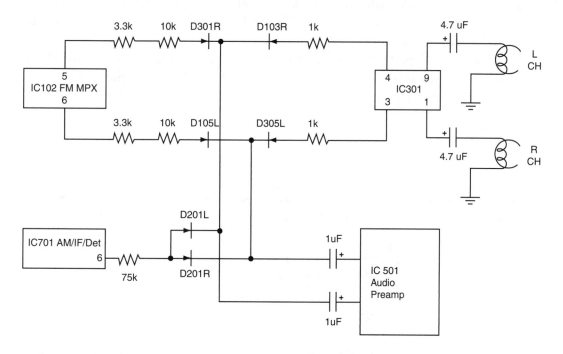

FIGURE 4-19.   Diodes D103R and D105L provide FM stereo switching to the input of the audio preamp IC501 stage.

When a stereo cassette is inserted into the radio-cassette player, the tape heads pick up the music from the tape, amplified by the tape-head preamp IC301. The amplified cassette music is fed out of pins 3 and 4 of IC301 to fixed diodes D301R and D305L. In turn, these diodes are connected directly to coupling capacitors C501 and C502 (1 uF) to the input preamp audio IC501 and output circuits. D201L and D201R block the tape music from entering the AM radio circuits. D103R and D105L block the tape music from entering the FM matrix stereo circuits, leaving the tape music to enter the preamp audio output circuits without any type of mechanical switching.

Check each diode with the diode test of the DMM, when crosstalk or interference is noted in the speakers. A dead stereo channel might result from a leaky or shorted silicon diode. A leaky diode might have a weak and distorted signal, while the other stereo channel might be normal. Of course, an open silicon diode can produce a dead stereo channel with a slight hum in the dead channel.

DECOUPLING VOLTAGE SOURCES

The decoupling voltage sources are found in the low-voltage circuits separating the AM and FM stereo circuits. A small resistor and a decoupling electrolytic capacitor might separate the FM and AM voltage sources within the battery voltage supply source. In **Figure 4-20**, S8 turns the radio-cassette player on and off, and S4-4 switches the battery voltage to the radio or tape functions. When S4-4 is switched to the radio circuits, R902 and C902 provide a decoupling voltage circuit.

R901 and C901 perform another voltage decoupling voltage circuit with D901 as a zener diode that provides a regulated voltage of 9.1 volts. Here, the 9-volt source is fed to either the AM or FM circuits by switch S5-2. C901, C902, and C918

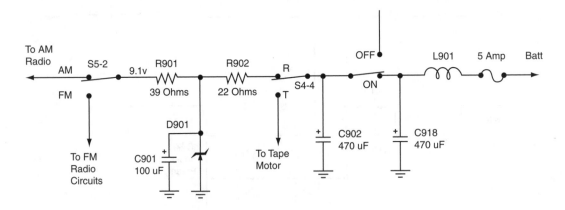

FIGURE 4-20. C901, C902, R901, and R912 provide decoupling voltage sources in the AM/FM radio circuits.

provide filtering action in the battery-supply voltage source of the radio-cassette player. In the latest cassette and CD power supply sources, you will find several regulated voltage sources and decoupling circuits separating the different voltage sources to each separate radio, cassette, and CD functions.

FIGURE 4-21. The whole front-end circuit of a Delco auto radio was exchanged instead of replacing unavailable module components.

MODULE REPLACEMENTS

Today, the AM/FM CD cassette player might appear in two or more containers. The AM/FM radio and preamp audio circuits might be found in the front dash unit, while the audio and power supply circuits are found in another unit mounted on the firewall. Some of the front-end circuits might appear in a modular form, and should be replaced if defective. Often the whole unit can be replaced if the modules are not readily available.

For instance, within a Delco 16046360 AM/FM radio, the front-end circuits were dead and were found to be a defective RF module (**Fig. 4-21**). Because the module was not available locally, the whole front-end unit was replaced and exchanged at a local Delco repair dealer. The time and cost of the module might have cost as much as exchanging the whole front-end radio. Sometimes these small modules might not be available at all, even at a manufacturers parts depot. Check with your local car dealer or service depot for special auto radio components.

AM ALIGNMENT

Although the new auto receiver might have fixed IF ceramic filter networks with a fixed 455 kHz frequency, the early AM alignment and tracking adjustments must be made for correct front-end alignment. Remember, the early auto receiver had an IF frequency of 262.5 kHz, while later AM auto radios had a 455-kHz IF frequency. The typical early IF alignment and tracking procedures are made with the AM signal generator, VTVM, or the AC range of the VOM or output meter. Connect the AM generator to the antenna terminal and the VTVM or VOM to a 4- or 8-ohm load at the speaker terminals.

Step	Adjust	Connect AM Generator	Connect Spk	Point of Adjust Dial
1.	IF	SG to Ant Connector	VTVM or VOM	262.5 khz T251, T252, T253, TT254 IF
2.	AM SG to Antenna	510 khz	Minimum	Maximum L152 Ant
		1650 khz	Maximum	Maximum T153 Mixer
3.	Repeat Steps 2 and 3			
4.	Tracking	AM SG to Ant Terminals	VTVM VOM	1400 KHZ CT152 & CT153 on Dial to Max on the Meter

Chart 4-1. The typical AM-IF and tracking alignment chart to align the early audio AM circuits.

Turn the AM tuning dial to a minimum and set the generator frequency at 262.5 kHz or 455 kHz for the IF frequency. Adjust the IF transformers (T251-T254) for maximum on the VTVM or output meter. Likewise, set the AM generator to 510 kHz and the set the tuning dial to the lowest on the AM dial and adjust L152 for a maximum reading on the output meter. L152 is the oscillator coil within the AM band. Now rotate the AM frequency to 1650 kHz with the tuning dial set at 1650 kHz. Adjust CT153 for a maximum reading on the meter. Repeat step two at 510 kHz and 1650 kHz, respectively.

Check the AM tracking with the AM signal generator connected to the antenna terminals and the frequency set at 1400 kHz. Make sure that the AM dial pointer is at 1400. Adjust CT152 and CT153 for maximum reading on the VTVM or output meter. CT152 and CT153 are trimmer capacitors located across the RF permeability tuning coils (**Chart 4-1**). Always follow the manufacturers alignment procedures found in the front of the service manual. After replacing the car radio back into the dash, readjust the antenna trimmer capacitor to a local AM station around 1400 on the dial, until the station is the loudest in the speakers.

A dummy antenna plug can be made from a couple of small 30 pF NPO capacitors and an antenna male plug. The signal generator is connected to the plug for alignment (**Fig. 4-22**). Another 30 pF capacitor is soldered from the center terminal to outside metal ground. The dummy antenna is plugged into the antenna receptacle when making RF adjustments. If the auto radio has a 455 kHz IF transformer, then set the signal generator to 455 kHz instead of 262.5 kHz.

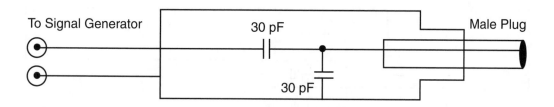

FIGURE 4-22. A dummy antenna can quickly be constructed with two 30 pF NPO capacitors and a male antenna plug.

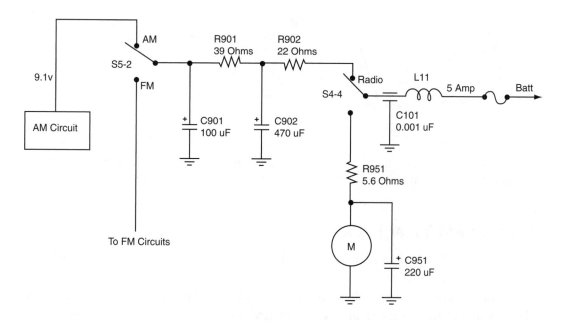

FIGURE 4-23. C901 caused a slight humming noise in a Sanyo FT1496 AM car radio.

Hum on AM/FM — Sanyo FT1496

A very slight hum was heard in a Sanyo AM/FM cassette-auto receiver with the volume turned all the way down. Both of the main filter capacitors were clipped with a 1000 uF 25-volt electrolytic with no results; the hum was still present. Because the hum was not heard when the cassette player operated, the hum must have originated within the AM/FM circuits. Upon checking the schematic, R901 and C901 formed a decoupling network to the AM/FM switch (S5-2). When C901 was shunted, the hum disappeared and was replaced with a 100 uF 25-volt electrolytic capacitor (**Fig. 4-23**).

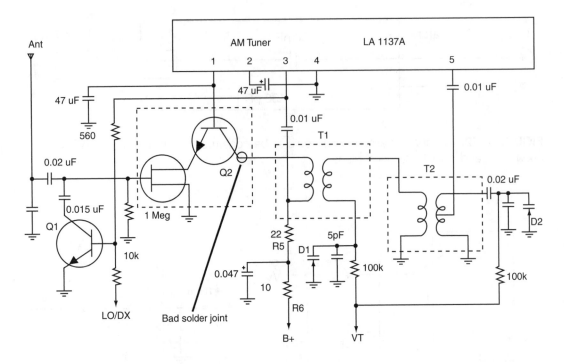

FIGURE 4-24. In an AM tuner section of a Radio Shack 12-1999 AM/FM compact disc car radio, the AM seemed dead and resulted from a badly soldered connection.

Defective AM Tuner Pack

In a Radio Shack AM/FM compact disc player, the AM radio was dead and the FM reception was good. Go directly to the AM tuner circuits tied to LA1137M IC and take voltage measurements. The voltage on varactor diodes D1 and D2 was checked; the voltage would change as the radio dial was turned, indicating correct tuning voltage was present. The voltages were fairly normal on the switching Q1 transistor. When voltage was measured on Q2, no voltage was found on the collector terminal.

A quick check on the schematic indicated that the B+ voltage to T1 was fed through resistors R5 and R6. The voltage was going into the primary of RF coil T1 and not out to the collector terminal. A badly soldered joint on the RF coil to the collector terminal of Q2 was located. Resoldering the badly soldered junction solved the defective AM tuner symptom (**Fig. 4-24**).

CHAPTER 5

Repairing the FM Circuits

In the latest AM/FM car radios, the front panel of the radio can be detached in order to discourage thieves from ripping the unit from the automobile. Simply detach the front cover and take the panel with you, as the car receiver now cannot be operated without the front panel. In some models, before detaching the front panel, be sure to press the power-off button first. Then press the side button and detach the front panel by pulling the cover toward you.

Be careful not to drop the panel when detaching it from the car radio. Sometimes a small case is provided to place the front panel into, so that it can easily be protected and carried inside the plastic case (**Fig. 5-1**).

THE FM RADIO

Removing the Car Radio from the Front Dash

The latest car receiver can be installed inside a metal sleeve that is bonded or bolted to the front dashboard. A metal sleeve is mounted into the auto's dashboard, and the car radio can now easily be pulled from the front, removing the radio out of the metal sleeve. Now you can take the auto radio with you so that it cannot be stolen.

FIGURE 5-1. To prevent auto radio theft, remove the front cover so that the car radio cannot be operated without it.

Some car radios have a flat metal key that can be inserted to unlock the radio assembly on one or both sides of the front cover, while other car radios have a slotted area at the top side of the front cover.

Simply insert the flat key and push upward to release a small spring-like lever to release the radio from the metal sleeve. A metal nut at the back of the radio must be removed from the flat rear strap support to free the car radio. Now pull out the car radio from the front of the dashboard. The auto receiver is easily removed for proper service or so it cannot be stolen.

Distance or Local Reception

In some early auto receivers you might find a DX or a local SPST switch that applied a strong distance radio signal directly to the front-end RF stage. A small PF capacitor coupled the long-distance radio station directly through SW-1 to the base terminal of the RF FM transistor. When a strong, local FM station was tuned in, SW-1 switched the picked-up antenna signal through a 680-ohm resistor to the base terminal. By switching the 680-ohm resistor in series with the antenna to the base of the FM RF transistor, you could reduce the strong FM signal (**Fig. 5-2**). Now the strong FM station does not swamp the front-end circuits.

You might find two or more resistors that can be switched into the front-end car radio to lower the strong FM radio signal for local operation in other front-end circuits.

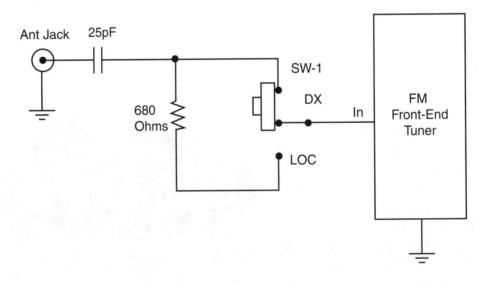

FIGURE 5-2. The distant DX local switching of the input signal from the antenna to the car radio circuits.

Not only was a low-ohm resistor switched in series with the antenna, but another resistor was also switched to a resistor tied to common ground, lowering the signal from the local FM station.

Block Diagram of the FM Circuits

The basic FM radio circuits consist of an RF FM amp, FM mixer, FM oscillator, plus the first IF, second IF, and third IF transistor stages, with a discriminator circuit of two fixed diodes. The early car radio FM stage consisted of bipolar or FET transistors. Likewise, the FM mixer and oscillator circuits consisted of a high-frequency or an FET transistor. The first, second, and third intermediate frequency (IF) stages were transistorized with two fixed diodes found in the discriminator circuits (**Fig. 5-3**). Here, the audio was switched or coupled to the first audio amplifier stage.

THE FM RF AMP STAGE

The FM radio signal is picked up by the car antenna and is connected to the first RF FM amp transistor through a small picofarads coupling capacitor. In the early auto receivers, the FM RF transistor might be a high-frequency or FET transistor. The RF signal is tuned in with either a small variable capacitor or with permeability coil tuning. The tuned FM frequency range is between 87.5 and 108 MHz, while the intermediate

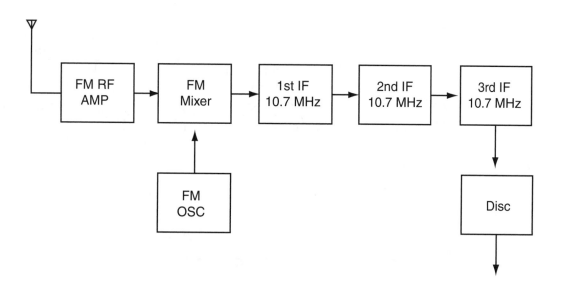

FIGURE 5-3. Block diagram of typical auto radio FM stages.

frequency is 10.7 MHz. Today, a small variable capacitor might still be found to tune the AM and FM circuits of a low-priced AM/FM/MPX auto radio (**Fig. 5-4**).

The field-effect transistor (FET) has very high impedance, like that of a vacuum tube. The FM RF stage of the auto receiver might have a bipolar or FET as the RF amplifier. The input or gate (G) terminal has high impedance to the incoming signal. The drain (D) terminal is comparable to the transistor collector terminal

FIGURE 5-4. The small variable capacitor can still be found in the low-priced auto radio-cassette player.

with a high positive voltage. The source (S) terminal is at ground potential.

The FM signal is coupled directly to the base terminal of a bipolar FM RF transistor with the emitter terminal connected directly to common ground, and the collector terminal connected through a small picofarads capacitor to the mixer stage. While in an FET FM RF transistor, the FM RF signal is coupled to the gate terminal through a 3-pF coupling capacitor and the drain terminal at ground potential. The source terminal of Q101 is coupled to a permeability tuning stage and into the FM mixer circuits (**Fig. 5-5**).

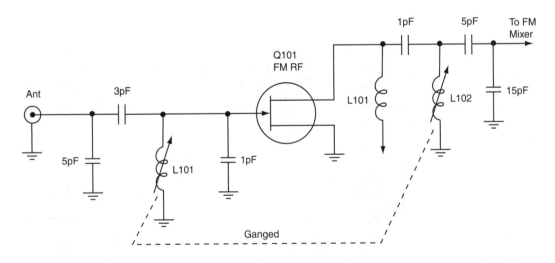

FIGURE 5-5. A typical FET FM RF stage with permeability tuning.

FM Permeability Tuning

Some of the lower-priced car radios still use permeability tuning in the front-end stages. Permeability tuning consists of moving a metal-ferrite rod inside a coil to properly change the inductance of a coil. The RF, mixer, and oscillator sections of the FM radio front-end circuits are tuned with permeability tuning. All three ferrite slugs or rods are found on a metal-ganged tuning assembly. When the front dial is rotated, the metal rods are fed in and out of the different coils at the same time, changing the inductors or tuning of each coil. Often, the metal-ganged assembly is rotated by a set of gears or with a dial cord assembly.

FM RF Amp Reception Problems

Most of the servicing problems related to the FM RF stage are a dead FM symptom or that you can only tune in one local FM station. Clean up the radio/tape switch for no FM reception, or check for an open RF permeability coil. Replace the open RF FM transistor for no FM reception or when only one local FM station can be heard. Check for badly soldered connections or replace a small picofarads capacitor from the antenna to the first FM RF transistor. Suspect near-zero voltage on the RF FM transistor with a leaky transistor. Resolder the base terminal of the FM RF transistor for no FM reception.

A source resistor broken loose on the FM RF transistor can cause a dead FM RF stage. Realignment of the padder capacitor on the antenna coil can cause poor FM reception and no FM stations tuned in. Poor FM reception can result from a bad antenna circuit. With no FM reception, clean up the AM/FM switch, and check for zero voltage on the source terminal of the RF FM FET transistor.

No FM reception can be caused by no or improper voltage applied to the FM RF transistor with a defective transistor voltage regulator and leaky diode. A defective voltage regulator IC, especially when the volume is turned up, can cause FM reception with a loud hum. Replace an open or leaky 1 uF 50-volt decoupling capacitor for no FM reception. No voltage found on the FM/AM section can result from a leaky voltage regulator transistor.

Suspect a leaky diode regulator on the PCB with no FM reception. A red-hot voltage supply resistor with a shorted or leaky 47 uF electrolytic within the voltage regulator circuits can also cause no FM reception, as can a bad zener diode on the PCB.

A dirty or broken AM/FM switch can cause intermittent FM reception. In this case, solder all board connections on the FM board. Intermittent FM can also be caused by a bad emitter terminal connection on the PCB. Simply moving the FM RF transistor with an insulated tool can cause the intermittent part to act up. A dirty or worn local AM/FM switch can also cause intermittent FM reception, as can a broken wire resistor. An intermittent RF FM transistor can cause only one station to be heard. A bad trimmer capacitor in the FM circuits can be another cause of intermittent FM reception.

A weak FM reception can result from a defective local distant switch. A defective FM tuner can cause a poor and weak FM reception. No voltage on the collector terminal of the FM RF transistor can cause really weak FM reception. A broken metal permeability slug can cause weak and intermittent FM reception, as can a leaky RF FM transistor. Suspect a shorted FET with a weak FM and distorted symptom.

FIGURE 5-6. A ganged-permeability tuning assembly is still found in the lower-priced auto radio-CD players.

FM Mosfet RF Stage

You might find a MOSFET RF stage in the latest auto receivers' front-end circuits. The MOSFET transistor has two different input elements (G1 and G2), a drain, and source terminals. The MOSFET is a metal-oxide field-effect transistor. Today, the MOSFET is found in the RF and power audio output transistor circuits of the auto receivers. Gate 1 of the RF FM amp connects to the input terminal, while gate 2 contains a bias voltage. The source (S) terminal is at ground potential and the drain (D) is the output terminal (**Fig. 5-6**).

The FM antenna picks up the FM signal and is capacity-coupled with several picofarads capacitors to the input gate terminal (G1). The FM RF station is tuned in by permeability coil L52. The source terminal of Q101 has a positive voltage applied between two 180-ohm resistors within the ground RF circuits. Drain voltage is applied through a 100-ohm voltage-dropping resistor in series with adjustable coil L51.

A 4 pF ceramic capacitor can couple the FM RF amplified signal to permeability tuning coil L53. A 6 pF padder capacitor helps trim up the output FM RF circuits. Coils L52 and L53 are ganged-tuned with the oscillator permeability coil tuning in the desired FM station (**Fig. 5-7**). Now the FM RF signal is capacity-coupled to the FM mixer circuits.

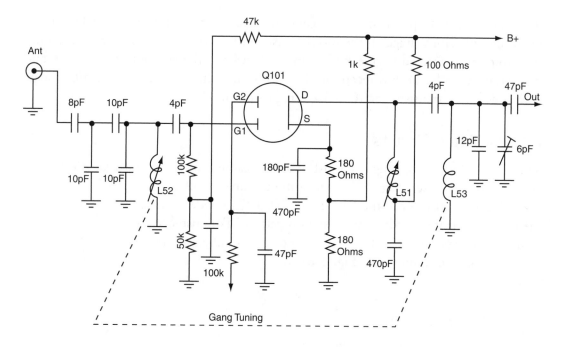

FIGURE 5-7. An FM Mosfet is found in the FM RF stage of the recent auto radio front-end assembly.

FM MIXER CIRCUITS

The FM mixer circuit combines both the FM RF signal and the local oscillator circuit. A 5 pF coupling-capacitor connects the tuned FM RF circuits to the base terminal of the FM mixer transistor Q102. The FM oscillator signal is coupled to the FM mixer transistor through a 5 pF ceramic capacitor to the base of Q102. Both the tuned RF signal and the local oscillator signals are mixed together to produce the intermediate frequency (IF).

In the early auto receivers, the AM IF frequency was 262.5 kHz, and today the IF frequency is set at 455 kHz. T101 is an FM IF transformer that amplifies the FM IF signal and is adjusted at 10.7 MHz.

The low-priced AM/FM auto radio cassette player might have only two IF stages, while the deluxe auto receiver might have three or more 10.7 MHz IF stages. The first FM IF stage is a transformer coupled to the next IF transistor.

Suspect a defective FM mixer stage when a rushing noise is heard and no FM stations can be tuned in. An open mixer transistor can cause a loud rushing noise with the volume turned up and no FM reception. A leaky mixer transistor might cause a really low collector voltage on the mixer transistor. Loss of FM or a dead FM reception symptom can be caused by a leaky FM mixer transistor. Replace both the FM RF and mixer transistors for no FM reception. Check for no voltage applied to the mixer stage with a defective transistor or IC voltage regulator circuit.

FM Oscillator Circuits

The FM oscillator stage might consist of a permeability tuned circuit that is capacity-coupled to the mixer stage. The permeability tuned coils are tuned with the same ganged tuning assembly as the RF circuits. TC101 provides a trimmer capacity adjustment for determining the correct station setting. For instance, tune in a local FM station and readjust TC101 to set the correct numbers of the known FM station frequency on the tuning dial (**Fig. 5-8**).

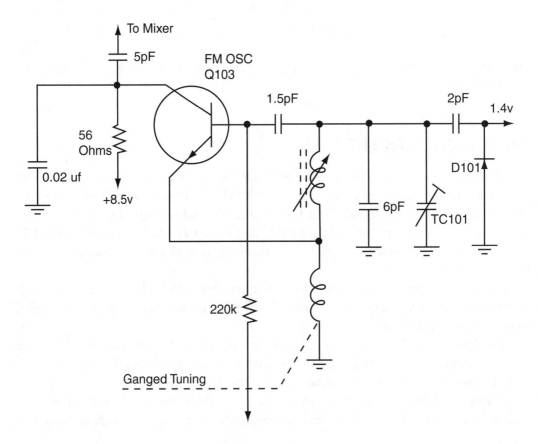

FIGURE 5-8. The FM oscillator circuit is connected through a 5 pF ceramic capacitor to the mixer circuits.

No FM stations can be heard when the local FM oscillator stage is not functioning. Suspect a defective component within the FM oscillator circuits when stations drift off frequency. A broken metal slug inside the oscillator permeability coil can quickly change the different FM stations on the dial at a different location. The FM station will drift off to another spot on the dial assembly.

A defective oscillator transistor can cause an FM station locked in at a certain frequency. The open oscillator transistor can cause a no-rush and no-station symptom. A leaky oscillator transistor can result in a no-FM symptom.

A no-FM MPX symptom can result from an open FM antenna or an oscillator coil. Lubricate the ground end of the variable capacitor when the FM band is noisy as the dial is rotated. Replace both the FM converter and RF transistor with no AM or FM reception. A roaring noise within the FM band can be caused by a leaky 68 uF tantalum 35-volt electrolytic capacitor.

Suspect a defective oscillator transistor when the radio comes on after sitting out in the cold weather. Suspect poor insulation between the trimmer padder capacitor when the FM stations cut out and are quite noisy; replace the damaged insulation with clear plastic insulators. Check for a bad padder capacitor mounted on top of the variable tuning capacitor for noisy FM and fading of FM stations. A defective voltage regulator IC can cause loss of FM with only a hissing noise. Suspect a bad oscillator stage with no FM action.

No AM — Normal FM

Only a rushing noise could be heard on the AM band with normal tuned-in FM stations in a Motorola 5FM485 auto radio. No AM stations could be tuned in and no cross-tuning noise could be heard in another car radio operating on the service bench. You can check the local AM oscillator stage by placing the car radio near another radio and trying to tune through the AM band. You should hear a squealing noise when the normal radio signal is tuned in at the same frequency, indicating that the oscillator circuit in the defective radio is operating.

A quick voltage test on the oscillator transistor Q9 indicated an increase in oscillator voltage. When a forward-bias voltage test was made between base and emitter terminals, only 0.26 volts was measured. The silicon NPN oscillator transistor should have measured a forward-bias voltage of 0.6 volts (**Fig. 5-9**). The leaky oscillator transistor was replaced with a GE-82 transistor.

FM Transistor Replacement

Visually inspect the FM transistor's mounting and soldered connections when a defective front-end transistor must be replaced. If the transistor elements are not marked directly on the PCB, locate the correct collector, base, and emitter terminals with a

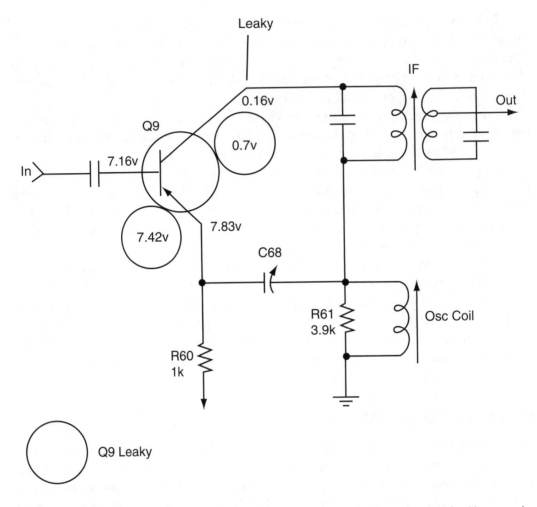

FIGURE 5-9. Leaky oscillator transistor caused a symptom of a loss of AM with normal FM in a Motorola auto radio.

transistor tester or the diode test of the DMM. The quickest test is to use the diode test of the DMM.

Remember that the base terminal is always common with a diode resistance test to either collector or emitter terminal. The resistance measurement between the emitter terminal and base are always higher than the test between base and collector terminals. The emitter terminal of an NPN FM oscillator is at ground potential. Also, the collector terminal voltage is higher than any other oscillator transistor terminal in determining the various unknown transistor terminals.

After removing the defective oscillator transistor, cut the lengths of each terminal the same length as the removed transistor. Mount the oscillator transistor replacement

Early Delco NPN	Function	GE	RCA	Sylvania
209247-0714	FM RF Amp	GE-17	SK3018	ECG108
2092417-0715	FM Mixer	GE-17	SK3018	ECG108
2092471-0716	FM OS C	GE-17	SK3018	ECG108

FIGURE 5-10. A typical transistor chart of the early universal transistors that can be replaced in the RF FM amp, FM mixer, and FM oscillator circuits.

in the same position as the removed transistor. Carefully solder up each transistor terminal. Use a pair of long-nose pliers to dissipate the excess heat of the soldering iron. Grip the transistor to be soldered into position about one-quarter inch from the soldered end.

Make sure that the transistor replacement lies in the same position as the original transistor. This means that the oscillator trimmer capacitor will only need a slight adjustment for correct frequency setting on the front dial assembly.

Choose the correct FM RF, oscillator, and mixer transistor from Sam Technical Publishing's Photofact auto radio series manuals or from the manufacturers service manual. Most original transistors found throughout the AM/FM auto receivers can be replaced with universal replacements transistors when original part numbers are not available (**Fig. 5-10**). Very few adjustments are required when using universal replacement transistors in the front-end section.

VARACTOR FM TUNING

Varactor diode tuning might be found throughout the latest AM and FM front-end tuning units. When a varying DC voltage is applied to the varactor diode, a different capacity is developed and tunes in the FM RF, mixer, and oscillator circuits. Separate AM and FM front-end circuits might be found within each tuning circuit. The varactor tuning system is found within the AM RF and converter stages.

The variable tuning voltage controlled by a controller circuit applies different voltage on the RF and converter tuning circuits, known as a VT supply voltage. Both the AM RF and AM converter transistors (Q201 and Q202) are FET-type transistors. Varactor diode (VD201) tunes RF signal coil L202, and VD202 tunes the oscillator or

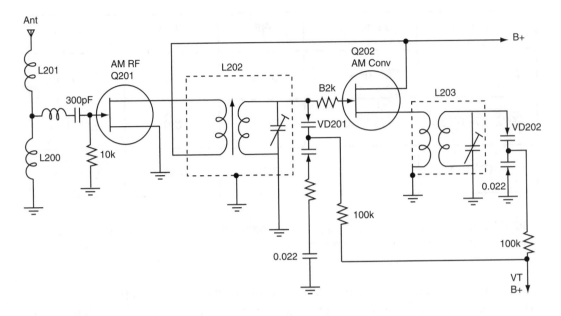

FIGURE 5-11. The AM RF and converter circuits tuned by varactor tuning diodes with a varying applied voltage (VT).

converter stage coil (L203). The B+ tuning voltage (VT) is supplied to both the RF and converter coil circuits (**Fig. 5-11**).

Besides varactor tuning, the AM RF coil (L202) can be adjusted for maximum reception in the RF stage. The trimmer capacitor (L203) of the AM converter stage can be adjusted to place the correct AM station at the exact frequency setting on the radio dial. The correct AM station can be received by varying the DC voltage on VD201 and VD202.

The FM stations can be tuned within the FM RF and FM oscillator circuits with a varying DC voltage. VD101 provides FM RF tuning within the RF stages, and VD103 tunes in the correct FM station. A common B+ tuning voltage (VT) is applied to VD101 through a 47 K ohm resistor to tune the FM RF stage. A varying DC tuning voltage (VT) is applied on VD103 to tune in the FM oscillator circuits (**Fig. 5-12**).

Suspect improper tuning voltage applied to VD103 when no FM station can be tuned in. Measure the voltage on VD103 as the FM dial is rotated. Check for a defective controller unit with no supply tuning voltage. Suspect a defective varactor diode (VD103) when a varying voltage is applied to the varactor diode with no station tuned in. Remember that this varying voltage is quite low, but will vary when the dial is rotated. Check the varying tuning voltage at the controller IC. Suspect that improper DC voltage is measured at the FM oscillator varactor diode (VD103).

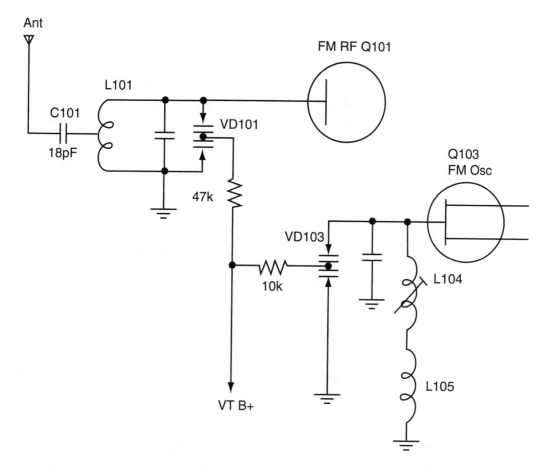

FIGURE 5-12. VD101 tunes the FM RF stage, while VD103 tunes the FM oscillator circuits.

FM IC CIRCUITS

Today you can find separate transistors in the FM RF, FM oscillator, and FM mixer circuits with IC components in the FM IF and FM multiplex circuits. You might find a transistor in the RF amp and oscillator circuits, IC IF amp, and IC MPX circuits. In fact, some of the latest FM circuits are entirely found inside one or two IC components (**Fig. 5-13**). Some of the latest front-end circuits might have both the AM and FM circuits inside one large IC component and another IC containing the multiplex FM stereo circuits.

Instead of regular IF transformers, the present auto receivers might have ceramic filters in the IF circuits. The ceramic filters are made up of piezoelectric material that is cut to the exact IF frequency. Each ceramic filter is cut at 10.7 MHz and does not

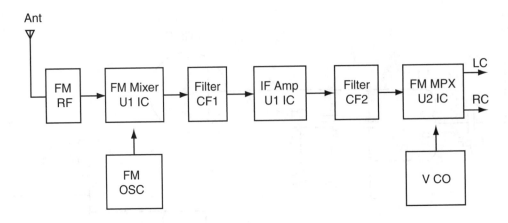

FIGURE 5-13. A block diagram of the FM IC circuits found in today's AM/FM/MPX car radio.

require any IF adjustments. In other words, the IF ceramic stage frequency is fixed at 10.7 MHz. You might find two or more ceramic filters in the FM circuits.

The ceramic filter is a component made up of piezoelectric material converting a mechanical stress into electrical energy or vice versa. The ceramic filter is made from slices of quartz, and when current is applied, a vibration results. The ceramic filter may have three or more terminals. Ceramic filters cause very few problems in the auto receiver.

In today's FM front-end tuner circuits you will find an FET FM RF amp transistor, FM mixer, and FM oscillator stages. All three stages are tuned with coils and varactor-type tuning. A variable tuning voltage is supplied through a 3.3 K ohm resistor controlled by a synthesized phase-locked loop (PLL) circuitry. The VT voltage is supplied between the two varactor diodes circuit to tune each coil. The varactor diode voltage is controlled by a PLL control IC (**Fig. 5-14**).

CIRCUITRY PROBLEMS

Defective FM Front-End Circuits

A defective FM RF, mixer, and oscillator transistor within the FM front-end tuning unit can cause loss of FM reception. A leaky or open FM oscillator transistor can cause no stations to be tuned in. Suspect an open or leaky FM transistor with only a strong local station that can be tuned in. The local FM station might be weak and distorted with a defective FM RF transistor. Leaky coupling picofarads capacitors within the

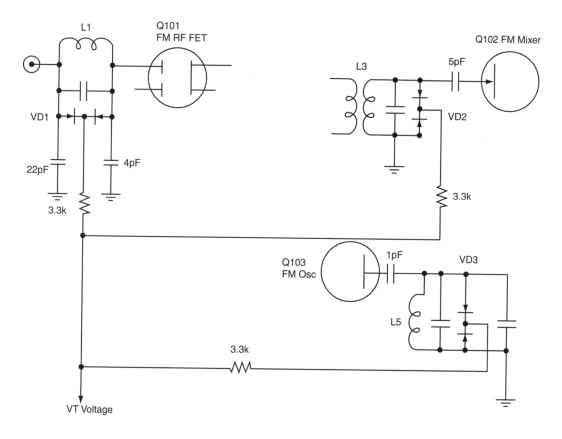

FIGURE 5-14. The FM FET RF stage is tuned by a varactor diode VD1, FM mixer is tuned by VD2, and VD3 tunes the FM oscillator circuits.

front-end circuits can cause loss of FM reception. Improper or no tuning in of the different FM stations can be caused by improper or no tuning voltage supplied by the PLL controller IC.

Suspect a defective PLL controller IC or improper supply voltage (Vcc) tied to the PLL IC. Rotate the radio dial and check the voltage that should vary on each varactor diode. If the voltage does not vary at either varactor diode, check the VT voltage at the PLL IC. Double-check the supply voltage applied to the PLL IC, and compare this voltage on the schematic.

Suspect a leaky PLL IC when the supply voltage is really low on the Vcc voltage pin terminal. The PLL circuitry is normal when the supply (VT) voltage varies on each varactor diode. Check for a poor or open FM coil winding or a defective varactor diode with normal VT tuning voltage. Do not overlook improper supply voltage from a defective transistor or IC voltage regulator circuitry.

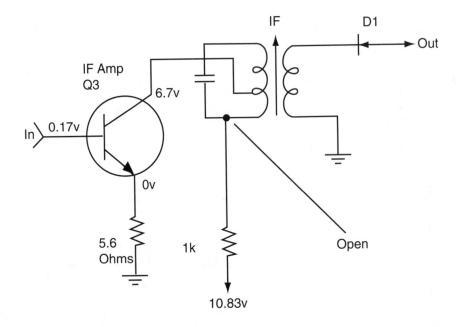

FIGURE 5-15. Open transistor (Q3) caused a dead front-end with no radio reception.

Dead Front-End — Ford

The front-end circuits of a Ford DAF-18806 auto radio were completely dead; when the volume control was turned up and down, a rushing noise could be heard, indicating that the audio stages were normal. Quick voltage tests on the RF and oscillator stages were quite normal. The collector voltage on the IF amp (Q3) was zero and should have been around 6.7 volts.

The primary winding of the IF transformer was open inside the metal case. The primary lead of the IF transformer was found broken off after removing the metal case, and it was repaired (**Fig. 5-15**).

Ceramic Filter IF Problems

Very seldom do the ceramic filter networks cause any service problems. Sometimes poor ceramic filter board terminals can cause intermittent FM reception. Resolder all ceramic filter terminals for possible intermittent FM reception.

Check the suspected ceramic filter with the low ohm range (1 Kohm) of the DMM. Measure the resistance of the input pin to common ground. No measurement should be seen at this point. Likewise, measure the output terminal to common ground. If a low ohm resistance is noted, remove the input and output pins from the PC wiring with

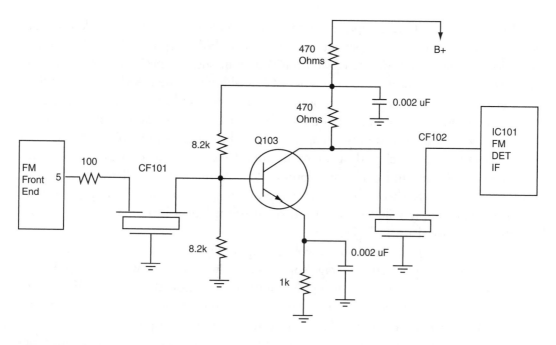

FIGURE 5-16. Check each ceramic filter (CF101 and CF102) with the low-ohmmeter range of DMM or on a crystal checker.

soldering iron and solder wick. Flick all unsoldered terminals with a knife blade or small screwdriver to remove each terminal from the etched pattern. Now take another measurement to common ground on all removed terminals. Replace the ceramic filter if any type of resistance is measured on the DMM (**Fig. 5-16**).

Another method is to test the ceramic filter network with a crystal tester. Remove the input and output terminals from the PCB. Clip the crystal checker to the input terminal and common ground. The ceramic filter should begin to oscillate and show a favorable reading on the crystal checker, if normal. Now do the same with the output terminal of the ceramic filter. Both input and output terminals should show a measurement on the crystal meter with a normal ceramic filter.

Defective IC FM Circuits

A defective IC component can cause a dead, weak, intermittent, and distorted FM reception within the auto radio front-end and IF circuits. A defective PLL Control IC can cause loss of AM or FM control. No tuner operation with the display showing only zeros can be the result of a bad IC in the timer circuitry. Replace the multiplex IC for no received FM stereo stations. A defective front-end IC tuner component can cause loss of AM or FM.

Replace the FM IC IF component with static found in the FM reception. Poor ground terminals on the FM IC component can cause noisy FM reception. Weak AM or FM reception can be caused by a defective front-end component. Check for poorly soldered connections on chip capacitors found off of FM ICs with intermittent AM or FM reception.

Locate the defective FM IC component with voltage and resistance measurements. Check the supply voltage (Vcc) terminal and compare it with the schematic. Suspect a leaky IC with really low supply voltage. Remove the voltage supply pin terminal from the PCB with solder wick and a soldering iron. Take a resistance measurement from the open supply pin and common ground. Replace the leaky IC if the resistance is under 500 ohms. Recheck the power supply voltage with the IC supply terminal removed. The supply voltage should return to normal or a little higher with the IC supply pin removed from the circuit.

Improper voltage supplied to the voltage supply pin can cause a dead or weak FM IC circuit. Suspect a defective voltage supply source with really low voltage applied to the IC. Check for a leaky or open supply regulator transistor, IC, or diode. A leaky or shorted decoupling electrolytic capacitor in the supply source can cause improper voltage. A leaky zener diode regulator can cause improper supply voltage. Check for overheating or a change in resistance in the power supply source for low or improper voltage applied to the FM IC circuits.

The FM front-end tuner circuit's supply voltage might pass through several switches, resistors, and voltage regulator circuits before reaching the FM circuits. L901 provides smoothing choke action with a 1000 uF electrolytic capacitor. The battery voltage is switched to the radio (RA) position with switch S4-4.

R902 and R903 with a 470 uF electrolytic provide additional filtering and a decoupling circuit to a zener diode (D901). D901 provides zener diode regulation with another 470 uF electrolytic, filtering and smoothing out the DC voltage for the FM switch S5-2.

C120 and R114 provide additional decoupling action before regulated DC voltage is supplied to pin 7 of the FM front-end tuning circuits. D903 and the 330-ohm voltage-dropping resistor provide voltage to the FM LED, indicating that the FM circuits are now operating.

Suspect a leaky electrolytic capacitor in the power supply for weak and distorted AM or FM reception. Replace an open electrolytic within the tuner section for no FM. Replace the voltage regulator transistor for either a weak AM or FM radio reception. Loss of FM or AM reception can result from an open 1 uF 50-volt electrolytic in the FM circuits. A shorted zener diode in the low-voltage source can cause loss of AM or FM reception. A leaky 47 uF decoupling electrolytic capacitor can cause loss of FM with a red-hot isolation resistor.

A defective voltage regulator IC can cause a hissing noise in the FM band. Suspect an open 50 uF electrolytic with a screeching noise when a station is being tuned in.

Tighten bad grounds on the main tuner board for noisy reception. Suspect a leaky voltage regulator IC when the volume is turned up and a loud hum is heard on both the AM and FM bands. A leaky decoupling electrolytic can cause distorted and low volume in the AM and FM bands.

PLL FM CIRCUITRY

Some of the latest auto receivers with a cassette or CD player use a digital phase-locked loop (PLL) circuitry to synthesize the AM and FM local oscillator frequencies. The FM mode PLL circuitry consists of a programmable divider, reference frequency divider, phase detector, and microprocessor included in IC501. Also included are a reference frequency oscillator, a fixed crystal (4.5 MHz), Q513, and Q514 as low-pass filter networks.

The 4.5 MHz crystal is used as a reference frequency. This crystal is connected to pins 24 and 25 of the PLL microprocessor IC501. The code determines "N," the divider that produces the required frequency for each FM station, spaced 200 kHz apart for FM stations. A 5-volt regulator circuit powers the PLL control (IC501).

When receiving a 98.1 MHz station, the VCO (Voltage Controlled Oscillator) generates a 108.8 MHz (98.1 + 10.7 IF MHz) signal. In the division of frequency, the Pulse Swallow system is used. The programmable divider (IC501) is used to produce 25 kHz from 108.8 MHz.

The 4.5 kHz reference oscillator is divided by N (=180) of IC501, resulting in another 25 kHz frequency. These two 25-kHz signals are fed to the phase detector, which is proportional to the phase difference between these two 25-kHz signals. This error voltage appears at pin 2 of IC501 and passes through the LPF. When the error voltage is integrated, harmonics and noise are filtered out. The resulting DC voltage is applied to the varicap diodes, whose capacity varies with the applied voltage (**Fig. 5-17**).

Now the output frequency of the VCO is corrected. With proper circuit design and correct adjustment, the VCO frequency is accurate and precise. Then the system is locked in, meaning that the phase detector senses no phase differences between the 25 kHz signals, and the VCO generates a frequency that is accurate and stable as the reference crystal oscillator. The reference oscillator of 4.5 MHz is divided by divider IC501. The FM band-N (=180)=4.5 MHz + 25 MHz. The low-pass filter (LPF) applies the DC voltage to the VCO circuitry, which is controlled from the output signal (25 kHz for FM), of the phase detector that is inside IC501.

The PLL Control (IC501) supply voltage (Vcc) is fed from a 5-volt voltage regulator IC. The locked signal is fed from IC501 to LPF transistors (Q513 and Q514) to the AM and FM front-end tuners. A variable tuning voltage is fed from pin 2 of the PLL IC501 to pin 4 of the FM tuning unit, to tune in the various FM stations with varicap or varactor diodes. The AM/FM oscillator, AM/FM signal meter, and FM

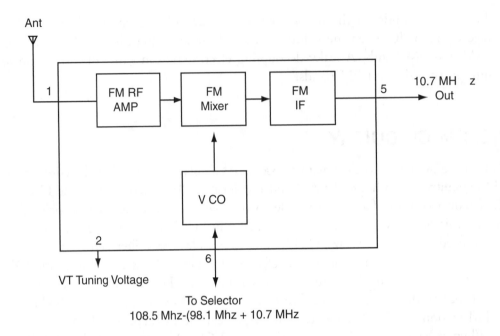

FIGURE 5-17. Block diagram of the PLL tuning that controls the front-end tuning of the FM RF, FM mixer, FM VCO, and FM IF circuits.

stereo indicator are controlled by IC501. The LCD display is connected to the PLL IC501 with the key matrix operation buttons (**Fig. 5-18**).

No or partial display can be the result of a defective IC501 or the display panel. Measure the variable DC tuning voltage (VT) at each varicap or varactor diode as the radio dial is rotated. Suspect a bad controller IC IC501 when no tuning voltage is measured on any varactor diode.

Check the PLL supply voltage (Vcc) at the supply pin of IC501. Suspect a leaky IC501 if the supply voltage is low. Do not overlook a defective 5-volt regulator IC for no or low-power supply voltage. Scope the 4.5 MHz signal and see if the crystal is oscillating. Replace IC501 if the supply voltage pin is low, you have no VT tuning voltage at pin 2, and no waveform on pins 24 and 25 of the 4.5-MHz crystal.

NO AM OR FM — COMBINED SERVICE PROBLEMS

Clean up the AM/FM switch for no AM or FM operation; spray the radio/tape switch with cleaning fluid.

Replace an open or leaky decoupling capacitor in the tuner voltage supply source for weak or low volume on the AM/FM band. A shorted zener diode and burned resistor

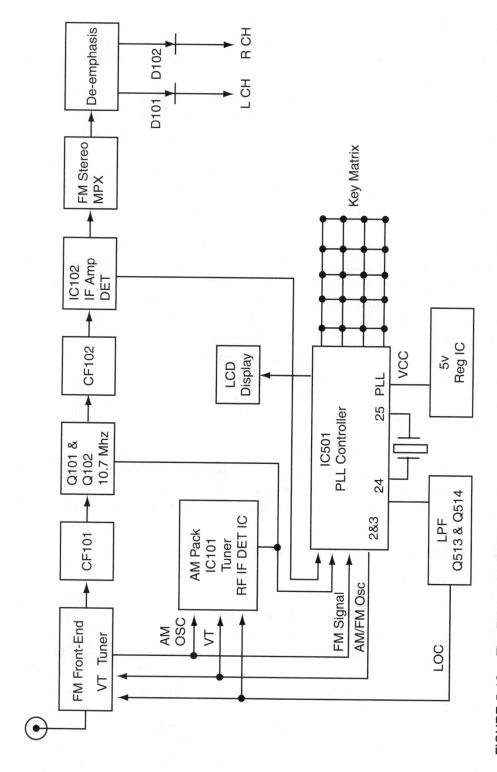

FIGURE 5-18. The PLL controller IC controls the VT voltage tuning, lock-in, AM/FM oscillator, FM signal meter, and FM stereo indicator circuits.

within the voltage source can cause loss of AM or FM reception. An open 1 uF electrolytic on the main tuner PCB can cause loss of AM/FM operation. Check for open decoupling electrolytics in the power source for weak and distorted AM and FM stations. Replace the voltage regulator transistor or IC for no AM or FM reception.

Also suspect a broken voltage source resistor terminal with a loss of AM or FM reception. Resolder the base terminal of a first IF transistor. Replace open FM RF transistors for normal AM and no FM reception. Check for a faulty FM RF transistor for only a weak FM local station and normal AM.

Check the FM oscillator transistor for only a rushing noise and normal AM reception. Replace the FM oscillator transistor when the FM station begins to drift with normal AM operation. Replace the FM oscillator transistor for intermittent or no tuner operation.

A leaky FM IC201 can cause loss of AM or FM. Resolder the antenna input terminal connection on the PCB for no AM or FM reception. Replace the jammed M1 switch when the car radio is locked on only one FM station. Replace the monitor switch assembly, with normal tuning and no sound. Check for a bad front-end IC board assembly with no AM and normal FM. Look for poorly soldered chip capacitors tied to the FM/AM tuner IC for intermittent FM reception.

Suspect a leaky trimmer capacitor within the collector circuit of the AM RF amp transistor for really noisy AM reception across the entire AM band. Realign trimmer capacitors on the antenna coil circuit for bad FM and AM stations on the lower end of the dial. Poor AM alignment can cause motorboating between 8 and 1 kHz on the dial assembly.

Only one FM and two AM stations can be found with an intermittent FM IF and a leaky AM RF transistor. Replace the fixed 4.5 MHz crystal for full loss of AM or FM reception. Suspect a defective PLL controller IC for loss of varactor diode tuning voltage for either AM or FM reception. Replace PLL IC, with no AM or FM control.

Noisy Chrysler 3501235 Auto Radio

A Chrysler car radio came in with a constant noisy and frying condition, and sometimes the radio would cut out completely. At first a noisy transistor was suspected of causing the frying noise. The noisy signal was signal-traced to the front-end circuits.

When a RF trimmer capacitor was bumped, the noise would begin and then quit. Water had seeped into the bottom metal chassis. No doubt that water had seeped into the trimmer capacitor, causing voltage arc-over between the metal plates. The suspected trimmer was connected from the secondary tuning coil to ground, with a voltage source found across the trimmer capacitor.

Sometimes a defective trimmer capacitor can be repaired if one is not available. Remove the top screw completely and spread out the metal plates to dry out any type of moisture. Replace broken or damaged insulation pieces with cut plastic or with one cut from the insulation of a power output transistor. Punch a hole in the center of the piece of insulation for the metal top screw. Realign the repaired trimmer capacitor for maximum station reception, tuned in the middle of the radio dial.

TROUBLESHOOTING FRONT-END CIRCUITS WITH A SIGNAL GENERATOR

Besides AM and FM alignment, you can use the RF signal generator to troubleshoot the FM and AM front-end circuits. Connect the AM signal generator to the antenna terminal and common ground. Set the volume, tone, and balance controls to the center rotation. Flip the local/DX switch to the DX position if one is found on the auto receiver. Rotate the signal generator signal to 1400 kHz with the dial setting at 1400 kHz.

With audio modulation turned on the generator, you should be able to hear a tone in the speaker if all AM circuits are normal. If not, inject the AM signal generator to the base or input terminal of the AM IC or transistor and common ground. Rotate the IF frequency to 450-455 kHz on the AM generator. You should hear a loud tone in the speakers if the IF and audio stages are normal.

Suspect a defective AM IF stage with no sound in the speakers. Slightly rock the 455 kHz frequency back and forth, because the IF frequency might be off to one side. Apply the IF frequency all the way up to the base of the last transistor or IC and common ground to signal trace the AM IF circuits (**Fig. 5-19**).

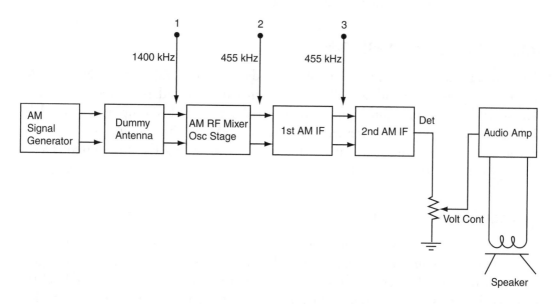

FIGURE 5-19. Inject the RF signal generator signal at points 1, 2, and 3 to troubleshoot the various auto radio circuits.

Likewise, signal-trace the FM circuits with the RF signal generator with no FM reception. Inject a 98.1 MHz RF signal (1 kHz-100 percent) modulation at the antenna terminal and common ground of the auto receiver. Set the volume, tone, and balance controls to center position. You should hear the 1 kHz tone in the speakers if the FM circuits are normal. If not, inject a 10.7 MHz IF signal to the base terminal of the FM mixer transistor and common ground. Suspect a defective RF or oscillator stage with no sound in the speakers. Proceed to the last FM IF stage to locate the defective FM circuit.

After locating the defective stage, take critical voltage measurements on the suspected transistor or IC component. When the signal is not heard at the collector terminal of a suspected transistor, check the condition of the transistor in circuit transistor tests. Improper collector, base, and emitter voltage might indicate a leaky transistor. Low or no collector voltage might point out a leaky transistor or improper voltage source. A leaky transistor might have low comparable voltages on all three elements or terminals. An open transistor might have no voltage on the emitter terminal and very high collector voltage.

The signal generator can locate a defective IC component by injecting an RF or IF signal at the input terminal. If no signal is heard at the input terminal and a 1 kHz tone is heard on the output terminal, suspect a bad IC or defective components connected to the front-end IC. Take critical voltage measurements on each IC terminal and compare them to the schematic. Check the resistance of each terminal to common ground for a possible leaky IC. Especially take a critical voltage measurement on the supply pin (Vcc) of the suspected IC component.

Intermittent Delco 1982 Series

A Delco radio came in with intermittent FM radio reception and normal AM. The FM signal was signal traced with a 10.7 MHz IF signal at pin 1 and 16 of the first IF IC1. The signal was present at pin 16 and had no signal at the number 1 input terminal (**Fig. 5-20**). Critical voltage measurements on the IC (DM-37) were quite low when the signal was in the intermittent state. Replacing DM-37 solved the intermittent Delco radio FM reception.

Typical FM Alignment

For FM IF adjustment, set the signal generator to 10.7 MHz and the dial on the radio to a minimum or under no signal from a broadcast station. Connect the FM signal generator to the FM antenna terminals and the output meter, VTVM, or FET-VOM connected to the speaker terminals (**Fig. 5-21**). Clip a 4-ohm resistor across the speaker terminals.

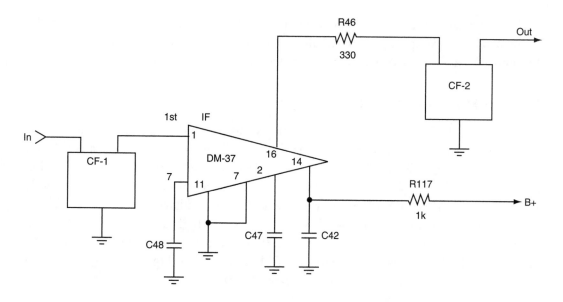

FIGURE 5-20. A defective IST IF IC1 caused intermittent radio reception in a Delco 1982 Series auto radio.

Adjusted Circuit	Input	Output	Frequency	Dial Settings	Adjust	VTV Scope Output Meter
IF	SG Ant FM Terminal	Across 4 Ohm Load	10.7 MHz	Near Min.	IF Transformer T101, T102	10.7 MHz Max
DET	SG FM Ant Terminal	Across 4 Ohm Load	10.7 MHz	Near Min.	T202	10.7 MHz
Tuner Tracking	SG FM Terminal	Across 4 Ohm Load	109 MHz	109 MHz	CT101	Max on Meter

FIGURE 5-21. A typical FM alignment chart showing the input and output instrument connections with alignment procedures.

Now adjust each IF coil or transformer to a maximum reading on the meter. No IF alignment is required with IF circuits containing ceramic filter networks. Check the tuning range coverage by setting the FM oscillator trimmer capacitor to a maximum reading on the output meter. The volume, tone, and balance controls should be set at the center of rotation.

Keep the gain control of the signal generator low enough to prevent overloading the input signal or causing clipping in the alignment procedures. Follow the manufacturer's AM and FM alignment procedures for correct alignment.

Tunable Hum — RCA RC1248A

The AM and FM radio had a hum in the speakers, and the hum seemed to be controlled by the volume control. At first the large filter capacitor was suspected, but when shunted with a good filter capacitor, the hum was still present. The FM tuning also seemed to change, because the FM tuning was accomplished with varactor diode tuning. The hum seemed to come from the 11.9-volt regulated power source. Upon checking the schematic, the 11.9-volt source was quite high at 21.5 volts. The 11.9 regulated voltage was provided by IC202. IC202 was replaced and appeared to be leaky with 21.5 volts feeding both the AM and FM circuits. The higher voltage and hum were removed after replacing voltage regulator IC202.

CHAPTER 6

Troubleshooting the Auto IF and MPX Circuits

Only two intermediate frequency (IF) stages are found in the early auto radio circuits. The first AM IF transformer is connected to the collector terminal of the converter or mixer transistor. The second IF transformer is transformer-coupled to another IF transistor. The second IF amp transistor is directly connected to the AM detector at the output of the second IF.

In the early AM IF stages, the 262.5-kHz IF transformers were connected directly to the collector and base terminals of each IF transistor. You might have found three or four separate IF circuits in the deluxe auto receiver. Today, the auto radio IF frequency is 455 kHz (**Fig. 6-1**).

FIGURE 6-1. Regular 455 kHz IF transformers are found in the low priced auto radio-cassette player.

Often the AM front-end circuits consist of an AM RF transistor and a converter transistor, which serves as both mixer and oscillator stage that connects to the FM IF circuits. The FM RF amp, FM mixer, and FM oscillator transistors are directly coupled to the FM IF circuits. The AM IF transformers might be in a series or parallel circuit with

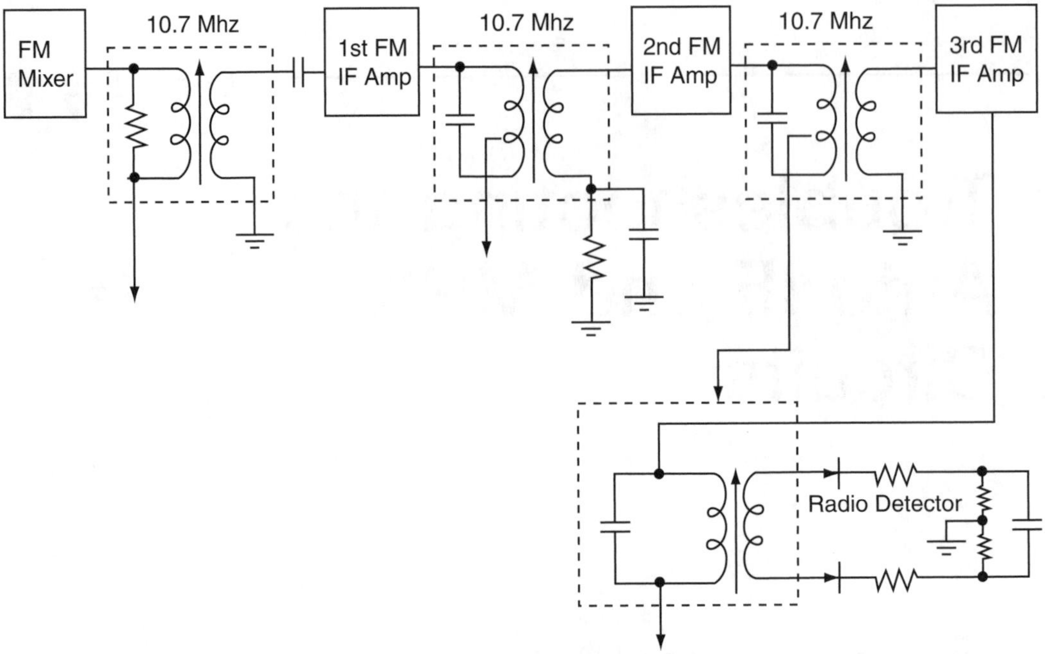

FIGURE 6-2. Block diagram of the FM IF circuits.

the FM IF stages. You might find that a couple of IF circuits serve both as AM and FM IF stages. While in other AM/FM circuits, the whole RF, mixer, oscillator, and IF circuits are broken into separate AM and FM stages throughout the whole IF circuit (**Fig. 6-2**).

The early AM IF circuits consisted of PNP transistors throughout the front-end and IF circuits. Usually the IF transformers were aligned at 262.5 kHz. Notice that the collector terminal voltage is zero or at ground potential. The base terminals of the PNP transistors are highly positive, operating directly from a decoupling capacitor and resistor network. Also, the bias voltage between the base and emitter terminals is positive, with a 0.3-volt reverse-bias voltage. The collector terminal voltage of a PNP transistor is highly negative compared to the base and emitter terminals.

Notice that in all IF transformers, the primary winding is tied to the collector terminal, and the secondary is connected directly to the next transistor-base terminal. The emitter terminal of each IF transistor is tied to ground, but is highly positive compared to the collector terminal.

AM IC IF CIRCUITS

Today, the AM circuits might contain transistors within the RF and converter circuits that are connected to an IC IF detector amp. Some auto receivers have one large IC

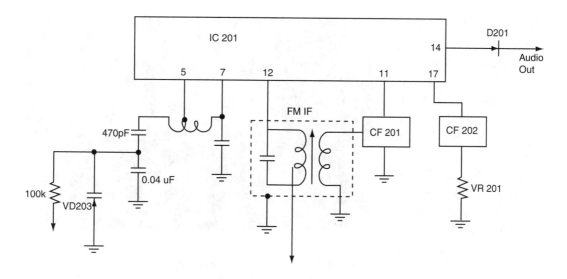

FIGURE 6-3. The AM IC IF circuits found in one IC component.

that serves as AM RF, oscillator, IF, and detector circuits. The oscillator coil (L203) is connected to pins 5 and 7 of IC201. The AM IF transformer is connected to pin 9, and feeds directly into a ceramic filter network (CF201). The amplified IF signal is out of pin 14, and is detected by D201 to an analog or audio signal connected to the audio switching circuits (**Fig. 6-3**). These AM IF stages operate at 455 kHz instead of the early auto radio 262.5 kHz frequency.

In either transistor or IC IF circuits, where an IC IF component might serve as IF amplifier for both AM and FM reception, a defective IF IC can cause weak, intermittent or dead AM and FM reception symptoms. Likewise, in older IF transistor circuits, where the FM and AM transformers are in series or cascade operation, in which one transistor serves as both AM and FM amplification, both AM and FM reception can fail with a defective common IF stage.

Go directly to the common IF circuits when both the AM and FM reception are dead or intermittent. Check the front-end circuits when only the FM reception is low or weak with normal AM reception. Suspect the AM front-end RF and converter circuits when the AM is weak or dead with normal FM reception.

TRANSISTOR FM IF OPERATION

The early FM transistor IF circuits were found with IF transformers tuned to a 10.7 MHz intermediate frequency. The FM IF circuits are still located in the low-priced AM/FM auto receivers (**Fig. 6-4**). In many cases, the AM and FM IF stages might include a FM transformer coupled to the mixer transistor, and the second IF stage directly coupled to a ceramic filter network. The 10.7-MHz IF signal is coupled directly to the base terminal of Q201, through a small ceramic capacitor, to the NPN IF transistor. The collector terminal of Q201 is tied directly to the ceramic filter (CF201) with a 330-ohm load resistor. The ceramic filter network is cut at a fixed 10.7 MHz frequency and directly coupled to another transistor IF circuit.

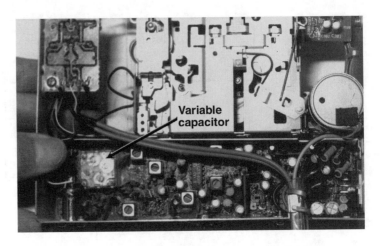

FIGURE 6-4. A variable capacitor and IF transformers are still found in today's low-priced auto radio-CD players.

Ceramic filter networks are quite common within the latest AM and FM circuits of the auto radio-cassette players. The first IF transformer stage might be coupled directly to the FM front-end circuits, with only a 10.7-MHz ceramic filter between the front-end and first IF stage. CF1 couples directly to the base of TR1, and another ceramic filter network (CF2) is tied directly to the filter network. The output of the second IF ceramic filter connects to input terminal 5 of the IF discriminator circuit. There are no IF transformers found within the IF circuits except T1, which is an IF discriminator transformer (**Fig. 6-5**). The audio output from IC1 couples through a 1-uF electrolytic to the FM switching circuits and onto the audio preamp or AF circuits.

FM Stations Drifting Off Channel

Within a Sanyo FT1490 car radio, the FM stations would drift off of the dial setting after the radio operated for several hours. The DC supply voltage to the FM oscillator collector terminal was monitored when the FM stations drifted off and did not have a change in supply voltage. R111 had increased in resistance and was replaced. Q103, the FM oscillator, was also replaced (**Fig. 6-6**). Replacing Q103 and base resistor (R111) solved the drifting of the FM stations.

FM IC IF CERAMIC CIRCUITS

Today, ceramic filter networks are found throughout the FM IF and AM IF circuits. Often the FM and AM front-end circuits are made up of transistors, with the IF,

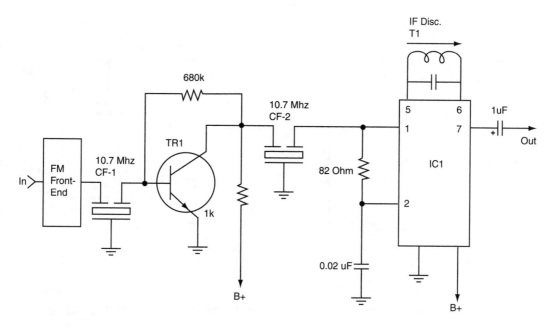

FIGURE 6-5. Two or more IF ceramic filters might be found in the auto receiver.

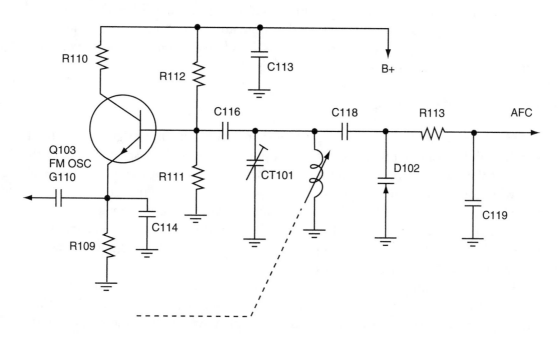

FIGURE 6-6. Replace R311 and Q103 FM oscillators for FM drifting in a Sanyo FT1490 auto radio.

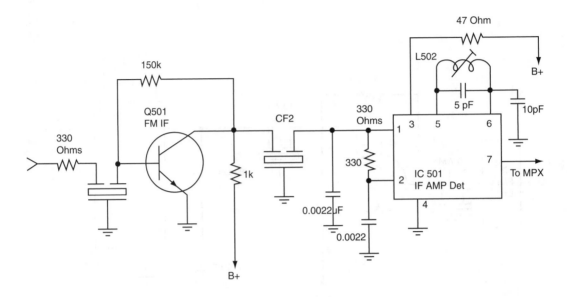

FIGURE 6-7. The IF amp detector IC might be located between the FM IF stages and the FM MPX circuits.

detector, and multiplex circuits possibly found in one or two different IC circuits. IC501 contains the second IF stage with a detector network. Within the mono FM discriminator stage, a ratio detector or discriminator circuit might be found inside IC501. The FM IF signal is amplified by IC501 and detected into an analog or audio signal switched to the audio input circuits (**Fig. 6-7**).

Intermittent IC IF Symptoms

To determine if the IC IF circuits are intermittent, check the input signal at the input terminal of the suspected IC. Monitor the output terminal of the suspected IC with the scope. If the stage becomes intermittent and there is no signal at the output terminal, but with normal signal at the input, then suspect the IF IC component.

Monitor the supply voltage (Vcc) with the DMM. Notice if the voltage changes when the intermittent radio signal occurs. Take critical voltage measurements on the suspected IF IC while the radio is operating. If the supply voltage drops or changes, suspect a leaky IC or a poorly regulated supply voltage.

Remove the supply voltage terminal pin and take another measurement. If the supply voltage returns with the supply terminal disconnected, suspect the IF IC. When the supply voltage remains low with the IF IC supply terminal disconnected, check the voltage regulator supply source.

IC3 combines both the AM input front-end circuits and the output signal from T2, and couples the IF signal to a 10.7-MHz ceramic filter (CF1). Besides the IC front-end

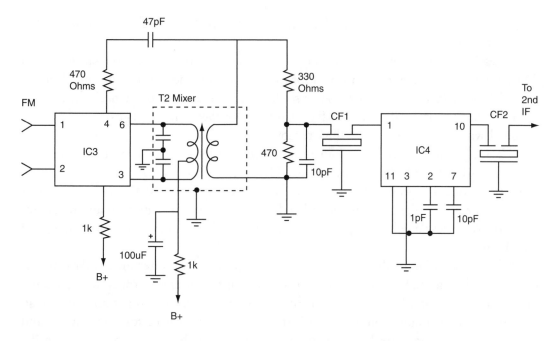

FIGURE 6-8. All of the FM IC ceramic circuits are found in two IC components.

circuits, all of the IF stages are amplified by IC components (IC3 and IC4) to the second IF stage. The multiplex FM IC circuits are tied to the third FM IF ceramic filter network. Here, the whole FM circuits are combined in three different IC components (**Fig. 6-8**).

If Stage Service Problems

The AM and FM IF transformers can cause many different service symptoms within the auto radio. Sometimes the small internal capacitors can become cracked inside the metal case and cause intermittent radio reception. An open primary or secondary winding can cause a dead AM or FM radio reception. Other times, the inside coil winding might be pulled too tight, and after time it can come loose, causing a dead or intermittent radio operation.

Do not try to adjust the IF transformer cores by hand, for they do not turn by themselves. You should only align IF transformers with a signal generator and output meter. You cannot simply touch up the IF core adjustment to improve the AM or FM reception; you can make the reception worse.

Replace the first IF transformer when only a local FM station can be heard. An open first IF NPN transistor can cause a dead AM or FM reception. No AM or FM reception, with only a rushing noise, can be caused by an open IF amp transistor.

Suspect a leaky first IF transistor when the radio begins to play after the inside of the car warms up.

A dead or distorted FM reception can be caused by a bad second IF transistor. No AM or FM can be caused by no or improper voltage on the collector terminal of the second IF transformer, due to a poorly soldered board connection. Check for a badly soldered connection on the second IF transformer for low or no voltage on the collector terminal with very weak AM or FM reception.

Weak AM or FM reception can result from a leaky first or second IF transistor. Weak FM sound can be caused by a faulty FM IF transformer. Poor RF or IF tracking can cause a weak FM sound. Improper or low voltage supply to the IF transistor circuits can cause a really weak sound signal with a defective decoupling capacitor and resistance network. A defective voltage regulator transistor or IC can cause improper voltage on the IF circuits.

An open winding of the second IF transformer can cause intermittent AM or FM reception with normal audio circuits. An open or intermittent IF transistor can cause intermittent AM or FM reception. Suspect an open second IF amp transistor for an intermittent Delco radio sound. Replace the second IF transistor for intermittent and noisy radio reception. Check for a bad connection inside the IF transformer when the FM reception is intermittent by moving the metal shield of the IF transformer. Realignment of the FM IF transformer is required when the FM sound is weak and distorted.

FM DISCRIMINATOR CIRCUITS

The early FM discriminator stages consisted of a limiter transistor with a 10.7-MHz transformer found in the input and output circuits of Q7. The last IF transformer serves as the FM discriminator circuits. The FM signal within the discriminator circuits is rectified by diodes X4 and X5.

The right channel audio is found at the cathode terminal of X4, and the left channel at the cathode terminal of X5. In the early chassis, the FM and cassette audio circuits were mechanically switched into the AF amp audio stereo circuits. The FM multiplex audio signal is now switched in with fixed diodes within each leg of either the left or right channels.

In the FM circuit employing a ratio detector, no limiter stage is found. The ratio detector has two fixed diodes, except that one of the two diodes is reversed, and the junction point of the load resistors are at ground potential. The ratio of the DC voltage output of this detector is proportional to the ratio of the IF voltages applied to the two diodes.

Replace the FM detector coil for loss of audio or intermittent FM reception. Realign the FM detector coil when the stations will not lock on to the FM band. Check to see if the FM discriminator coil is out of alignment if the FM stations drift off frequency. Tighten the ground screws on the main FM board for noisy FM reception. Resolder the negative sides of both stereo coupling electrolytics for loss of FM reception after a few minutes of operation.

Check the last IF transformer for a bad winding connection, for intermittent FM reception. Test the second FM IF transistor for weak and intermittent FM operation. Check the readjustment of the FM IF transformers with distorted FM music. A bad IF detector IC can cause weak and distorted FM audio. Measure the supply voltage terminal of the AM/FM detector IC for poor radio reception.

AM DETECTOR CIRCUITS

The AM detector within the early auto radio was a fixed diode of a 1N34A or 1N295 diode. An ECG109 universal replacement can replace either AM detector diode. The fixed diode is found at the tapped winding of the last IF transformer winding and is capacity coupled to the first AF transistor. The detector diode separates or rectifies the audio signal from the IF signal so that it can be amplified and heard in the speakers. A leaky detector diode can cause a distorted audio sound, while the open diode has no audio sound. A bad diode connection inside the secondary winding of T102 can cause intermittent AM radio reception (**Fig. 6-9**).

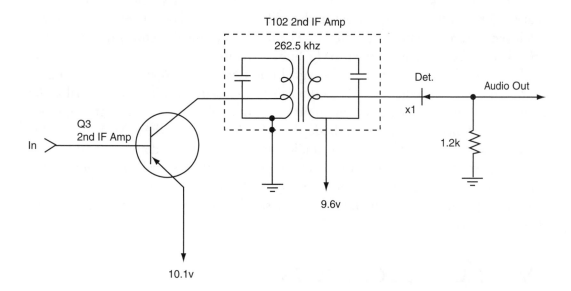

FIGURE 6-9. The AM crystal detector might consist of a 1N34A, 1N295, or a universal replacement ECG109.

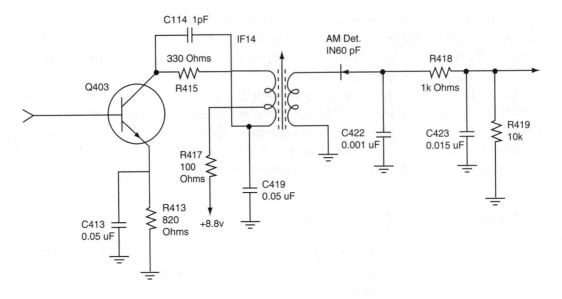

FIGURE 6-10. Connect the external audio amp to the crystal detector to locate a dead detector circuit.

Weak AM Detector Signal

The amplitude modulator (AM) detector is a device that senses a signal and indicates that it's there with the audio output in the AM car radio. A diode detector recovers the intelligence from a demodulator circuit. The function of the detector is to demodulate the RF wave before it is coupled to the audio circuits.

The IF and detector stages can be signal-traced with the scope and a demodulator probe. After the ratio detector, the audio can be signal-traced with the external audio amp. Simply connect the audio amp input to the output terminal of the detector diode (**Fig. 6-10**). Check the suspected diode with the diode test of the DMM.

An open detector diode has no audio output. The leaky detector diode might produce hum, distortion, and a weak radio signal. A badly soldered connection on the diode can cause intermittent sound. A really weak audio sound can be caused by an open primary or secondary winding of the last IF transformer that is coupled to the detector diode.

FM IC MULTIPLEX CIRCUITS

The frequency modulated (FM) stereo uses a multiplex method to receive stereophonic music in the FM channel. The multiplex FM stereo employs multiplexing techniques

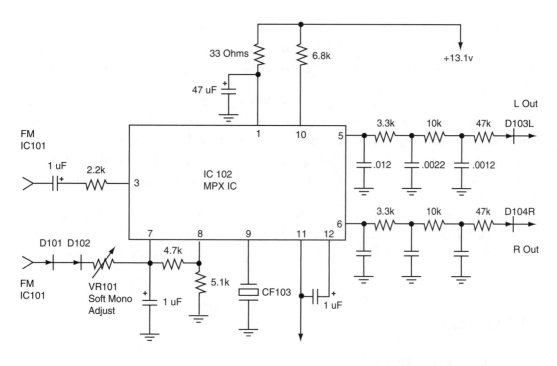

FIGURE 6-11. The FM MPX circuits of the AM/FM/MPX stereo auto receiver.

to broadcast both channels of a stereophonic program on a single carrier wave. The right and left audio channels are fed from a multiplex stereo circuit.

The FM detector and IF signal are fed into pin 3 of the MPX IC102, and out terminals 5 and 6. D101 and D102 with VR101 provide soft mono FM sound at terminal 7 of IC102. The left audio channel is fed from pin 5 through several resistors and the capacitor network to the left channel output diode D103L. Likewise, the right channel audio is fed from pin 6 through three resistors to the left discriminator diode D103R. Both the right and left channel multiplex stereo audio are coupled through a fixed diode switching arrangement (**Fig. 6-11**).

FM Multiplex Service Problems

Check for the MPX IC running red-hot with only a local FM station picked up. Replace the MPX IC with no FM stations received. A bad FM multiplex IC can cause weak and distorted FM reception. The defective MPX IC can cause loss of FM sound in the speakers.

Suspect a defective band selector switch for no FM MPX reception. Resolder all capacitors on the matrix board for no FM stereo after 20 minutes of operation. Replace the third IF FM transistor for the symptom of an FM station drifting off and finally having no sound. Check for a bad board connection in the MPX circuits, for intermittent FM on the left channel. Slightly push down upon the board to see if the sound cuts in and out.

Measure the supply voltage (Vcc) at the FM MPX IC for weak and distorted FM stereo. Suspect a defective voltage regulator transistor or IC with shorted diode for low voltage applied to the FM MPX terminal. Check for an open electrolytic decoupling capacitor or an open multiplex diode for no right stereo audio. Check all small electrolytic capacitors within the multiplex circuits for open conditions with the ESR meter.

Cracking and noisy FM stereo music can result from a defective stereo multiplex IC. Check for a defective ceramic filter network within the multiplex IC circuits for weak and distorted FM. A leaky switching diode can cause distorted FM music in the right channel.

INDICATORS

FM Stereo Indicators

The FM lamp stereo indicator lights up when the FM music is played in the car radio with stereo sound out of the left and right channels. The early stereo indicator consisted of a pilot lamp bulb that came on when stereo music was played. Today, the FM MPX stereo light is an LED stereo indicator. Here, the stereo LED509 indicates stereo operation at pin 7 of the FM MPX IC103. WR102 adjusts the stereo voltage applied to LED509 (**Fig. 6-12**). Often, WR102 is adjusted at the point of lighting up the stereo light, and when the stereo music is played, the stereo light indicates the stereo system is in operation.

Besides changing FM mono music into a stereo system, with a left and right stereo channel output, pins

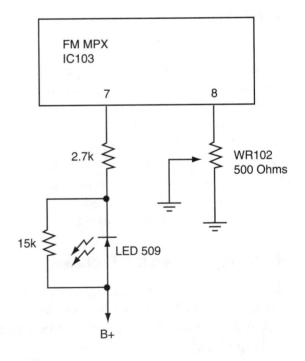

FIGURE 6-12. LED509 indicates when the FM stereo music is being played.

4 and 5 of IC502 provide diode switching of the audio channels. The FM music is coupled through a 1-uF electrolytic and a 15K-ohm resistor to pin 2 of the FM MPX IC. The FM stereo lamp is excited at pin 1 and with transistor Q501 to the FM radio switch. Adjustment of the preset controls can be made for proper adjustment of the FM stereo indicator lamp or LED.

A defective lamp or an open electrolytic on the MPX board can cause improper lighting of the FM stereo lamp. Check all small electrolytics within the FM MPX stereo circuits with the ESR meter. An open stereo lamp or LED can cause loss of FM indication with the stereo music at both left and right channels.

Suspect a defective FM MPX IC when the stereo light stays on all the time. Check for a defective MPX IC with loss of FM but working AM, and the stereo light being on all the time. A defective FM MPX IC can cause loss of left channel FM music. Replace the FM MPX IC when both channels are dead and there is no FM stereo light.

AM/FM Indicators

A simple SPST switch arrangement can turn on the AM or FM indicator lights. When the auto radio is switched to FM operation and the music is played in stereo, the FM stereo indicator should come on. The FM MPX IC101 controls the FM stereo indicator light. The radio/tape switch applies a DC voltage to either the radio or tape circuits. When the tape is played, a tape indicator lamp is turned on. Most of the early FM MPX indicators were turned on with a small lamp bulb. Today, LEDs are found throughout the entire indicator board.

Check each lamp or LED with a resistance or continuity measurement. Set the DMM to the 200-ohm scale and check the continuity across the lamp bulb terminals. Check each LED with the diode test of the DMM. The normal LED will show a resistance measurement in one direction, and with reversed test leads will show an infinite reading. A leaky LED will show a low ohm measurement in both directions. Often the LED indicator will last the lifetime of the car radio.

AM/FM Radio Switching

The output audio signal from either the AM or FM radio circuits are switched separately into the audio AF or volume control circuits of the car radio. By placing fixed diodes in both the AM and FM audio output circuits, the audio can be switched with manual switches.

D352 and D351 pass the FM signal audio circuits to the volume control of the AF circuits. D401 and D402 are placed in the AM output circuits; with the cathode terminal connected to the audio circuits, the FM signal is blocked out from entering the AM radio circuits. Likewise, D401 and D402 pass the AM radio signal to the audio

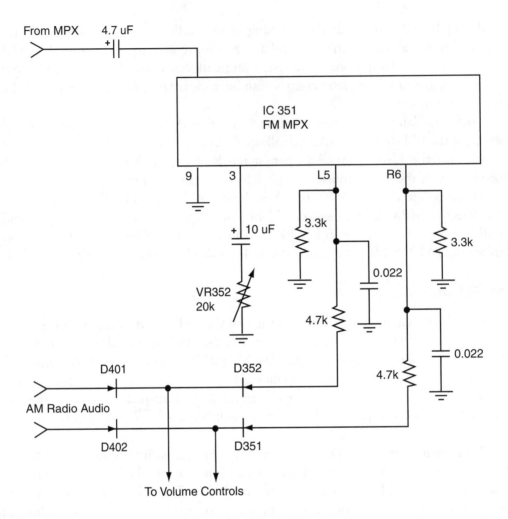

FIGURE 6-13. The AM and FM radio signals are switched into the audio circuits with D401, D402, D352, and D351.

amp circuits, and it is blocked from entering the FM circuits by D351 and D352. The FM or AM radio circuits are only turned on when the AM/FM switch is turned to either the AM or FM position (**Fig. 6-13**).

FM VOLTAGE SOURCES

The early front-end AM radio circuits had a large voltage-dropping resistor and a large decoupling filter capacitor to common ground. A 12.6-volt source was fed to the single audio output transistor, while the 10.3-volt source fed the radio circuits. Another large

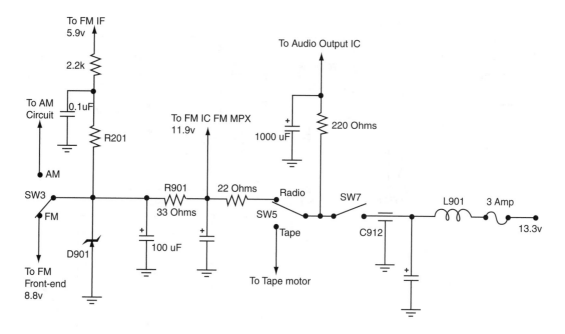

FIGURE 6-14. The FM voltage sources with resistance and electrolytic capacitor filtering.

resistor and decoupling filter capacitor fed the 9.5-volt source to the driver or AF amp transistor.

The early car radio with a cassette player might have several voltage-dropping resistors and electrolytic capacitors to provide a different voltage source to the FM front-end, FM IF IC, FM MPX IC, and the audio output ICs. Zener diode D901 provides a regulated 8.8-volt DC source to the front-end circuits, while the FM IF transistors have a 5.9-volt DC voltage applied. The FM IF and FM MPX IC have an 11.5-volt source found between R901 and R902, with a 470-uF filter capacitor tied to common ground. The audio output right and left ICs operate through R912 and are filtered with a 1000-uF electrolytic (**Fig. 6-14**).

The latest AM/FM MPX auto receiver with a tape player might have a transistor-diode regulated system that provides a different voltage source to the FM MPX IC, FM IF DET IC, FM IF transistor, and front-end FM circuits. SW1 turns on the DC power to the collector of the voltage regulator Q601. A 10.8-volt regulated voltage is applied through the voltage-dropping resistor placing 9.7-volt source to pin 1 of FM MPX IC102. A 33-ohm resistor and a 470-uF electrolytic supply another 9.7 volts to pin supply terminal 12 of the FM IF detector (IC101).

The FM IF transistor (Q102) has a 6.7-volt source provided by two 470-ohm resistors and bypass capacitor (0.0022 uF). Last but not least, the FM front-end circuits, which include the FM RF, FM mixer, and FM oscillator transistors, have a

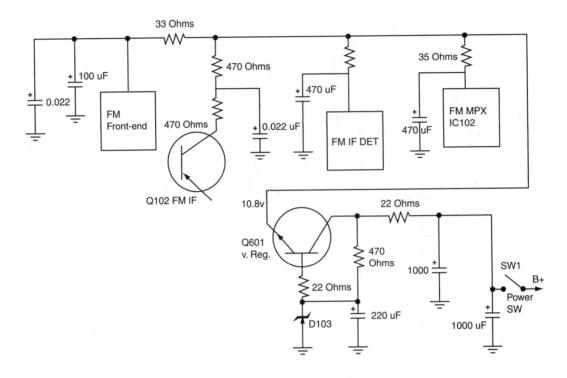

FIGURE 6-15. Q601 provides 10.8 regulated voltage to the FM circuits.

voltage-dropping network of a 100-uF electrolytic and a 3-ohm voltage-dropping resistor (**Fig. 6-15**).

Distorted Right Channel

Check for proper FM IF alignment for a distorted sound symptom. A leaky MPX IC can cause a weak and distorted right channel. Replace MPX IC with no audio in the left channel. Suspect a multiplex IC when there is no sound from the right channel and the indicator light is on all the time. A poorly soldered right channel terminal connection on the IC can cause a poor or distorted right channel.

Check for leaky coupling electrolytics for a distorted left channel. A leaky switching diode can cause distortion in the left channel. Signal-trace the distorted left channel from the MPX IC output terminal to the switching diodes with the external amp (**Fig. 6-16**). Suspect the MPX IC with distortion in the right channel but with a normal left channel. Readjustment of the MPX alignment controls within the multiplex can cause distortion in both channels. Do not overlook leaky bypass capacitors from MPX IC to ground for weak and distorted sound.

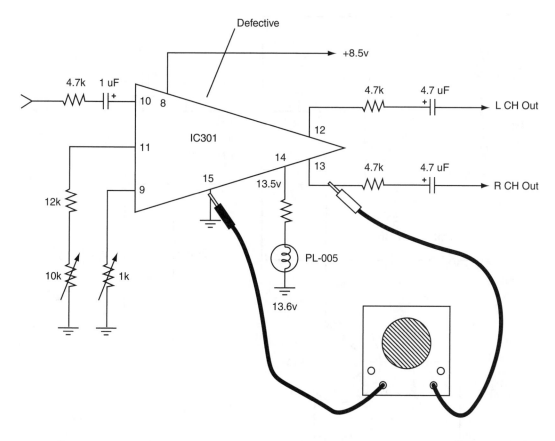

FIGURE 6-16. No sound was found with a defective IC301 in the right channel of an Aiwa CTR-2050Y radio.

TROUBLESHOOTING AUTO RADIO CIRCUITS

Checking FM Circuits with Voltage Tests

A quick voltage test on the supply voltage terminal of each FM IC component and on the collector terminal of transistors to common ground can indicate a dead or defective stage. Go directly to the FM MPX circuits with no stereo light and a dead right channel symptom. Take critical voltage measurements on the supply terminal 1 of IC502. Suspect an improper supply voltage or a defective regulator circuit with low or no voltage. Low voltage or no voltage at the collector terminal might indicate an open Q501 (**Fig. 6-17**).

Compare all voltage measurements with those found on the schematic, if one is handy. Take critical voltage measurements on each IC terminal and compare to the schematic. If the change in voltage is quite great, suspect a defective component close at hand.

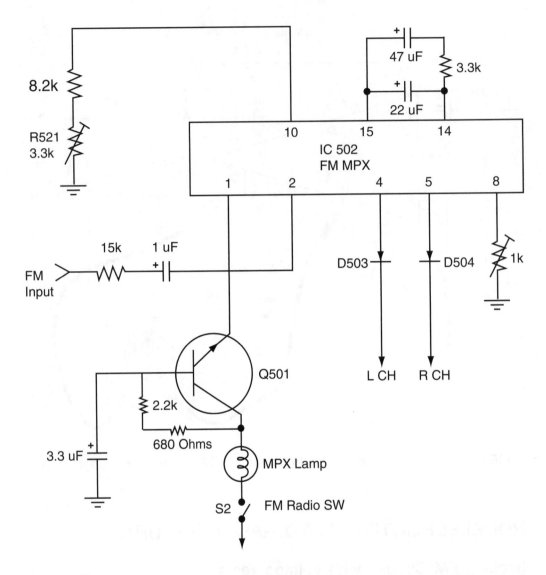

FIGURE 6-17. Low voltage at Q501 might indicate a defective MPX lamp indicator circuit.

Suspect the FM MPX IC or an FM IC component if there is no FM stereo signal out of either left or right channel. Critical voltage measurements on the supply terminal of the FM IF IC detector stage might indicate a leaky or open FM IC or a bad voltage source. A quick voltage test on the supply terminal of the FM IC or the collector terminal of the FM IF transistor can indicate a defective component nearby. Double-check all voltages on each terminal of the FM IC and compare to the schematic. Remember that the highest voltage found on an unmarked IC is the voltage supply terminal (Vcc).

When a FM station cannot be tuned in or is not heard in either channel, look for a defective front-end stage or tuning unit. Check the supply voltage from the power supply source to the tuning front-end. Low or no voltage might indicate a defective component within the front-end circuits or a defective power supply.

Quickly check the power supply source on another FM IC component with the same voltage source. Check for an open supply resistor or for one that has increased in resistance. Check each decoupling capacitor within the same circuit for leakage or ESR problems. Critical voltage tests on the supply voltage source can quickly locate a defective stage or voltage source.

Signal-Tracing the FM Stages with a Noise Generator

By using the speaker as an indicator, you can signal-trace the auto radio from the antenna to the audio output circuits with a noise or tone generator. Start at the volume control and go either way to determine which stage is defective. If the tone or noise is normal at the volume control, you may assume that the audio circuits are normal. Proceed toward the front end of the radio with the generator. Go from base to base of each transistor or from output to input of each FM MPX IC to the FM IC component (**Fig. 6-18**).

The volume should increase as each stage is signal-traced toward the front-end circuits. Suspect a defective stage when the generator noise ceases or becomes weaker.

FIGURE 6-18. Using a tone pencil generator to locate a dead AM or FM IF stage.

Check both the base and collector terminals of each transistor for the bad component. Suspect a defective IC, or a component tied to one of the IC pins, when the signal is weak or dead at the input terminal and sound can be heard at the output terminal.

Sometimes a defective transistor will test open or leaky within circuit tests, and when removed, the transistor tests normal. Replace the suspected transistor. Excessive heat from the soldering iron can sometimes restore the junction of the defective transistor. Other times, just moving a transistor terminal can make it act up or return to normal.

Check all terminal voltages on a suspected IC component. When the voltage is really low or has no voltage compared to the schematic, check each component tied to the same pin terminal. A resistance check to common ground of each pin terminal can indicate a leaky or shorted capacitor. Test all electrolytics tied to the various IC pin terminals with the ESR meter. Determine if the IC component is defective or if a bad part is tied to a certain IC pin terminal. Make sure that the IC is defective before removing it from the PCB.

Signal-Tracing FM MPX Circuits with External Audio Amp

The audio signal from the input of the FM MPX IC can be signal-traced all the way through to the audio circuits with an external audio amplifier. Start at the input MPX IC terminal and notice if the audio is normal to this point. Often, the FM audio signal is capacity-coupled to the input of the FM IC. Now check the audio at the left and right stereo output terminals. Go directly to the right output channel when the symptom is a weak or low audio output right channel. Check the left channel audio to see if both channels balance up (**Fig. 6-19**).

The FM audio signal tested normal at the input terminal 3 and was very weak at pin 6 in the right channel of the FM MPX IC102, on the external audio amp. A quick voltage check on pin 1 was really low, at 1.7 volts. Either the IC102 was leaky or improper DC supply voltage was supplied to the Vcc supply pin 1 terminal. The voltages on the IC were not all the same compared to the schematic.

To eliminate the power supply circuits and determine if IC102 was internally leaky or shorted, pin terminal 1 was removed from the PCB wiring. Excess solder was removed from pin terminal 1 with the soldering iron and solder wick. A quick resistance test between pin 1 of IC102 was quite high, indicating the service problem could be in the voltage supply source. The 33-ohm voltage-dropping resistor was running quite warm, with very little voltage measured after the terminal was removed.

A resistance measurement between the supply voltage of the PC wiring pad and common ground was less than 10 ohms. The 47-uF electrolytic was checked with the ESR meter and showed leakage. Replacing the 47-uF capacitor solved both radio output channels and restored the 9.7 volts to pin terminal 1.

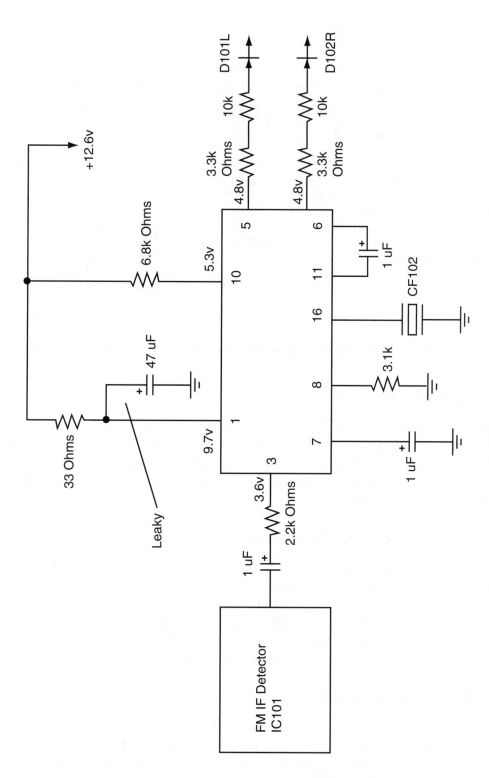

FIGURE 6-19. A leaky 47-uF electrolytic caused weak audio in the right channel of the FM MPX IC102.

Step	Generation Connection	Generator Frequency	Dial Setting	Adjustment	Output Connections	Remarks
FM IF	"	98.1 Mhz 1k Modulation	98.1 Mhz	Adjust Coil L101	Spk Terminals	Adjust Max.
FM V CO	"	90.1 Mhz	90.1 Mhz	L104 Osc Coil	TP103	Adjust max.
FM IF	"	10.7 Mhz	Quiet Point on the Dial	Adjust IF Transformers	Spk Terminals	Adjust Max.
FM MPX	"	87.9 Mhz 1k% Modulation	87.9 Mhz	VR351 in MPX Circuits	Frequency Counter to TP301	Adjust to 76 khz
FM MPX	"	98.1 Mhz 1k% Modulation	98.1 Mhz	VR 352 in MPX Circuits	Across the Spk Terminals	Separation for Max - More Than 30 DB

FIGURE 6-20. A typical FM alignment chart with the different FM generator frequencies.

Intermittent Left Channel

The left channel in a Sanyo FTC26 auto radio was intermittent and would sometimes fade in and out. All electrolytics and small bypass capacitors were moved around, and they tested okay with the ESR meter. When one end of IC301 was pushed downward, the sound began to cut out. Resoldering all terminals on IC301 solved the intermittent left audio channel (**Fig. 6-20**).

Typical FM RF and MPX Alignment

Connect the FM signal generator to the antenna receptacle and an output meter, VTVM, or FET-VOM across the speaker terminals. Clip a 4- or 8-ohm 10-watt resistor across the speaker terminals. Set the volume, tone, and balance controls at center rotation. First make the IF, RF, and MPX adjustments in that order (**Fig. 6-21**). Keep the signal generator output signal as low as possible to prevent overloading and clipping. Follow the manufacturer's alignment procedures found within the service literature.

FM Drifting—Sanyo FT415

In this case, even the local FM station sometimes would drift off frequency, with a roaring noise in the speaker. All voltages on IC1 were quite normal, and diodes D1 and D2 were good. A touch-up of the discriminator coil (T2) cured the FM drifting problem (**Fig. 6-22**).

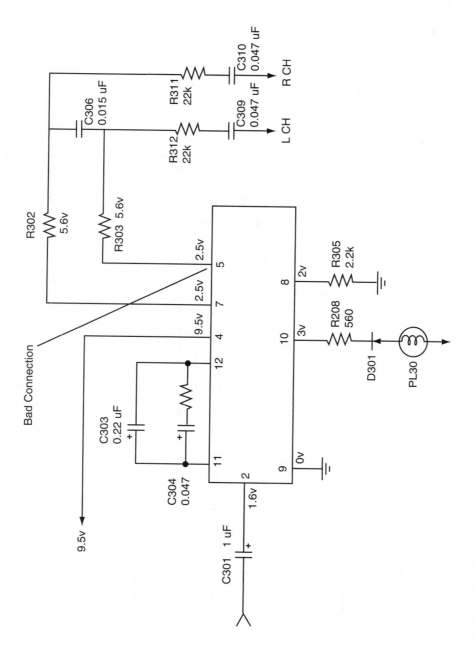

FIGURE 6-21. A bad pin 5 terminal of IC301 caused an intermittent left channel in a Sanyo FTC26 auto receiver.

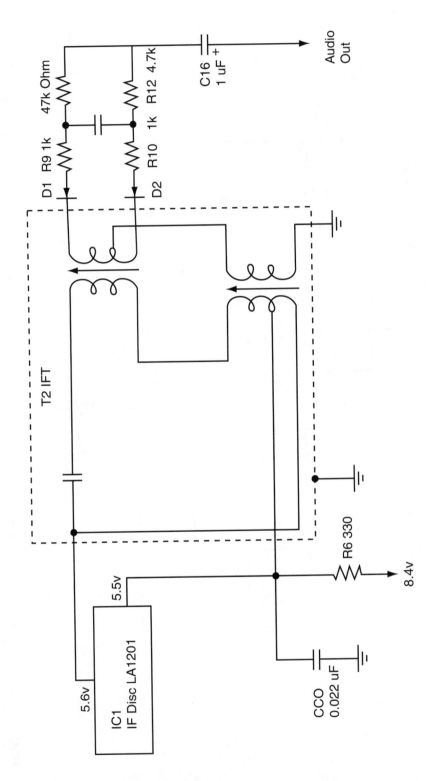

FIGURE 6-22. A touch-up of the discriminator coil (T2) solved the FM drifting symptom in a Sanyo FT415 auto receiver.

CHAPTER 7

Servicing Auto Preamp Audio Circuits

The preamp audio circuits are found ahead of the main amplifier to boost the weak input audio signal. Usually the preamp circuit is the first transistor or IC amplifier stage that feeds the amplified input signal to the volume control and audio output circuits. The audio equalization stage might also be called the preamp audio circuit.

A preamp audio stage is found to boost the AM/FM radio signal to the AF or power output circuits. The first audio stage connected to the audio cassette-tape head can be called the preamp circuit. The preamp stage might be called the first audio circuit within the auto CD player. Often, the preamp stage operates ahead of the volume control circuits.

The audio frequency (AF) stage is often found to amplify the audio signal after a preamp circuit. AF circuits operate within a frequency range that one can hear between 20 Hz and 20 kHz. The audible or analog signals are found as the first stage in the audio amplifier circuits. The AF audio circuits might operate ahead of or after the volume and tone-control circuits (**Fig. 7-1**).

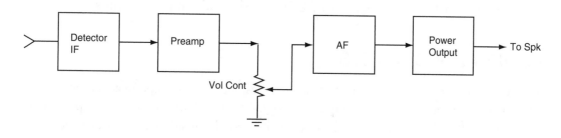

FIGURE 7-1. Block diagram of the IF detector, preamp, and AF amp circuits found in the auto audio amplifier.

THE EARLY PREAMP CIRCUITS

The early audio preamp stage consisted of a triode tube to amplify the audio signal within the car radio. Later on, the transistor replaced the radio tubes as the first preamp circuit. The transistor preamp circuit in the auto radio amplified the audio after the signal detector of the IF circuits. The preamp transistor input audio signal might be switched into the base circuit with a manual switching method or with silicon and switching diodes. A preamp radio stage is usually found where a very weak signal must be amplified before the audio stages, such as a tape head, or magnetic pickup in the record player.

A typical early record player and changer contained a crystal cartridge with a sapphire needle or diamond stylus. The crystal cartridge developed a higher output voltage than the magnetic cartridge. Very little voltage was developed with the magnetic cartridge, which must have a preamp stage to boost the audio signal to the audio circuits from the record. A variable-reluctance phono pickup is sometimes called a magnetic cartridge.

Likewise, the audio signal developed by a tape head within the auto tape player must have a preamp circuit to amplify the audio signal to the AF or driver circuits. The early transistor tape-head preamp circuits in the auto receiver might have a single transistor as the preamplifier stage.

Later on, the preamp tape-head preamp circuits connect to a directly coupled transistor circuit. Here, two different transistors are found to amplify the tape head signal with the collector terminal of the first preamp transistor tied directly to the base of the second preamp transistor (**Fig. 7-2**). No resistor or capacitor is found between the two transistors.

A 68K-ohm load resistor provides collector and base voltage to Q501 and Q502. Notice that the direct coupling of the collector terminal of Q501 is connected directly to the base terminal of Q502. Q502 emitter bias resistor of 1K ohms provides a bias voltage of 0.6 volts between the base and emitter terminals. The weak tape-head signal is capacity-coupled to the base terminal of Q501, and the preamp amplified output signal is coupled from the collector of Q502 to the audio output circuits.

IC PREAMP CIRCUITS

Today, a single or dual IC component amplifies the cassette tape-head audio signal. The dual IC amplifies the weak tape-head signal with a 1-uF electrolytic capacitor to pin 1 of IC201. The weak signal is amplified internally inside the IC and is coupled to the audio stages through a 4.7-uF electrolytic capacitor (**Fig. 7-3**).

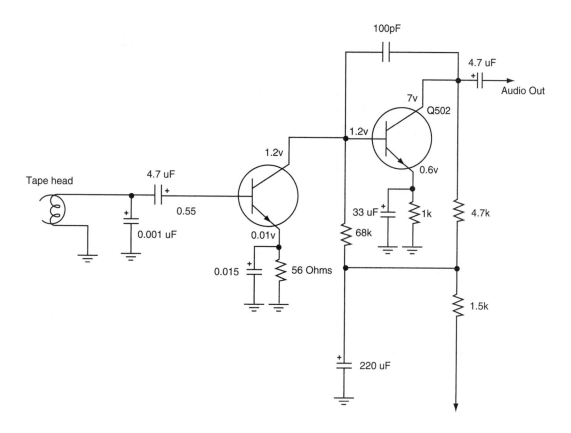

FIGURE 7-2. The directly coupled tape-head preamp circuits within the auto cassette player.

Both the left and right tape heads are capacity-coupled to the preamp IC. One half of the preamp IC might contain a single or dual IC component of the audio amplifier circuits. Likewise, a dual IC might contain both preamp tape-head circuits.

Notice that the tape-head preamp circuits are ahead of the mechanical or diode switching circuits. Often the preamp audio output signal is switched into the audio output circuits of an AF or driver amplifier circuit when the AM/FM radio signal is switched out of the audio circuits. The tape-head IC circuit is switched into the cassette audio circuits when B+ voltage is applied to the cassette amplifier circuits as the cassette is inserted.

The dual-tape heads are switched into the stereo circuits of a dual-IC amplifier component with two different 4.7-uF electrolytic capacitors. S1-1 and S1-2 switch the left and right dual-tape heads (CH1 and CH2) into the stereo input circuits. IC701

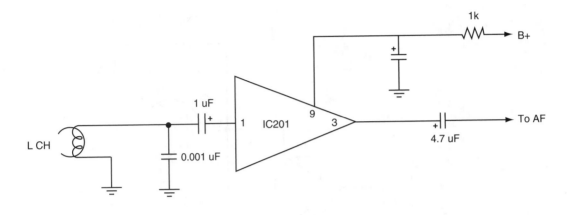

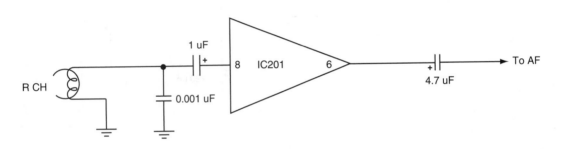

FIGURE 7-3. The left and right tape heads are connected to terminals 1 and 8 of a dual IC201 preamp circuit.

amplifies the tape-head signal and the output is found at terminals two and six. Again, the amplified tape-head signal is capacity-coupled to the audio driver or AF circuits (**Fig.** 7-4).

Tape Head-Radio Switching

When the cassette is inserted into the auto receiver-cassette player, the tape-head circuits are switched in by applying DC voltage to the tape-head preamp, audio, and cassette motor circuits. The battery voltage is switched to the tape-head circuits and out of the radio circuits. SW2 switches the tape-head audio circuits into operation by applying a DC voltage back to the radio circuits. When the cassette is out of the auto

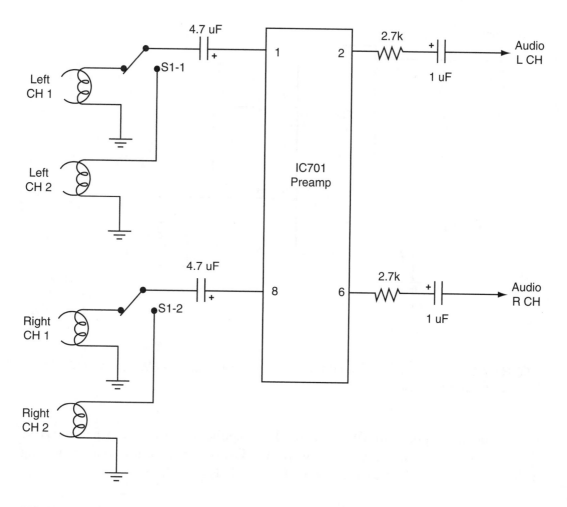

FIGURE 7-4. Left channels 1 and 2 and right channels 1 and 2, connected through a 4.7-uF electrolytics to the input stereo terminals of preamp IC701.

radio cassette player, SW2 switches the DC voltage back to the radio circuits. Now when the car radio's on/off switch is turned on, the radio circuits are activated.

AM/FM Radio Switching

The radio AM circuits become alive when the radio AM/FM switch controls SW1 is switched to the AM circuits of the tuner front-end circuits. Oftentimes the front dial radio circuits are activated with the FM and AM push buttons. The regulated DC

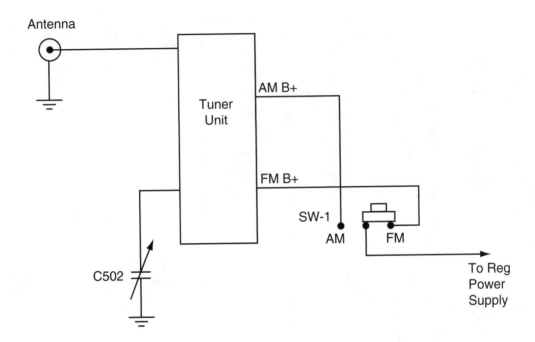

FIGURE 7-5. The AM/FM regulated power-supply switching with SW1 to the AM and FM supply voltage circuits.

battery voltage is applied to SW1 and switched to either the AM or FM B+ voltage of the tuner front-end circuits. A dirty or worn AM/FM switch can cause erratic switching or loss of switching of either the AM or FM band (**Fig. 7-5**).

RADIO OR TAPE-REGULATED VOLTAGE CIRCUITS

The radio or tape B+ voltage supply circuits might have transistor or IC voltage regulator circuits providing a regulated voltage to either circuit within the battery supply source. In **Figure 7-6**, the B+ AM/FM voltage source has a transistor-regulated voltage source supplied by TR101. SW-2 switches in the radio AM/FM supply voltage to the transistor-regulated circuit of TR101, and the regulated voltage is supplied to the AM/FM radio switch SW-1. SW-1 switches the radio B+ supply voltage to either the AM or FM radio circuits.

Audio Switching

In the early auto radio circuits, the audio was switched from the radio into the audio switches with a sliding manual switch arrangement. Later on, the radio and cassette circuits were

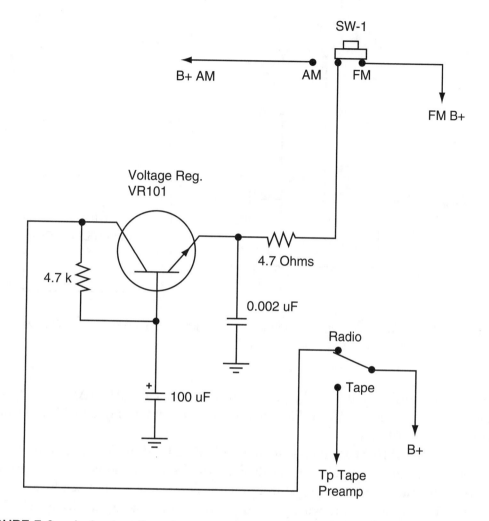

FIGURE 7-6. A simple radio voltage regulator circuit supplying regulated B+ voltage to the AM or FM radio circuits.

individually switched into the audio circuits with fixed silicon diodes. Because the AM radio circuits are manual, both AM switching diodes were tied together with the anode terminals, and were separated to the stereo circuits at the cathode terminals.

Likewise, the FM stereo circuits are silently switched into the audio circuits with fixed diodes. By placing the cathode terminals together in each stereo channel, the FM or AM circuits cannot feed back into each other's circuits.

Here the FM diodes D1 and D2 feed the FM signal into C101 and C102, which couples the FM radio signal to the input terminals 8 and 9 of the preamp or driver IC101 circuits. Diodes D4 and D5 feed the AM signal into C101 and C102 and onto

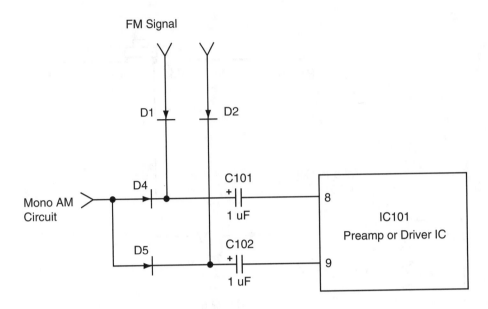

FIGURE 7-7. D1 and D2 silently switch in the FM radio band while D4 and D5 switch on the mono AM radio circuits to preamp IC101.

the input terminals 8 and 9 of the preamp IC101. The radio AM/FM switch applies a DC voltage to either the AM or FM circuits, which in turn is only applied at the input terminals of IC101 (**Fig. 7-7**).

In the early AM/FM radio cassette mono player, a fixed diode is found in the AM or FM output circuits, applied through fixed diodes to the audio volume control. Likewise, the monaural audio signal from the cassette player is amplified by a preamp transistor or IC, and then switched into the monaural radio circuits at a 4.7-uF electrolytic.

D101 applies the FM monaural signal to the volume control circuits, while D102 connects the AM radio signal. D103 applies the early monaural cassette audio signal to the 4.7-uF electrolytic and volume control circuits (**Fig. 7-8**). When the stereo signal is found in the FM MPX and stereo tape-head circuits, a fixed diode is found in each audio input circuit.

You might find that in some early audio circuits, the AM and FM audio signal was fed directly into the preamp audio circuits of electrolytic capacitors. Either the AM or FM stereo audio is switched into the dual-stereo tape-head circuits. Here the preamp tape-head audio circuits are functioning as the preamp or AF radio circuits.

A separate stereo radio AM or FM switch applies B+ voltage to the AM or FM radio circuits. The radio circuits are switched out of the audio preamp circuits when the

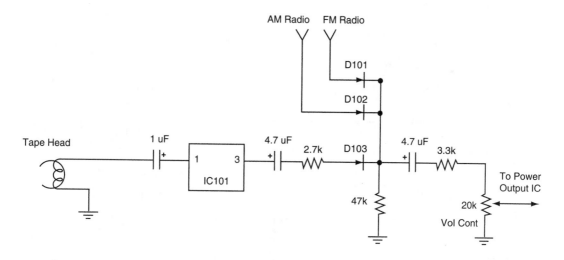

FIGURE 7-8. D101, D102, and D103 silently switch in the AM, FM, and tape circuits to the volume-control circuits.

cassette is inserted into the cassette player. Thus, only one radio signal is applied to the preamp or AF audio circuits (**Fig. 7-9**).

PREAMP CIRCUIT PROBLEMS

Weak Audio in a Sanyo FTC26 Auto Cassette Player

Very weak audio was heard in the left channel of either the tape player or radio circuits. Because both the AM and FM signals were weak, the defective component must have been in the audio circuits. Also, the left channel was weak when music was played on the audio cassette player of a Sanyo car radio. A loud hum was heard when the center terminal of the volume control was touched with a small screwdriver blade. This meant that the audio output circuits were fairly normal.

Hum was noticed in the speakers when the volume control was wide open and the DMM voltage probe touched the collector terminal of the AF transistor Q701. All voltage tests on Q701 were normal. Very little voltage was found when the negative side of C710 (4.7-uF electrolytic) was touched.

After inserting a 1-kHz audio signal at the positive terminal of C710, it was fairly normal but very weak on the other terminal of the coupling capacitor (**Fig. 7-10**). Replacing the open C710 with a new replacement solved the weak sound from the Sanyo radio-cassette player.

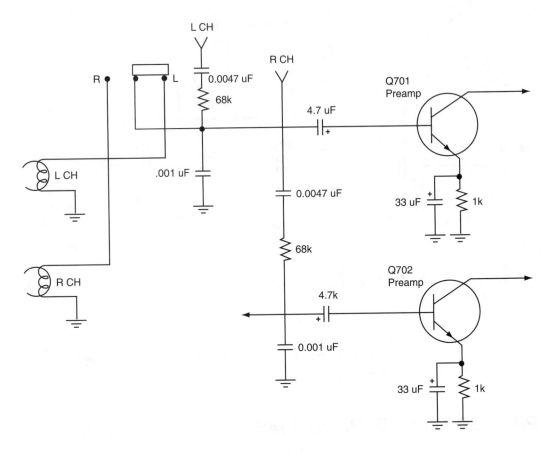

FIGURE 7-9. Here, preamp transistors Q701 and Q702 also provide AM/FM radio amplification as well as the stereo tape-head circuits.

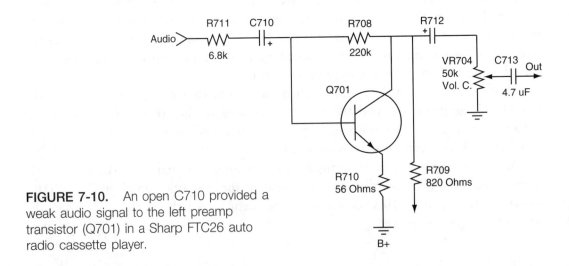

FIGURE 7-10. An open C710 provided a weak audio signal to the left preamp transistor (Q701) in a Sharp FTC26 auto radio cassette player.

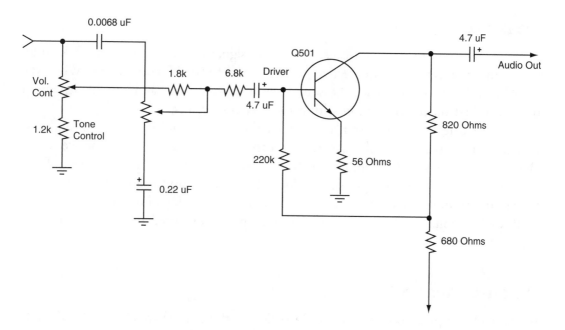

FIGURE 7-11. The tone and volume controls might be found in the base circuits of driver transistor Q501.

The single-driver audio transistor can be coupled to the volume and tone controls with a 4.7-uF electrolytic. Driver transistor (Q501) amplifies the audio signal and is coupled to the output circuits through another 4.7-uF capacitor (**Fig. 7-11**). The 50K-ohm volume control controls the audio signal that is applied to the base element of the driver transistor. Another 50K-ohm tone control is found within the base circuit of Q501. Today you will find both transistor and IC driver circuits within the auto-cassette receiver.

Loss of Audio in the Preamp Circuits

Replace the open coupling capacitor for a loss of audio in the speaker. Check for an open preamp transistor for loss of audio. A no-audio trouble might result from a badly soldered joint at the preamp transistor or voltage regulator transistor. Replace the open 1-uF 50-volt electrolytic coupling capacitor for loss of AM or FM reception.

Suspect an open preamp IC for a loss of audio in the tape-cassette circuits. A defective voltage regulator transistor or IC might cause loss of audio with a slight noise in the speaker. The no-audio symptom can result from a open 0.1-uF capacitor off of the input terminal of the audio preamp IC.

Weak Audio Signal

A leaky coupling capacitor can cause low audio signal in one channel. A weak audio signal might be caused by a defective preamp transistor or IC component. When the auto hits a bump in the road, the sound becomes very weak and can result from a broken or bad board connection within the preamp circuits.

Suspect a leaky preamp transistor when the left channel is very weak and distorted. A bad 1-uF electrolytic coupling capacitor can cause a low or weak left channel. The weak right channel can be caused by an increase in resistance of the emitter resistor of a preamp transistor. An open or dried-up bypass capacitor tied to the emitter terminal of a preamp transistor can cause a weak audio symptom. Suspect weak or low volume when pin 5 of the preamp IC is shorted internally to the B+ supply voltage.

Distorted Preamp Problems

Replace a leaky preamp transistor for distortion in the right channel. A leaky, directly coupled preamp transistor can cause tape distortion in the left channel. Check for a leaky coupling electrolytic when the audio slowly fades away and distortion sets in. When the sound becomes distorted after warming up, suspect a preamp IC and check the voltage on the preamp. A distortion in the left channel at high volume can result from a bad bias resistor in the preamp audio circuits.

Suspect a bad voltage regulator transistor or IC when the audio gets distorted after several hours of operation. A bad preamp transistor can cause distorted audio after operating for a few minutes.

Noisy Preamp Circuits

Replace the first audio preamp transistor for a microphonic audio chassis. Suspect a noisy preamp transistor when a static noise is heard with the volume turned way down. A low hissing noise with a slight hum can be heard with a defective preamp transistor when the volume is turned down. A noisy preamp transistor or IC can cause a noisy left channel. The open preamp transistor can produce loss of audio with hum in the speakers. A noisy preamp transistor can cause noise in the left channel that could not be balanced out with the balance control.

Check for a defective preamp IC with extreme noise in the left channel. Suspect a leaky 4.7-uF electrolytic coupling capacitor with noise in the right channel and the volume control turned down. A bad decoupling electrolytic in the preamp circuits can cause hum in the audio.

AF OR DRIVER CIRCUITS

The audio (AF) amplifier and driver circuits operate within the frequency range of 20 Hz to 20,000 kHz. The AF transistor or IC is the first audio amplifier connected to the preamp or volume control circuits. In the early car radios, the AF transistor was connected to the signal detector from the second IF stage. Today, the audio frequency amp might be a single IC component or within one large power output IC. The AF directly coupled stage is found in the radio, cassette, and both amplifier channels.

The driver audio circuits provide audio driving power to the audio output circuits. The driver stage provides power input signal or voltage onto the next stage. In the early auto receivers, the driver transistor provides audio to two different transistors in a push-pull output circuit. The driver transformer couples or transfers the driver signal to the audio output circuits. The power amplifier stage, whether a transistor, tube, or IC component, supplies signal to a higher audio-output power circuit.

Instead of two AF or driver transistors directly coupled, you might find one dual-IC component performing audio amplification to both stereo circuits. The IC AF circuit might contain more than one transistor in each stereo channel within the IC component. You will find AF IC and driver circuits within the latest auto receivers, cassette, and CD players.

TYPES OF DRIVER PROBLEMS

Music Fading Out in Chevy 986771 Model

When the Chevy car radio was first turned on, the music would fade out and become really weak in the speaker. The audio was still weak when a screwdriver blade was touched to the center terminal of the volume control. A quick voltage measurement on the first AF amp indicated higher-than-normal voltage found on the schematic of the collector terminal (10.7 volts) and zero voltage on the emitter terminal.

The DS-46 AF transistor tested normal with the diode test of the DMM within circuit transistor tests. Either the emitter resistor was open or the first AF amp transistor was intermittent. The emitter bias resistor (R2) was found open within the emitter circuit to common ground. When a 560-ohm resistor was subbed for R2, the music returned to normal (**Fig. 7-12**). R2 was replaced with a fixed 560-ohm resistor.

In **Figure 7-13**, the AF or driver IC1 provides signal-driving power to the audio output IC circuits. A 10K-ohm volume and a 100K tone control are coupled with a 1-uF electrolytic coupled to pin 1 of IC1. The auto radio IF detected signal is greatly amplified before being coupled to a 3.2-ohm oval speaker. Here the driver audio circuit is found within one large power output IC component.

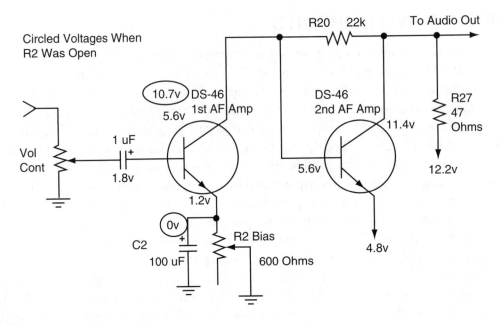

FIGURE 7-12. The music faded out in a Chevy 986771 auto radio with an open bias resistor R2.

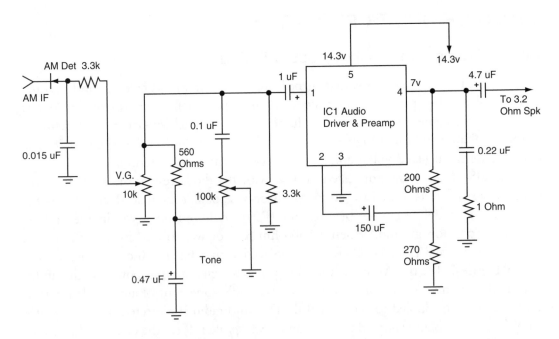

FIGURE 7-13. The AM radio detector is connected through a 3.3K-ohm resistor to the volume control and driver IC1.

AF or Driver IC Problems

The defective AF or driver transistor and IC might become leaky, shorted, or open. Extreme distortion with weak tape audio is heard with a leaky or shorted IC. An open front-end IC might have weak or no audio reception.

When the IC becomes leaky or shorted, the supply voltage will decrease with a slight change in voltage on the other IC terminals. The voltages on all IC terminals might match those on the schematic when the suspected IC opens up. Critical voltage measurements on each IC terminal and an input and output signal-tracing method can help to locate the defective IC component.

A loss of left-channel signal in the AF or driver transistor stage can be caused by a badly soldered connection on the collector terminal of the AF transistor. An open driver transistor can cause a no-audio symptom in the right channel. Suspect a shorted driver transistor with a weak and distorted right audio channel. A leaky driver transistor can cause a weak left channel with slight distortion.

A defective driver transistor can cause really weak or no volume in any channel, as well as cause a dead car radio with no sound. Suspect really weak or no audio output with an open AF transistor.

Replace the AF IC for no audio out of either channel. Check for a bad driver IC with no audio output at the speakers. For no tape or tuner audio, replace the input or AF driver Audio IC. Suspect a shorted 0.1-uF bypass capacitor off of the input driver IC for no audio in either stereo channel.

Noisy or Distorted AF or Driver Stage

Replace a leaky driver transistor for a distorted audio right channel. Check for a leaky driver transistor for a weak and distorted left channel. When the sound gets distorted after warming up, replace the audio driver transistor or IC. Check for a change in a bias resistor of the AF or driver transistor when the sound is distorted and the volume control is advanced halfway off of rotation. Replace both driver and output transistors with distortion in the left channel. Check for a defective driver IC with distorted audio.

A defective AF IC can cause a noisy high-pitched sound in the speaker. Suspect a noisy driver transistor with static noise in the speakers with the volume turned down. A noisy AF IC can cause the frying noise in the speakers.

A bad input amp IC can cause a noisy left channel. A slight hum and a noisy left channel can result from a noisy driver transistor or IC. The noisy right channel with the volume turned down can result from a defective 4.7-uF electrolytic coupling capacitor in the driver circuits. An open AF amp transistor or IC can cause a humming noise without audio or any stations tuned in.

Noisy Pace XMC-3763 Left Channel

The audio was really noisy with a hissing sound when the volume was turned down on a Pace auto receiver. The hissing noise was present in both the tape and radio left channel. Of course, this indicated that the noisy component was in the audio circuits. An external audio amplifier was brought to the scene to try to locate the noisy stage.

Starting at the output circuits and signal-tracing toward the front of the audio circuits, the noise was still present at pin 10 of the left IC4 AF amplifier. When the external audio amplifier was connected to the volume control, the music was normal; no hissing noise could be heard. No noise was found at the input terminal 6 of IC4. IC4 was suspected of being noisy and was replaced, solving the noisy left channel (**Fig. 7-14**).

Intermittent AF or Driver Circuits

Suspect a badly soldered terminal connection on the AF transistor when the audio quits or cuts out. A leaky driver transistor can cause the right channel to cut out. Check for an open bias control on the first AF amp transistor when the radio fades out. Suspect poorly soldered contacts on the driver transistor with intermittent volume in the speakers. Poorly soldered connections on the regulator transistor or IC can cause intermittent volume.

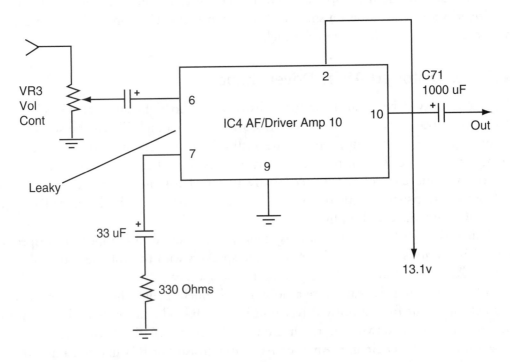

FIGURE 7-14. AF driver amp IC4 in a Pace auto radio caused a hissing sound in the speaker.

Simply prod or move the coupling capacitor suspected of causing the intermittent tape and radio sound. Sometimes just touching the suspected component can cause it to act up. The radio signal might also come and go with badly soldered connections on the volume control terminals.

Voltage Control Problems

Suspect a bad volume control when the audio cuts up and down in the speaker. Loss of volume or audio can be caused by an open volume control. Check for a bad ground connection on the volume control when you cannot turn the volume down. Replace or clean up the volume control with a noisy rotation when the radio is first turned on (**Fig. 7-15**). Check for a bad volume control when the audio cuts out at a high volume setting.

Check for a loose ground wire on the volume control when the volume cannot be turned up or down. Look for a bad volume control connection on the PCB with intermittent audio and hum created with the volume wide open. Check for a crack in the foil pattern on the volume control if the volume becomes erratic or intermittent when the car hits a bump in the road. Intermittent and noisy volume control can result from a bad or worn control. No sound in the left channel can be caused by a defective volume control. A worn volume control can cause the volume to cut in and out at a certain control setting.

Poor or no volume balance can be caused by a bad or worn balance control. Suspect a bad balance control when one channel is weaker than the other or when stereo balance is poor. Audio that will not balance up can result in an open volume control in the right channel. Clean up or replace the noisy tone control. Intermittent or erratic tone operation can be caused by a poorly soldered joint or connection on the tone control.

SPECIAL CONTROLS

When one channel becomes weaker than the other channel, the balance control must be rotated toward the normal channel to provide equal volume out of the stereo speakers. The left channel might be louder than the right when the balance

FIGURE 7-15. Clean up the dirty and erratic volume control with cleaning fluid applied inside of the control.

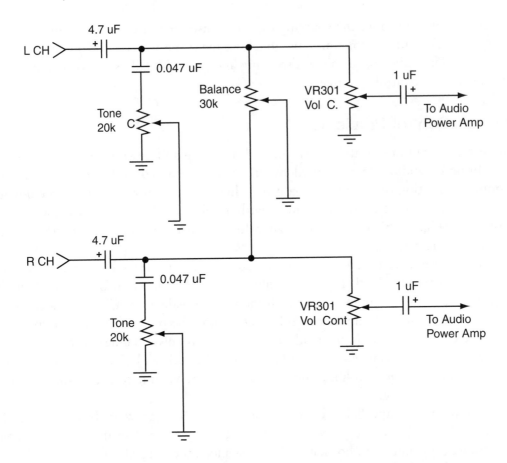

FIGURE 7-16. The tone and balance controls of the auto receiver are generally found in the volume control circuits.

control is not set at zero, thus indicating a weak right channel. The stereo channel will not balance up when the audio channel is a lot weaker than the other channel. The weak component must be located so the audio circuits will balance up.

The balance control with the tone control is usually connected ahead of the volume and tone control. The stereo signal from the left and right AF or driver stages are applied to the outside terminals while the center terminal (wiper) is grounded (**Fig. 7-16**). Special volume, treble, bass, dual, and balance controls must be obtained from the manufacturer.

When the stereo amplifier channels will not balance up, suspect a weak audio channel. A defective balance control might not balance the stereo channels with an open or broken control. Check both terminals of the control for poor or broken wire connections. Inspect all wiring and board connections. Make sure that the center terminal of the balance control is grounded. Check for correct resistance of

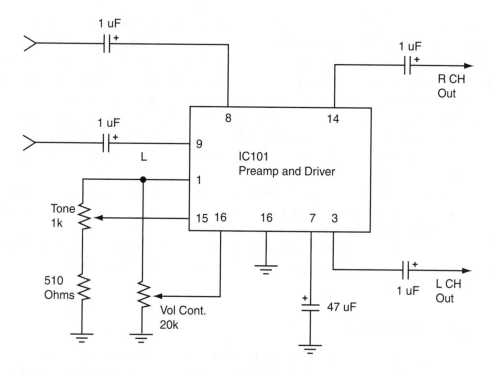

FIGURE 7-17. Here the tone control is found at pin 1 of the stereo driver IC101.

the balance control. Replace the defective control if open or erratic with a linear output taper.

Replace all volume controls with an audio taper. The balance or tone control should have a linear taper-type replacement. Dual controls should be replaced with the manufacturer's original part number.

The tone control might be located in the same circuits as the volume control. The tone control might be located ahead of or behind the balance control. In the auto stereo driver IC circuit, the tone control is found at pin 1 and ground of IC101 (**Fig.** 7-17). The tone control might be a single mounted control and be rotated, or might have a sliding type control (Ton), located on the front car dash panel beside the balance (Bal) and volume (Vol) controls.

OTHER PREAMP PROBLEMS

Hum PickUp

A slight hum pickup might exist in the preamp and AF circuits with a defective coupling capacitor. Clip a known electrolytic across the suspected coupling capacitor and observe correct polarity. Rotate the volume control to minimum. Notice if the hum has disappeared. If not, remove the subbed capacitor from the chassis.

Check for a preamp or AF amp transistor or IC for hum in the audio. A shorted internal base terminal to the collector element of the AF transistor can cause motorboating and a humming sound. Suspect an open tape-head winding or a torn connecting wire with a loud hum and a rushing noise with no music. Clip a known coupling capacitor across the suspected one for a really low humming noise and no-audio symptom. Measure the resistance of all base and emitter resistors of the preamp and AF transistors for incorrect resistance.

A slight humming noise might be heard with an poor or open ground connection of the volume control or cracked foil around the control PCB. Check for a leaky voltage regulator transistor feeding the AF or preamp voltage supply circuits for hum, especially when the volume control is advanced. A noisy preamp or AF transistor can produce a microphonic noise with a slight hum in the speakers. Poor PCB grounding can cause low hum in the speakers.

Defective PC Boards

Poor PCB terminal connections of the preamp or AF transistors can cause a dead, intermittent, weak, and distorted audio circuit. Push up and down on the PC board with an insulated tool. Sometimes poorly tinned transistor terminals might produce an intermittent connection under a large blob of solder. Resolder all transistor and IC terminals on the PCB to help locate the poor foil connections.

Poorly soldered PC board connections of a function or tape-switching contact can cause intermittent sound or noisy operation. If the switching contacts are still intermittent after cleanup with cleaning fluid, check for poorly soldered contacts on the PCB. Double-check for poor connections from the switch contacts to the foil connections with the ESR meter. The ESR meter should sound off with poorly soldered connections.

Sometimes just moving components around with an insulated tool can cause a resistor or capacitor terminal to act up in the AF circuits. Pushing up and down on the PCB around suspected components or a tie-wire can uncover a poor connection. Sometimes, tightening up the mounting screws of the preamp or AF PCB can solve intermittent and noisy audio problems. Placing a star washer over the PCB screw or bolt can make a better ground connection and eliminate a low hum pickup problem.

Loud Hum and Rushing Noise—Cassette Preamp

In a Sanyo FT415 auto radio-cassette stereo recorder a loud hum and rushing noise was heard with no music, when the volume control was turned up in the left channel.

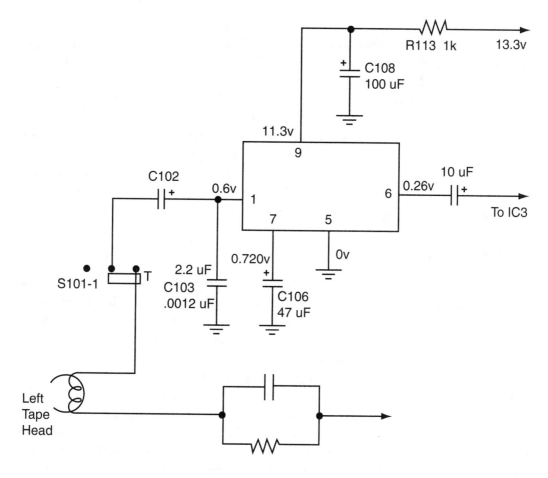

FIGURE 7-18. C102 (2.2-uF) electrolytic was replaced in a Sanyo FT415 car radio for a loud rushing noise in the speaker.

Because an open or defective tape head can cause this problem, both the tape-head continuity and connections were checked without any results.

All voltages were checked on the preamp IC1 (TA7137) and were fairly normal. When the red voltage probe touched C102, the rushing noise quieted down. Replacing coupling capacitor C102 (2.2 uF 10 volts) solved the hum and rushing noise (**Fig. 7-18**).

Distorted Preamp Circuits

Although most audio distortion is found in the audio output circuits, leaky components in the preamp can also cause distortion, as can a leaky preamp transistor or electrolytic coupling capacitor. The weak and distorted audio symptom might result from a leaky preamp IC. Suspect a leaky AF transistor for a weak and distorted audio problem. Check for a bad or dirty radio-cassette switch with distortion and garbled music. Slight distortion in the right channel might be caused by an open AF transistor, but that tests normal when tested in-circuit. Replace the suspected transistor, if in doubt.

A change in resistance of a burned bias resistor of the preamp or AF transistor can cause a distorted symptom. A dirty tape head can cause distortion in the recorded sound. The right channel in a tape player can be distorted with a defective tape head.

Replace the preamp or AF transistor for a microphonic-distorted noise. Apply several coats of coolant on each transistor or IC to locate the one that produces the microphonic sound. Sometimes just touching the top of the AF transistor or IC can create the microphonic ring in the speaker.

Weak Preamp Circuits

Check and shunt electrolytic capacitors within the preamp circuits for a weak audio symptom. The leaky AF amp transistor can cause weak reception. A leaky preamp or AF transistor or IC can cause a weak audio stage. An open preamp transistor might cause loss of audio in that channel. A defective preamp transistor might cause a very weak and intermittent-sound symptom.

Shunt electrolytic capacitors in the emitter circuits for weak sound in the AF circuits. Do not overlook a dirty tape head in the cassette player for a weak audio channel. A weak preamp circuit might result with an improper voltage source.

You can quickly signal-trace the preamp circuits by taking critical waveforms throughout the preamp circuits with a 1-kHz test cassette that is operating within the cassette player. Start at the volume control and work toward the tape heads in the preamp or AF circuits. The test cassette should show a waveform on the scope if the preamp stages are normal. Check the weak waveform against the normal audio channel at the volume control, coupling capacitors, collector, and then base terminal of the preamp or AF transistor to locate the defective component.

After locating the weak stage, check each transistor with in-circuit transistor tests. Measure the bias resistor and check the resistance with the schematic or color code. Sometimes the resistor might overheat and change the color bands on the body of the resistor. Remove one end of the resistor for a correct measurement. An increase in resistance of a bias or base resistor can cause a weak audio signal with some distortion.

Dead—No Audio in Preamp or AF Circuits

The dead chassis with no audio is much easier to service than a weak or intermittent symptom. No audio in the preamp or AF circuits can be signal-traced with an external amplifier or scope. A clicking test is a quick method to check the front-end circuits.

Turn the volume control wide open and listen for a humming sound. Touch the tape-head ungrounded wire, tape-head coupling capacitor, or base terminal of the first preamp transistor with a small screwdriver blade. A clicking noise or a hum should be heard in the speaker, but no click, no hum, no operation, and no dead stage.

Another method is to use a screwdriver blade across the front of the tape head and listen for a thumping sound in the speaker. Each time the metal screwdriver blade passes across the small gap in the tape head, a thump is heard with a normal tape-head and preamp circuit. Suspect a defective tape head, broken tape-head wires, open coupling capacitor, or preamp transistor with no thumping noise heard in the speaker. An open preamp or AF transistor might cause the no-audio symptom. A leaky preamp transistor can cause a no-audio sound. A dirty or worn input-switching terminal can cause a no-audio symptom. A burned or open emitter resistor can cause a loss of sound. The broken ground connection of the tape head in the cassette player might produce a dead chassis.

An open preamp IC with a normal voltage source can cause dead front-end circuits. Improper or no voltage applied to the preamp or AF circuits can produce a dead chassis. An open voltage regulator transistor in the low voltage power source can cause the dead preamp stage.

When both channels are dead, suspect a leaky preamp IC or shorted decoupling capacitor in the voltage source that is feeding the preamp circuits. A leaky AF or driver transistor might result in no audio with hum in the speaker. A dead auto receiver with a low humming noise can result from a leaky decoupling capacitor in the voltage source of the preamp circuits.

Defective Tape-Head Circuits

The dirty tape head within the cassette player might cause weak or distorted music. A heavy-packed tape head with oxide dust might be dead or might produce weak and distorted music. Broken tape-head wires might result in a loud rushing or howling noise in the speakers. A loud rushing noise with no music indicates a broken ground wire on the tape head.

Suspect an open tape-head winding with a broken wire when moving one of the tape head terminals, and when the music cuts in and out. A worn tape head can cause missing high notes of the recorded music. Replace the worn tape head. A packed-oxide tape head might cause a weak right channel (**Fig. 7-19**).

Replace the defective tape head when it will not record in the left or right channel. A weak and rushing noise in the recording might result from a defective tape head. Sometimes the recording will come to life when pressure is applied from an insulated tool on the tape head. Moving the tape head might remove the distorted recording in the right channel.

A broken ground wire on the erase head might cause a jumbled recording. The open erase head will leave a jumbled or mixed-up recording. Excess oxide on the erase head can cause a jumbled recording; the previous recording might be partially erased.

Most weak and distorted cassette music can be cured by thoroughly cleaning up the tape head. Clean up the front of the tape head with alcohol and a cleaning stick. An old toothbrush dipped in alcohol can clean up the packed tape head.

FIGURE 7-19. Clean up the auto stereo tape head to remove packed tape oxide that can create weak and distorted sound.

Check the tape head for broken connecting wires or poorly soldered connections. You can check the suspected tape head by clipping the external amp across the tape-head terminals. Now play the tape head music through the external amp. Suspect a bad head if one side of the tape-head music is quite weak compared to the other stereo winding.

Measure the continuity and resistance of the tape head. A typical cassette tape head has a resistance from 200 to 850 ohms. A defective stereo tape head might have a different resistance measurement than the other tape-head winding (**Fig. 7-20**). For instance, if any one of the tape heads measures 323 ohms and the other 325 ohms, the stereo head is normal. If one side measures 323 ohms and the other winding measures 450 ohms, replace the defective tape head.

SIGNAL-TRACING TAPE-HEAD CIRCUITS

The tape-head circuits can be signal-traced with a 1-kHz test cassette or with the signal from an external audio signal generator. Solder a piece of flexible hook-up wire across the two ungrounded terminals on the tape head, or clip the two head wires together with a shorting clip wire. Now connect the audio signal generator to the ungrounded wire and the ground clip of the generator to the grounded lead on the tape head.

Set the external audio generator to 1 kHz and then scope the preamp, AF, or driver transistor circuits with the scope. Likewise, scope the input and output of the preamp circuits with the scope probe. Notice the gain of each channel as you proceed through the front-end circuits.

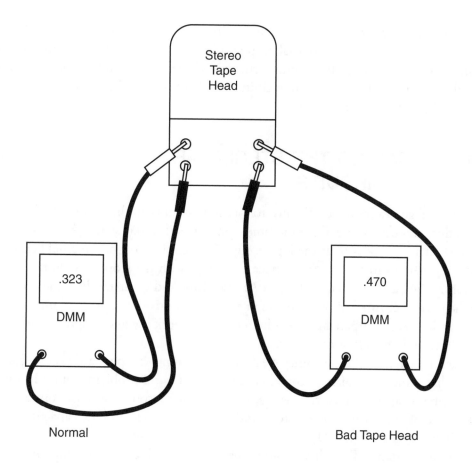

FIGURE 7-20. Comparing the tape head resistance of each stereo channel with the DMM.

Check the signal at the base terminal of the preamp transistor or the input terminal of preamp IC. When the signal stops, or becomes weak or distorted, check that stage with voltage and resistance measurements. Compare the voltage measurements of the normal channel with the bad stereo channel. Check those voltages on the preamp transistors with the exact schematic, if handy.

Measure the collector load resistor, base, and emitter resistors for correct value. Likewise, take a critical resistance measurement from each IC terminal to ground and compare with the normal stereo channel. The dead, intermittent, or weak preamp or AF circuits can be located with audio signal tracing methods, correct voltage, and resistance measurements.

Another method is to insert a 1-kHz cassette test tape and then either scope the audio signal or signal-trace the audio with an external audio amplifier. A regular music tape can be used to signal-trace the tape-head circuits when an external amp is used.

Check the signal at the base of each transistor and notice the gain of each stage. Monitor each stage with the external amp and leave connected at that point until the intermittent music acts up. If the music cuts out at the speaker and is still heard in the external amp, proceed to the volume control to determine if the intermittent stage is in the final preamp or AF circuits.

SIGNAL-TRACING THE STEREO AF OR DRIVER CIRCUITS

Clip the signal generator to the AF transistor or input terminal of the preamp IC. To signal-trace both stereo channels at the same time, connect both input stereo channels together after the switching diodes or at the input coupling electrolytic of the AF amplifier. Adjust the signal generator to 5 kHz. Signal-trace each stage with the scope as indicator.

Start at the base terminal of the first AF transistor and compare the same waveform with the normal channel. Proceed through the AF and driver circuits, and notice which stage is really weak compared to the other channel. Check the audio signal on each side of an electrolytic coupling capacitor.

When the signal becomes quiet weak in one channel, suspect a defective component in that same place of the normal channel. Most weak problems found in the AF or drive circuits are caused by transistors, ICs, electrolytic coupling capacitors, a change in resistance of emitter or base resistors, and bad bypass or electrolytic capacitors across the emitter resistor (**Fig. 7-21**).

The intermittent AF or driver audio channel can be located with the same signal generator connected to both input stereo channels. Instead of the scope as indicator,

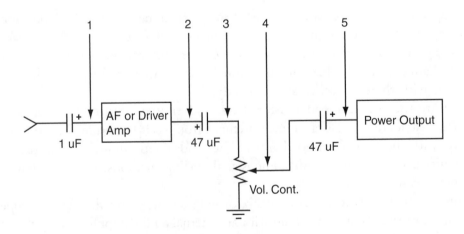

FIGURE 7-21. Connect the external amp by the numbers to locate a weak or intermittent stage in the audio circuits.

connect the external amp to troubleshoot the various stages. You do not have to watch the scope for an intermittent problem; the audio amp begins to act up when the intermittent or erratic stage acts up. Most intermittent components within the AF and driver circuits are transistors, ICs, cracked resistors, internal capacitor breakage, and poor foil wiring or component terminal connections.

Weak Left Channel—Pioneer KP3800E

The left channel in a Pioneer auto radio was really weak in both AM/FM/MPX and tape player. The left channel was still weak when the external amp was connected to the volume control. The balance control was rotated to one side to balance up the weak left channel. Of course, this meant that the weak audio stage was ahead of the volume-control circuits. The signal at the base of AF transistor (Q1) was normal in both channels. When the audio amp was connected to the emitter circuits, the left channel was really weak.

A quick voltage test indicated a higher voltage on the emitter and base terminals of Q1 than on the right AF transistor; Q1 was tested with an in-circuit transistor test and showed that Q1 was leaky. When Q1 was removed and tested out of the circuit, the AF transistor tested good (**Fig. 7-22**). Q1 was leaky and was restored to normal when heat was applied from the soldering iron. Replacing Q1 solved the weak left channel.

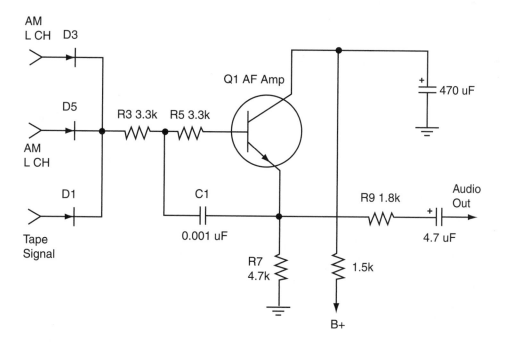

FIGURE 7-22. Leaky AF amp Q1 caused a weak left channel in a Pioneer KP3800E auto radio.

CHAPTER 8

Repairing Auto Radio Sound-Output Circuits

The early auto radio circuits consisted of a driver, AF, and push-pull circuit of transistors. Later on, the push-pull stages were replaced with a single output transistor with a tapped-output transformer stage. Today, the audio output circuits might consist of power output transistors and IC components. The high-powered output transistor amplifier might have 10 or more power output transistors (**Fig. 8-1**).

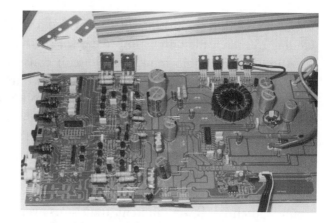

FIGURE 8-1. The high-powered amp might have more than 10 audio output transistors.

OUTPUT CIRCUITS

Early Auto Transistor Circuits

In the early auto circuits, the audio signal from the last 262.5-kHz IF stage was detected by a fixed diode (1N295) and fed to the volume control. The center terminal (wiper) of the volume control coupled the controlled audio to the base of a driver transistor. PNP transistors were found in these audio circuits. An inter-stage transformer coupled the collector primary winding to the secondary and into the base of the single-ended audio

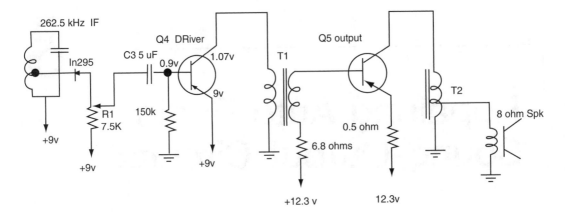

FIGURE 8-2. The coil tap on T2 equals the impedance of the auto radio speaker.

output stage. Auto transformer (T2) provided a speaker load with a tapped transformer winding to match the output impedance of the 8-ohm PM speaker (**Fig. 8-2**).

Another early audio transformer circuit consisted of an AF stage directly coupled to a driver transistor. The driver transistor was transformer coupled with two separate transistors in a push-pull audio output circuit. NPN transistors were found in this type of a driver circuit. The collector terminal of the driver transistor was coupled to a primary winding of an inter-stage transformer. The secondary circuits of T1 consisted of two separate windings feeding directly to separate audio output transistors.

Two separate audio output circuits were found in the stereo audio circuits. Here, the two push-pull transistors were PNP types with the collector terminal of Q4 at ground potential. Two different transistors were found in the base circuits of the transformer windings. Transistors Q4 and Q5 amplified the audio, and capacity-coupled to the 8-ohm speaker circuits with a large electrolytic capacitor. A speaker fader circuit provided the correct volume to either the front or rear PM speakers (**Fig. 8-3**).

Single-Ended Transistor Output Circuits

In the early transistor car radios, the audio output stage used a single power output transistor as the load. The base terminal was driven directly and sometimes with a resistor found between the AF stage and the audio output transistor. In the single-ended outputs, the audio base circuit was driven with an inter-stage transformer. The emitter terminal might be connected to a DC source through a bias resistor.

The power output transistor was a germanium PNP-type power transistor that was mounted on a heat sink or on the metal car radio chassis. The collector terminal was the outside shielded area of the power output transistor and insulated away from the

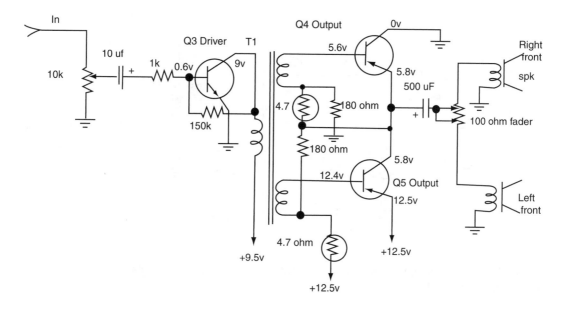

FIGURE 8-3. The 100-ohm fader control provides a louder volume at the front or rear of the automobile.

heat sink with insulating material. The collector load component was the power output transformer and speaker. The PM speaker was connected at a tap on the output transformer, or the transformer served as a load with the speaker connected to the collector terminal of the power output transistor.

Within the power output transistor circuits, the emitter terminal is always more positive than the collector terminal, which is connected to a common ground. The base and emitter bias resistors are always very low ohm-type resistors. The emitter bias resistor measures less than 1 ohm and is sometimes called a fusible resistor. Often, when the power output transistor becomes leaky or shorted, the fusible resistor will open up. The defective power PNP transistor could be replaced with either a GE-3, SK3009, and ECG104 universal replacements (**Fig. 8-4**).

Audio Output Transistor Problems

In this example, the audio output transformer had impedance from 12 to 25 ohms with a tapped secondary matching an 8- or 10-ohm speaker. A very low voltage on the collector terminal might have been from 0.7 to 1.5 volts that was also applied to the output transformer. In some single-ended output circuits, a 10-ohm speaker connected directly to the collector terminal with 1.5 volts applied, and the transformer inductance served as the load to common ground. The tap on the autotransformer measured a different ohm value, even when it matched either an 8- or 10-ohm PM speaker.

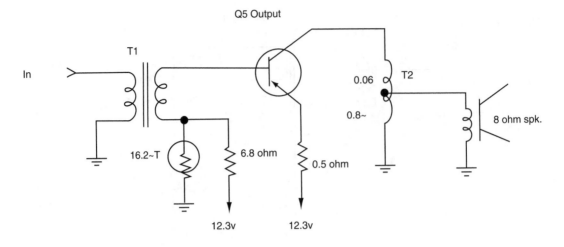

FIGURE 8-4. Universal audio output transistors can replace the single-ended power output transistor in the early car radios.

When the audio output transistor became shorted, and no emitter bias was found in the emitter circuit, the transformer might have begun to smoke and become damaged if the car radio was left on too long. The outside covering on the output transformer could have become burned and the transformer winding enameled wire become overheated and destroyed. Both the audio output transistor and transformer were damaged. Usually, the PM speaker is not damaged unless the transformer ground wire opens up.

Although the audio output single-ended transistor can be replaced with universal transistor replacements today, the power output transformer is not available. Either the damaged transformer must be repaired or a sub must be found on another old car radio, using the same type of output circuits found in a junked car radio. Repairing the tapped transformer is not as difficult as it may seem; just wind on a new coil.

Remove the old output transformer by drilling out the chassis rivets with a hand drill and bit. Remove the outside cover of the burned coil winding, cut it loose, and pull out the burned winding. These output transformers are wound with fairly large 22- 36 enameled magnet wire. After the burned winding has been removed, place a layer of masking tape over the inside transformer core to insulate it against the metal transformer framework.

Determine what size of wire is used with a wire gauge or by comparing the size of the enameled wire. Leave about four inches of enameled wire to make the connection to the collector terminal. Most singled-ended audio output transformers have a 0.4 to 0.8 ohm winding before the tapped joint. The bottom half of the coil winding is 0.8 to 1.5 ohms. If a Sams Photofact Radio Series manual is handy, check the coil resistance on the schematic.

For instance, if the top half of the old transformer measures 0.8 ohms, remove three to six feet of wire from the magnet spool and scrape off the enamel for an ohmmeter reading. Use an accurate low-ohm scale of the DMM to check for the correct resistance. Do not cut the wire until the complete coil is wound on the transformer core. If the three feet measures only 0.5 ohms, proceed to another spot on the enameled wire and scrape off enough enamel to get a good measurement. Keep winding the coil until the winding is around 0.8 ohms. An accurate DMM is needed for this measurement. Now cut the wire (**Fig. 8-5**).

Start feeding the cut end into the transformer area and wind each coil winding, side by side until the area is full, and then wind on another layer to continue the coil winding. Try to wind each winding side by side, and try not to kink the wire up when the area is looped through the core area. Wind the coil very tight to the correct resistance and then tin the enameled ends for a tapped connection. Hold the winding in place temporarily with a piece of cellophane or masking tape.

Now take another piece of the same-sized magnet wire and cut it to the correct resistance of 1.2 ohms shown on the schematic. Of course, the bottom half of the coil will have a longer piece of wire compared to the top half of the transformer winding.

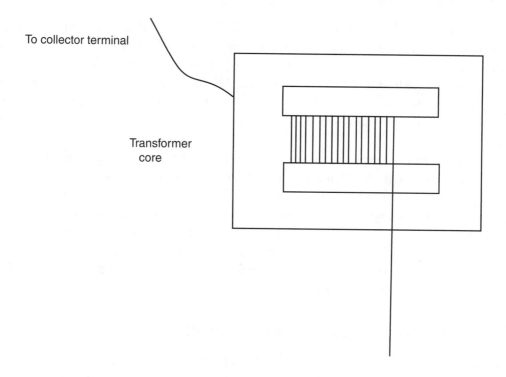

FIGURE 8-5. The burned output transformer can be repaired by winding the core with new enameled wire.

Wind the coil wire in the same direction. Solder a four-inch length of tapped wire at the junction for the speaker connection.

All three wire ends are soldered neatly together to contain a tapped connection, and are taped up. Now finish the secondary coil winding on the same core area. Wind the final winding in the same direction. Keep the soldered junction as small as possible, and cover up with masking or electrical plastic tape.

Place a larger layer of masking tape over the final winding with spaghetti insulation pushed over the enameled ends. Tin all three ends after scraping off a layer of enamel for a good connection. Apply a coat of solder paste over the wire ends to make a good-tinned connection. Now measure the total resistance of the entire winding, which should be around 2 ohms.

The tapped wire goes to the speaker connection, the wire with the least resistance (0.8 ohms) to the collector terminal, and the 1.2 ohm wire soldered to the common ground terminal. If a schematic is not handy to determine the transformer coil resistance, wind the top half at 0.8 ohms and the bottom half at 1.2 ohms for a DS503 or DS501 output transistor load impedance.

Early Directly Driven AF Amp Circuits

Within the early AF amp transistor circuits with singled-ended audio output stages, several audio transistors might be wired in series or cascade to another AF transistor. The collector terminal of the first AF transistor (NPN) is connected directly to the base of the following AF amp transistor. A second AF amp transistor (NPN) is connected directly to the base of the first PNP power output transistor. The collector terminal of the power output transistor is tied directly to a 10-ohm PM speaker and load output transformer.

Often, when the output transistor becomes leaky or shorted, the directly driven AF or driver transistor is also damaged. The driver transistor might go open and cause damage to the connected output transistor. When either transistor goes open or becomes leaky, the voltage on the directly driven transistor will also change. Check all three directly driven transistors with the diode test of the DMM with in-circuit tests.

Check for burned or open bias resistors within the base and emitter terminals of the audio output transistors. Replace all bias resistors that have changed resistance within the directly driven AF transistor circuits.

Push-Pull Audio Output Transistor Circuits

For greater power output in the audio output circuits, you might find two power-output transistors in a push-pull output circuit. The AF amp might consist of AF amp IC or a module driving the AF transistor and output circuits. Notice that the AF amp and both audio output transistors are directly connected to each stage. The

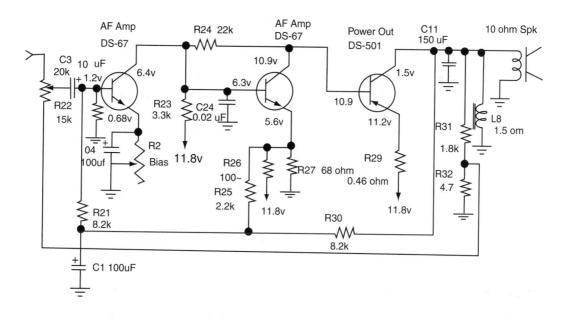

FIGURE 8-6. Directly coupled AF and power output transistors are found in the early car radio output circuits.

collector terminal of the AF stage is connected directly to the base of one power output transistor (**Fig. 8-6**).

Q102 provides bias voltage for the two audio output transistors. Q103 collector terminal is connected directly to the car battery source, and the collector terminal of Q104 is at ground potential. The power output audio signal is coupled to the speaker's fader control through a 1000-uF electrolytic. The fader control provides speaker balance to the right front or right rear speakers. The same audio system is found in the left channel power output circuits.

Winding Auto Choke Coils or Transformers

A burned choke coil winding in the early car radios power supply can result when a shorted electrolytic or output transistor will not blow the main fuse. If a fuse is temporarily installed that is larger than the correct one, a short in the load of the power supply can cause damage to decoupling resistors and choke transformers. Usually the coil-enameled wire becomes burned and the winding becomes shorted, causing hum in the power supply.

Suspect a burned choke coil or transformer when a loud hum is heard in the speakers and when shunting all electrolytics within the power supply does not remove the hum noise. The outside cover of the choke coil or transformer is often burned or scorched, indicating that the transformer has been running quite warm (**Fig. 8-7**).

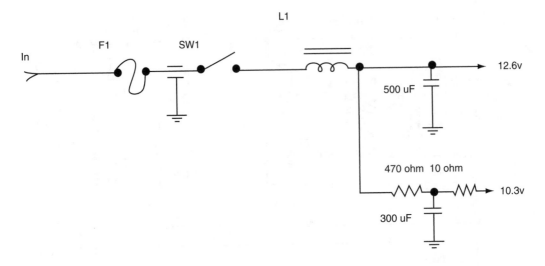

FIGURE 8-7. Burned choke transformer (L1) can cause a low humming noise in the speakers.

Most of the early choke transformers have a resistance of 0.3-0.6 ohms or an inductance around 1.5 to 10 MHz. If possible, sub a removed choke from a junked car radio and mount it in the place of the defective one. Most choke coil transformers can be interchanged with one another when found defective in the early car radios.

If one is not available, simply wind the choke transformer after removing the burned winding. Determine the size of the wire in the original transformer by using a wire gauge. Most wholesale electronic firms carry different sizes of enameled magnet wire. Simply wind the bobbin or core area until the desired resistance or inductance has been reached. Check the wire every several feet with an inductance meter or range of the DMM.

The magnet wire can be cut at several long lengths to feed through the open area and to wind in layers on the metal core area. Splice the wire junction with another piece of enameled wire until the required resistance or inductance has been reached. Tape up each coil splice with plastic tape. Slide a piece of spaghetti over each wire lead before soldering into the power supply circuit.

IC Power Output Circuits

The low-powered IC output circuits consist of a small IC component for each stereo channel. A larger IC component can be found as dual right and left audio output circuits, all in one component. One large power output IC might include all of the

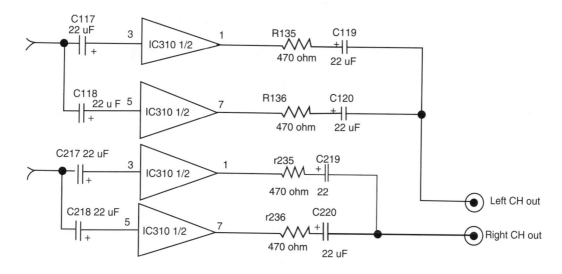

FIGURE 8-8. Two different power output ICs are found in the AM/FM/MPX auto CD player.

stereo circuits and audio circuits within the auto radio, while two different ICs might be found in each channel (**Fig. 8-8**).

A single power output IC can be found in each channel of the auto radio-cassette player with stereo sound. The input audio signal is fed into pin 8 of IC6 and out at pin 12. Likewise, the left input audio signal is fed into pin 8 and out of pin 12 of the left power output IC5. Pin 1 of either IC is the voltage supply terminal. Two separate main power ICs are found in a Sanyo radio-cassette player (**Fig. 8-9**).

One IC component might include both the left and right power output ICs driving several speakers. The right and left audio channels are fed to a set of different speakers. Remember that some of the Delco auto car radios' output circuits are directly coupled to the speakers and are not grounded. Be careful here, as the power output IC can be destroyed if one side of the speakers is grounded. A small power IC might be found in the audio circuits of a radio-cassette player.

Dual-IC Stereo Audio Output Circuits

One large power output IC might be located within the car stereo tape player. The left input audio signal appears at terminal 13, and the right input terminal on pin 2. The left audio output is found on pin 8 through a 470-uF electrolytic to the left speaker. The right audio output is at pin 7 of IC703 and coupled to the speakers with another 470-uF electrolytic capacitor (**Fig. 8-10**). Terminal pin 14 connects to the dual power output IC supply voltage.

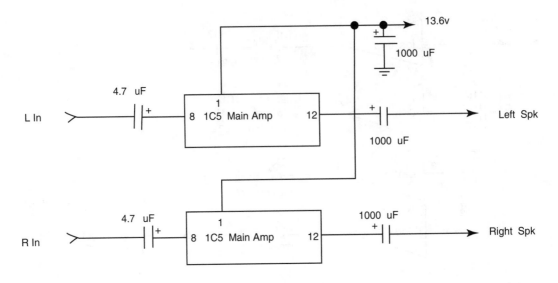

FIGURE 8-9. Two separate power output ICs are found in a Sanyo car radio-cassette player.

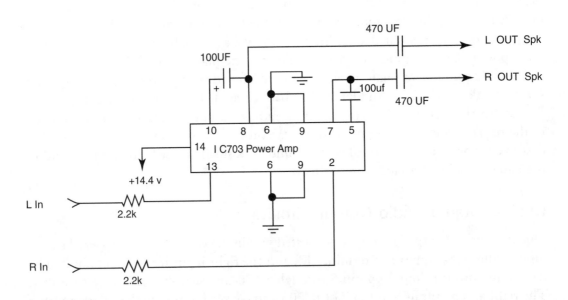

FIGURE 8-10. Two large 470-uF electrolytics couple the left and right channel speakers to the power output dual-IC.

Fader-Control Circuits

The fader-control circuits consist of a low-ohm fader or speaker control that provides a greater volume within either the front or back speakers. The fader control in amplifier circuits can attenuate the output circuits to fade out one signal and fade in another. The speaker fader control has a low-ohm resistance (35 to 100 ohms) with a 2-watt rating. The fader control can be found in each left and right output speaker circuits (**Fig. 8-11**). The fader control has a higher wattage rating and it is a wirewound type of control.

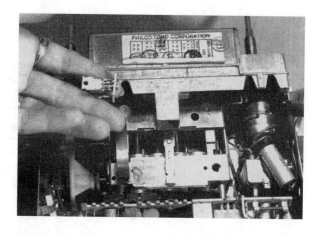

FIGURE 8-11. The fader circuit can apply more volume or can fade out the sound in the rear and front speakers.

Make sure that the fader control is not grounded within some Delco car radios. Some of these power output IC circuits have a common speaker terminal above ground with the power output tied directly to the fader and speaker terminals. The common wire of all speakers are fed back to the power output IC and leave a positive voltage on the speakers and IC terminals. If one side of the speaker is accidentally grounded, the power IC can be damaged.

In a Delco 1982 car radio series, the fader and speaker terminals have a DC voltage applied that is tied directly to the power IC without a coupling or blocking electrolytic. Make sure the common speaker terminal of the left and right return wire or cable is not grounded. Notice that a 6.6 DC voltage is found on the fader control and speaker voice coil (**Fig. 8-12**). Do not accidentally ground the speakers.

SOUND-OUTPUT PROBLEMS

Dead—No Power Output Signal

Replace the power output transistor with a dead chassis and no power output. Suspect a burned bias resistor and shorted power output transistor with a dead/no-sound symptom. Check for a shorted output transistor with open bias resistors and a dead chassis. For a dead left channel, check for a bad bias diode and output transistor. A shorted audio output transistor can cause the dead chassis with a blown fuse. Suspect a shorted output transistor, burned bias resistor, or diode with a dead chassis.

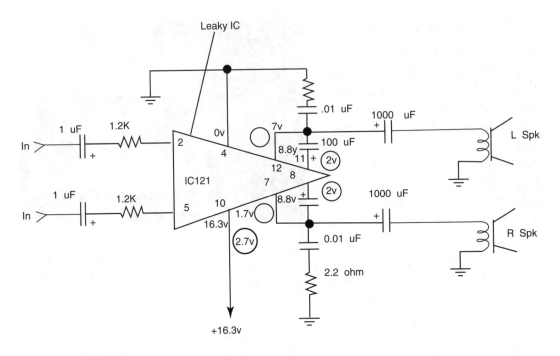

FIGURE 8-12. A leaky power output IC can cause weak and and distorted music with lower IC terminal voltages.

A no-sound symptom can be caused by a poorly soldered connection on the output transistor terminals. Replace both the output and the driver transistors for loss of audio and a blown fuse. A base-to-collector short of the audio output transistor can cause loss of power output. Suspect that the battery is connected up backward, or is charged up wrong, when the audio output transistor blows the fuse after the car radio is turned on. Replace both power output transistors with a blown fuse.

Check for a defective power output IC with no right or left audio channel. Inspect the power output IC for a hole or chip blown out of the body of the IC with no sound in the speakers. Suspect loss of power output with a bad driver and audio output power IC. Replace both output ICs with both dead output channels. Check the dual-output IC with no sound in either channel. Replace the power output IC when the IC is running red-hot with a dead chassis. After replacing the overheated power IC, check for poor foil terminal PC wiring connections.

Besides power output transistors and IC components that can cause a dead chassis, check corresponding components tied to the power output circuits. A badly soldered B+ jumper wire can cause no voltage to be applied to the power output circuits. The dead left channel can be caused by a bad PC board connection on the supply voltage source.

Suspect a bad voltage regulator transistor or IC for loss of output audio or sound. A leaky IC regulator can cause a slight scratching noise with no audio. Loss of audio output can result from a shorted or leaky electrolytic in the voltage supply source. A badly soldered joint on the voltage regulator can cause a no-audio symptom. Do not overlook a shorted tweeter speaker that repeatedly destroys the audio output transistor or IC.

Dead—Only a Humming Noise

No sound was heard from an Audiovox AU929 model auto receiver. Even when the volume control was rotated back and forth, no scratching noise or hum could be heard, indicating service problems in the audio output circuits. A quick voltage test on the audio output IC indicated lower voltages than normal.

No sound could be heard when touching the input terminal of the LA4445 output IC. The output IC was replaced with an ECG1707 universal replacement. Double-check the PC wiring where the power IC terminals are connected for possible damage in removing the power output IC.

Weak Audio Power Output Circuits

Replace the dual-audio output IC for really weak sound. Make sure that the speaker connections are not connected to ground in a Delco car radio when the power output ICs are found shorted. Replace the leaky audio power output IC for a weak and distorted left channel. Check both output power ICs in each channel for low volume in both channels. Really low volume on all stations can be caused by a leaky power output IC. Check for an open or leaky power IC with a weak audio output right channel. A shorted output IC component can cause a weak or low audio output (**Fig. 8-13**).

Check for an open or leaky output transistor for weak and distorted audio. Replace both output transistors for low audio sound in the speakers. Replace the audio output transistor for a weak left channel. Both high-powered output transistors can cause a weak and distorted audio symptom. Replace the right channel output transistors for low and distorted sound.

An open bias resistor in the transistor output circuits can cause a weak sound and no click in the speaker. A badly burned bias resistor and shorted output transistor can cause weak radio output signal. Check for an open or burned resistors in the output transistor circuits for low volume.

Distorted Power Output

A distorted right channel can result from a leaky output transistor. A shorted or leaky output transistor can cause low volume and distortion. Replace a bad power output transistor for distorted audio. One leaky output transistor and the other output open can cause a distorted right channel. A shorted output transistor with a burned bias

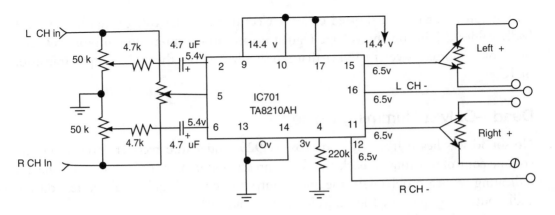

FIGURE 8-13. A leaky power output IC can cause weak and distorted music with low IC terminal voltages.

resistor can cause a weak and distorted sound output symptom, as well as cause excessive distortion. Low distortion in the audio can result with the open output transistor. Replace the suspected transistor after the car radio has operated for several hours and the sound becomes distorted.

A bad or leaky output power IC can cause excessive audio distortion. Improper voltage on one of the pin terminals of the power output IC can be caused by a shorted or leaky bypass capacitor, causing distortion in both output channels. Internal shortage of the B+ voltage inside the power output IC can cause distortion in the car radio speakers. Replace the power output IC and bias resistors for loss of audio output or distorted sound. Check the power output IC and ring cracks on the PCB for distorted audio in both channels.

Replace the defective voltage regulator transistor or IC with distorted sound after the radio has warmed up. A bad voltage regulator transistor can cause distorted audio after several hours of operation. A leaky voltage regulator transistor can cause audio distortion. Check for a leaky diode in the voltage regulator circuit for the audio distortion symptom. Distorted audio in both channels can result from ring cracks around the power output components.

Intermittent Audio Power Output Circuits

Replace the power output transistor for intermittent sound. A power output transistor can cause intermittent sound and low distortion (**Fig. 8-14**). Remove the power output transistor if there is a pop in the speaker and the audio goes dead after operating for only a few minutes. Replace the defective power output transistor and bias resistor when the audio does not come on for over a half an hour.

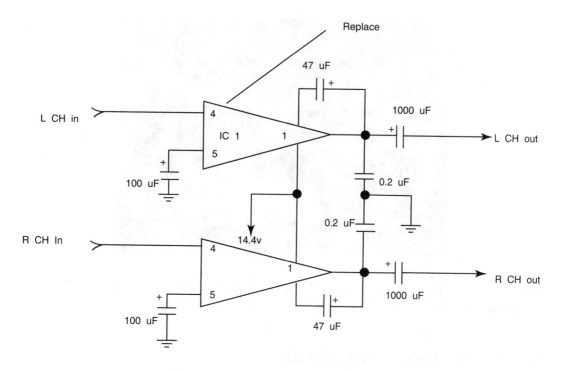

FIGURE 8-14. An open or leaky output transistor in a push-pull power circuit can cause intermittent and distorted sound.

Check the mica insulator on the power output transistor when the fuse opens intermittently. Check both power output transistors with a loud motorboating sound with the volume turned way down.

Replace the stereo audio output IC with an intermittent left audio channel. Suspect a leaky or shorted power output IC when the fuse keeps blowing. Check the power output IC for intermittent shutdown. Replace the power output IC when the speaker cuts in and out at maximum volume.

For the intermittent left channel, check the electrolytic coupling capacitor from the power IC to the speaker. The intermittent right channel can result from an intermittent coupling capacitor to the power output IC. A bad power output IC can cause a noisy and intermittent frying sound in the speaker. A poorly grounded heat sink can cause intermittent motorboating sound in the left power output IC amp. Check for a bad plug connection for an intermittent left channel.

A defective mute transistor can cause the audio to cut in and out at the speakers. Intermittent popping noise can be caused by a defective muting transistor. A bad mute switch mechanism can cause the audio to become intermittent and to cut out. Do not overlook a defective protect circuit, connection, or component in the output stages for intermittent and no audio output.

FIGURE 8-15. The left channel was intermittent in a Radio Shack 12-2103 AM/FM auto radio-cassette player with a defective IC701.

Intermittent Audio—Radio Shack

The sound was intermittent in the left channel of a Radio Shack 12-2103 AM/FM stereo cassette player. The audio acted up in both the AM radio and the cassette player, indicating problems within the audio output circuits. The right channel appeared normal. Sometimes the radio played for days without acting up.

The external audio amp was connected to the left volume control (50K ohms) and the sound played for several days. When the sound became intermittent, the audio was traced to the input terminal (2) output IC701. When the external amp was connected to output terminal pin 15 of IC701, only a low humming sound could be heard. Replacing IC701 solved the intermittent left channel (**Fig. 8-15**).

Noisy Power-Output Circuits

An ungrounded heat sink of the power output IC can cause hum and motorboating. Replace the noisy power output IC for a crackle and popping noise in the speaker when the car radio is first turned on. A bad power output IC can cause a static or frying noise in the speaker. A leaky power output IC can cause a motorboating sound in the speakers.

Replace the left power output IC that is running red-hot and that has no audio except a loud howling sound. A bad power output IC can cause a noisy, intermittent, and cracking sound in the speakers. Replace the power output IC for noise in the left channel. Distorted sound with a popping noise can result from a defective power output IC.

FIGURE 8-16. Push up and down on the car radio speaker cone to see if the sound cuts out.

Replace both power output transistors for noise in the speaker. Suspect the audio output transistor when the car radio pops off and on, and then goes completely dead. A leaky output transistor can cause a noisy and distorted symptom. Replace both output transistors with a noisy right channel and when the noise gets louder as the radio heats up.

A dead channel with only a slight humming noise can be caused by a red-hot power IC. Replace the overheated power IC with a low channel hum in the audio. The shorted power output IC can cause a hum in the sound, and can cause smoke to appear from the car radio. Replace the audio output IC with a popping and cracking noise in the speaker and a humming noise in the background (**Fig. 8-16**).

Look for a bad plug connection from the main amp to the line amp with intermittent hum in the audio. Check for DC voltage on the speaker terminals with a humming sound and a leaky IC or transistor. A bad ground heat sink can cause a motorboating sound in the left channel with a bad power output IC.

Burned and Shorted Output Components

Suspect a shorted or leaky power output transistor when the line fuse is burned black inside the glass area. Sometimes if too large a fuse has been inserted or tin foil has served as the emergency fuse, the "A" power cord might be burned clear up to the on/off switch. Go directly to the audio output circuits for shorted components that may blow the main fuse.

Check each power output transistor with the diode test of the DMM or with a transistor tester. Notice that a normal high-power output transistor will provide a lower resistance measurement than an RF, mixer, or oscillator transistor. Often, the power output transistor might become shorted between collector and emitter terminals. Double-check all power output transistors that may be open or leaky. You can easily spot the overheated transistor by the appearance with a dark-gray body.

Check each transistor bias resistor for a possible change in resistance. Make sure the low-ohm emitter resistors are not overheated or burned. Measure each bias resistor with the leaky transistor out of the circuit. Check directly driven transistors that tie to the base terminal of the output transistor for open or leaky conditions. Make sure that all bias diodes in the base circuits are normal. Remove one end for a good diode measurement.

Locate the suspected power output IC when the main fuse is blown in the car radio. Check the supply voltage terminal to common ground for a low-ohm reading. It is possible to have a dead right channel and a normal left channel with a defective dual-IC power output.

Measure each terminal to common ground for a leaky output IC. An overheated power output IC body might turn gray with burned marks and might even have a chip out of the IC component. Because the voltage supply terminal of a power output IC is connected directly to the battery source, it is a likely candidate for blowing the auto fuse.

Fuse That Keeps Blowing

Go directly to the power output transistors or IC components when the main fuse keeps blowing after replacement. Notice if the fuse opens before the on/off switch is turned on. If so, double-check the bypass capacitors and metal feed-through components for possible grounding. Sometimes a transistor or IC might arc over and then return to normal; now the radio works fine.

You might locate a shorted feed-through capacitor where the battery voltage enters the metal radio cabinet. Simply remove the "A" lead and solder it directly to the on/off switch. Cut the old capacitor wire in two. Solder a 0.001-uF bypass capacitor to the power lead and ground at the power switch. The internal grounded on/off switch can blow the radio fuse. Check for a leaky polarity diode within the power supply and power lead that could possibly blow the main auto radio fuse.

Fuse That Keeps Blowing—Kraco

In this example, the 4-amp fuse (F401) kept blowing as the radio was turned on in a Kraco KCB-2390 auto radio. A quick peek at the output ICs indicated that the left power IC had become overheated with a gray-looking body. The voltage supply pin 1 of IC303

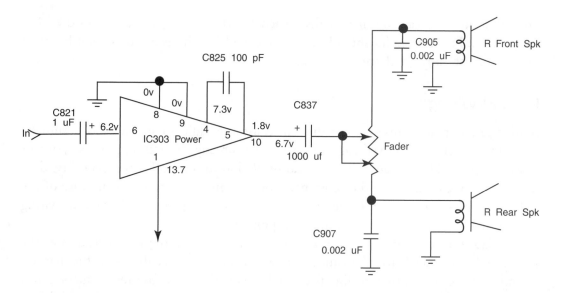

FIGURE 8-17. Leaky IC303 kept blowing the fuse in a Kraco KCB-2390 model.

in the left channel was unsoldered, leaving the right channel connected. This time the fuse did not open and a radio station was heard in the right channel. No doubt IC303 was leaky and was replaced with a universal ECG199 replacement IC (**Fig. 8-17**).

Dead—Fuse Open

Sometimes the main auto fuse opens without any explanations; nothing appears shorted or leaky. Simply replace the fuse and hope that it never happens again. At times, a transistor or IC component might flash over for an instant and blows the fuse. Make sure that the correct size fuse has been replaced. A lower amp fuse inserted instead of the correct one might hold for a while and then open up when the radio is turned on. For instance, if a 4-amp fuse is inserted instead of a 5-amp fuse, the car radio might blow several miles down the road.

Dead Fuse—Normal Operation

The auto receiver might play normally after the main fuse opens up. The auto radio appears normal and the new fuse holds. Make sure that the correct fuse has been inserted. Sometimes when inserting a disc or cassette, the main fuse might blow with a dead radio. Remove the tape and notice if the tape is wrapped around the capstan of the cassette player. The cassette might not be fully inserted, the tape pulled from the cassette, and the excess tape wrapped around the capstan. Extra pressure might be pressed against the cassette or CD disc when inserted, and the unit stopped with a blown fuse. The auto radio should be removed and bench-tested for several hours of operation, after a good cleanup.

Burned Wiring

A shorted component inside the radio can cause the radio to smoke, and eventually cause the power lead to overheat and burn. The directly shorted output transistor or power output IC could cause the "A" lead insulation to burn off if the main fuse does not open. Quickly measure the resistance from the fuse lead to the metal case of the radio for a low-ohm reading. If the fuse is too large or wrapped with tin foil, the wiring can get red-hot with a shorted output component.

Replace the fuse harness with a new set, or just replace the power lead after the repair has been made. Select the same size of wire cable or replace the whole fuse harness assembly. Select the correct in-line fuse holder with the right fuse replacement cable. Replace frayed wiring connections and damage two-way flat connectors. Several different types of in-line fuse holders are available for ATC, MAXI, and screw type fuses.

Hum in the Speakers

Most hum problems heard in the speakers are caused by dried-up electrolytic capacitors. Low hum problems can also be caused by bad decoupling electrolytics. Check all capacitors within the low-voltage power supply with the ESR meter. Discharge each capacitor before testing with the ESR meter. Clip another electrolytic of known value across the suspected electrolytic capacitor and notice if the hum is gone. Be sure to observe correct polarity of electrolytic capacitors.

A change in resistance of a filter choke transformer, or one that is burned, can prevent good filtering action, resulting with hum in the speakers. Notice if the outside cover of the choke coil is scorched or burned, indicating that the receiver is pulling too much current. Often, with a direct short in the radio, the filter choke transformer or coil becomes hot when the main fuse does not open. Remove a portion of the outside cover and take a peek. Accurate resistance readings might not tell the tale with low-ohm choke coils.

Hum in the speakers can be caused by a leaky or shorted output transistor or IC with a DC voltage measured across the speaker voice coil. Hum in the speaker can result from a leaky output component.

Usually the hum noise heard in the speakers might be coupled with a distorted or weak audio symptom. A defective dual-power output IC can cause hum in both channels. The shorted or leaky output transistor might cause a hum in only one channel. A leaky driver transistor can cause the output transistor to become shorted and produce hum in the speakers. Humming noises in the audio circuits can easily be signal-traced with the external audio amp.

Distorted Sound After Warming Up

A leaky audio driver or output transistor can cause distorted sound after the car radio has operated for several hours. A leaky power output IC can become overheated and cause a breakdown after warming up. The leaky zener diode has been known to cause the radio to shut down after the auto receiver has operated for a few hours.

Suspect a defective voltage regulator transistor or IC if the sound becomes distorted after warming up. Loss of audio with a slight buzz in the sound can be caused with a bad voltage regulator IC. Check for poorly soldered terminals on the voltage regulator for poor or distorted sound. Measure the voltage output of the voltage regulator component for an increase or decrease in applied voltage, indicating a leaky regulator.

A leaky coupling or decoupling electrolytic can cause the audio to slowly fade away and become distorted. A defective muting transistor can cause intermittent popping in the speakers. The defective power switch can cause distortion in the audio after operating for several hours. A leaky zener diode can cause noise in the speaker and then shut down the chassis after warming up.

Speaker That Cuts Out at High Volume

Replace electrolytic capacitors in the audio output circuits when the audio fades out and becomes distorted. A defective voltage regulator can cause the sound to shut off with a really high volume setting. A leaky coupling capacitor can shut the volume off when the volume is tuned up halfway (**Fig. 8-18**). A bad or worn volume control can cut out the audio with high volume applied. Check for a bad volume control board connection for intermittent sound and hum in the speakers with the volume wide open.

Check for open bias resistors when the left channel cuts out at high volume. A bad bias resistor can cause a distorted left channel when high volume is applied to the speakers. Suspect an increase in resistance or a change of bias voltage when the sound is distorted with the audio level turned three-quarters of the way up. Replace the power output IC when the speaker cuts out at maximum volume.

A low static noise with the volume turned down can result from a noisy driver transistor or IC. Noise in the right channel with the volume turned down can be caused by a leaky 4.7-uF electrolytic in the audio circuits.

A bad or worn volume control can cause the volume to cut out at a low control setting. When the volume control cannot turn the audio down, suspect an ungrounded volume control. Low volume and a distorted symptom can result from a bad electrolytic in the power supply. When the radio cuts out with the volume turned down low,

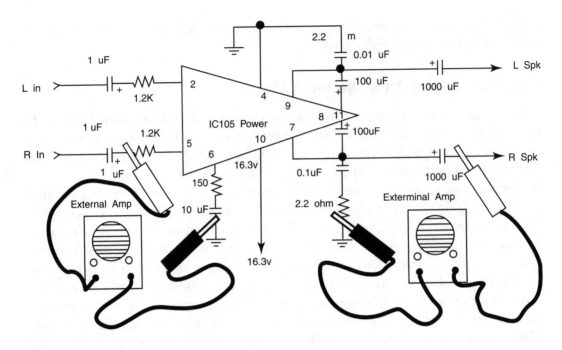

FIGURE 8-18. Check the input and output circuit of IC105 with the external amp to determine if IC105 is defective.

suspect an 0.047 bypass capacitor in the RF or audio circuits. Check all electrolytics with the ESR meter.

Popping and Cracking Noise—Automatic DC6500

In this example, when the Automatic car radio was first turned on, it began to crack and pop in the speakers. At first the noise was really loud, and then slowly the noise would quiet down. The audio seemed to be normal at the input terminal 6 of IC804 in the right channel with the external audio amp. When the external amp was connected to coupling capacitor C837, the noise could not be heard. Replacing the audio output IC804 with a universal IC replacement (SK3243) solved the noise in the speaker (**Fig. 8-19**).

Bad On/Off Switch

No lights or sound can be heard with an open on/off switch. Check for a possible defective power switch if the radio shuts off after several hours of operation. A dirty or worn on/off switch can cause intermittent auto radio operation. A bad power switch can cause a dead car radio with no power or dial lights with an arcing sound.

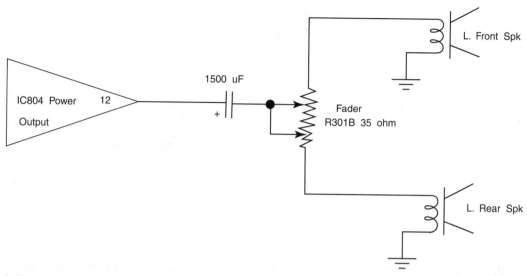

FIGURE 8-19. Replacing IC804 in an Automatic DC6500 auto receiver cured a cracking noise in the speakers.

Suspect a bad power switch when the radio will sometimes turn on and not at other times. A defective power switch can cause audio distortion after the car radio plays for several hours. Replace the on/off switch when the power will not stay on. Suspect a poorly soldered connection on the switch terminals for intermittent audio operation. Replace the power switch if a noisy sound is in the speakers when the car radio is first turned on.

Only Hum in the Speakers

An Audiovox 80-CT-TPX auto radio came in with hum in the speakers. The local radio station audio was traced up to the input terminal 6 of the power output IC202. Only a hum could be heard in the external amp when connected to pin 12 of the output IC. A quick voltage measurement on pin terminal 12 indicated only 2.5 volts. The same 2.5 volts were found on the speaker terminals. Replacing a leaky 470-uF (C223) speaker-coupling capacitor solved the loss-of-radio problem (**Fig. 8-20**).

SIGNAL-TRACING, REMOVING, AND REPLACING

Signal-Tracing Audio Circuits with External Amp

The external audio amp is used in signal-tracing the distorted, weak, or intermittent audio within the audio amplifier. The audio amplifier can have a PM speaker or

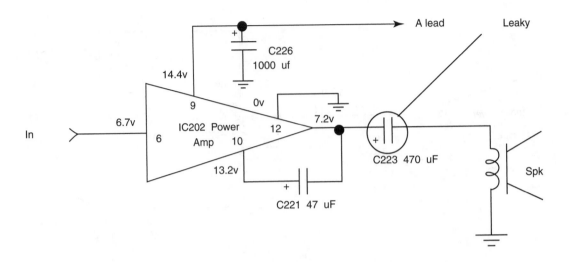

FIGURE 8-20. A shorted coupling electrolytic (C223) caused a DC voltage to be found on the speaker voice coil in the Audiovox auto radio.

earphones as indicator with the radio signal or audio signal generator at the input. The audio signal can be traced from the preamp or AF circuits to the audio output speaker terminals with the external audio amplifier. Start signal-tracing at the volume control and proceed either way to locate the defective circuit.

Insert the external amplifier into an isolation transformer for test equipment and product protection. When the audio is really weak in the AM radio, tune in a local broadcast station as the signal source. Insert a test cassette to signal-trace the distorted sound of the cassette player. A test disc can be used when signal-tracing the audio output circuits in a CD player.

Clip the output of an audio or function signal generator to the input terminals of the high-wattage amplifier for a signal source, and signal-trace the audio output circuits with the external amp. Rotate the signal generator to 1-kHz frequency for audio tests. Keep the output signal of the generator as low as possible to avoid overloading the audio stages.

Go from base to base of each succeeding transistor, from the preamp or AF to the output circuits, to check audio signal with the external amp (**Fig. 8-21**). Clip the ground terminal to common ground of the car radio and place the test probe on the base of each transistor. Check the audio signal at the volume control. When the defective component is in the output circuits, then begin at the volume control and check each output transistor or IC. Test for audio signal at the input and output terminals for a suspected IC with the external audio amp.

When the audio amp becomes weak, intermittent, or distorted within the external audio amp, you have located the defective circuit. Remember, the audio signal should

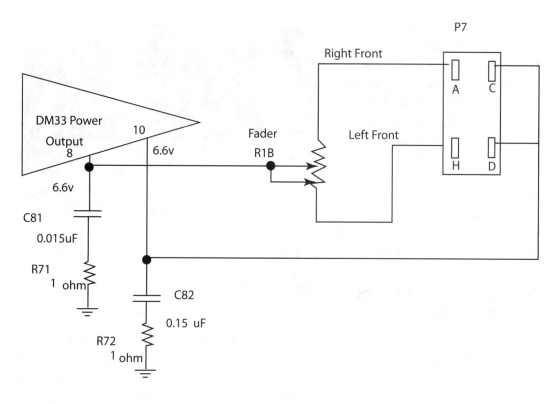

FIGURE 8-21. Check the input and output terminals of DM33 with the external amp.

become stronger as you proceed to the audio output circuits. The external amplifier can monitor the defective stage for weak or distorted reception. You can pinpoint the weak audio component with the external audio amp. Then, after finding the bad stage or circuit, take critical voltage measurements. Test each suspected transistor or IC with in-circuit transistor tests.

Removing Several PC Boards

Servicing the auto radio symptoms is somewhat like removing boards within a TV/ VCR chassis to get at certain audio output components. Several boards and components might have to be removed to get at the defective component. Some of the boards might be held in place with small screws, while others might have several different soldered connections to the metal radio chassis (**Fig. 8-22**). Sometimes removing the required boards takes up more time than finding the defective component.

If there is no board assembly chart available on removing and replacing the different boards, write down each component or board as they are removed. You might be called away or have the car radio half exposed at the end of the day. Mark down each ground wire, clip, and screw, and the spot from which they were removed. Some of these screws may be longer than others and they can short out parts below if inserted in the wrong place. After replacing the defective component, the boards and parts can be reversed in removal, so they are connected and mounted in the right sequence.

FIGURE 8-22. Several screws and soldered connections may have to be removed to find the defective component.

Replacing Power-Output Transistors and ICs

After locating the defective power output transistor, the part must be removed and replaced with a new output transistor. Before removing the output transistor, unsolder all terminals and suck up the excess solder. Make sure the new transistor terminals are the correct length and that the power transistor is mounted in the same spot. Remove the mounting screw that may hold the power transistor to the metal shield. Some power output transistors might be held in place with a small metal bar.

Notice if a piece of insulation is found between transistor and heat sink. Often a new piece of insulation comes with the power output transistor. Smear heat sink compound or grease on both sides of the mica insulator. Choose the clear-type compound, as a white grease can soil the bench, car radio, fingers, and clothing.

Now mount the new replacement in the same spot as the original. Sometimes the terminals have to be cut off to fit in certain areas. Universal power transistors can be replaced if the exact part is not available. Now solder up the transistor terminals to the PCB. Clean off any solder residue or paste and lopped-over soldered connections.

Replace the power output IC in the same manner. Usually the power IC does not have an insulator between the body and heat sink. Coat both the heat sink and metal body of the power IC with heat sink compound or grease. Make sure all IC terminals are located in each hole and that no terminals are bent over.

The small power IC might have only one screw mounting the IC against the metal heat sink or frame of the car radio. Large power ICs are mounted with a screw on each side of the component. Unsolder all IC terminals and suck up the excess solder from each terminal on the PC board. Flick each terminal with the blade of a pocketknife or small screwdriver to make sure the IC terminals are loose and not still stuck to the PC wiring. Lift the power IC out of the terminals from the heat sink.

Most car radio ICs, microprocessors and transistors can be replaced with universal replacements. Look up the universal replacement within the Auto Radio Series (AR) service manuals or in a transistor replacement guide. Check the old part number to a universal replacement within the Sylvania ECG semiconductor manual, NTE replacement guide, RCA SK Series replacement guide, or Sams Technical Publishing transistor replacement cross-reference book.

Take a low-resistance measurement from each transistor element or IC component to common ground with the DMM. Make sure that there are no poorly soldered connections or blobs of solder over two different terminals. Check each terminal to common ground for a possible short or a leaky component. Measure the resistance between any two terminals for possible leakage on the low-ohm range of the DMM. When a low resistance measurement is found, check to see where the terminals are soldered or connected within the service manual or schematic.

CHAPTER 9

Troubleshooting the Auto Cassette Deck

THE AUTO CASSETTE DECK

The auto cassette deck begins to play as a music cassette is inserted into the dash slot. The cassette triggers an on/off switch to turn on the cassette motor and amplifier circuits when the cassette is inserted. Most auto cassette decks consist of tape heads, transistor or IC preamps, a motor, and motor speed circuits. A deluxe cassette deck may have an auto-reverse circuit that operates with transistors within the motor circuits.

The early auto cassette player consisted of only tape heads, a preamp, and a cassette motor. Later on, the stereo 8-track cartridge player with up to four audio outputs lasted for several years. Today, the cassette player is still found within the auto radio receiver chassis. Now you will find a radio receiver with a cassette and CD player. Just lately, a digital music MP3 player capable of playing up to 10 hours of digital music has been placed within the auto receiver.

Tape-Head Mechanism

The early cassette player consisted of a stereo tape head, drive motor, capstan/flywheel, pinch roller assembly, idlers, pulleys, cassette reels, and a motor drive belt. The tape motor is designed to play forward, reverse, and a fast-forward operation. The tape motor and amplifier circuits begin to operate when the cassette is inserted. The early auto cassette player did not have any type of speed circuits, and the motor was reversed by changing the voltage polarity to the drive motor. A motor drive belt rotated the capstan flywheel in many models, and in some of the recent players they are gear driven.

The left and right stereo tape heads are located in one tape-head assembly, and the output is capacity coupled to the preamp transistors or IC circuits. The tape-head

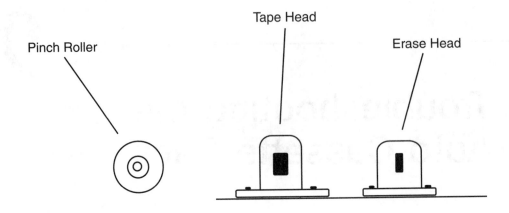

FIGURE 9-1.   A magnetized tape head can cause a noisy and distorted music.

azimuth adjustment screw levels the tape head with the moving tape for optimum performance.

Improper adjustment of the azimuth screw might cause the loss of high frequency reproduction on the tape. The worn or magnetized tape head can reproduce a loss of high frequency with some signs of distortion and noise (**Fig. 9-1**).

A motor drive belt rotates the capstan/flywheel that pulls the tape through the capstan drive and pinch roller assembly. A bad or worn motor drive belt can cause improper speed problems. Both the capstan drive area and pinch roller can accumulate tape oxide and cause slow or improper tape speed. The pressure pinch roller rotates against the drive shaft with the cassette tape riding between the two, pulling the tape across the tape head in playback modes.

The take-up reel rotates in a forward mode to pick up the loose tape as it comes by the tape head. A supply reel holds the tape and feeds it into the capstan/flywheel drive and pressure roller. In fast-forward mode, the take-up reel pulls the tape at a fast rate of speed. And when in reverse mode, the supply reel pulls the tape backward at a high rate of speed. Several pulleys and idlers might be used in the fast-forward and reverse modes.

Tape-Head Cleanup

Clean the tape head with alcohol and a cotton swab or stick. The tape head and pinch roller should be cleaned up every month, if used quite regularly, or after every 100 hours of service. Tape oxide can build up on the tape head and cause the tape to slow down with distorted sound. If packed oxide is found on the tape head and is difficult to remove with alcohol and cloth, use a flat piece of plastic to remove the stubborn oxide. Finish the job with alcohol and cloth. Always clean up the tape head after repairing the cassette player (**Fig. 9-2**).

Do not use a screwdriver blade near the tape head as it might magnetize the tape head. A magnetized tape head can cause hissing noise in the background. Use a cassette tape demagnetizing cassette or tool that can be passed over the tape heads. Pull the demagnetizing tool away from the

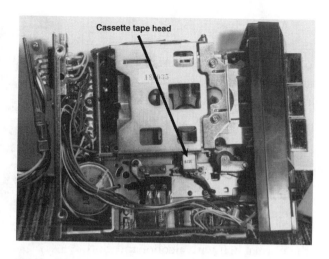

Cassette tape head

FIGURE 9-2. Clean up the front of the tape-head surface to remove tape oxide and dirt.

tape head and then turn off or unplug the tape demagnetizer tool. You can find a cassette cleanup and demagnetizing cassette at most electronic music departments. The demagnetizing cassette can be used to clean up the tape heads without removing the cassette player from the automobile.

Lubrication

Usually, most auto cassette players do not require lubrication until after many years of operation. Do not just place oil on rotating surfaces when the cassette player is in for repair. The capstan/flywheel bearing might require lubrication if it begins to squeak as it rotates. Sometimes idlers and pulleys can become noisy, and a touch up of light oil can make then run smooth again. A drop of oil on a noisy bearing can do wonders, but too much oil can drip down on to rubber moving parts and result in speed problems. Wipe up the excess oil from parts to prevent speed problems. Smear a light coat of grease on the capstan bearing if the flywheel bearing is dry and noisy when rotated.

Belt Replacement

Most auto cassette players have several belt drive systems within the cassette mechanism. In lower-priced models, one motor drive belt can drive the flywheel capstan assembly. Another drive belt may also be used with the drive motor to drive the idlers and reel assemblies. Most of the drive belts are flat, although a thin square belt can be found to rotate the capstan, idlers, and reel assemblies (**Fig. 9-3**).

A defective motor drive belt can be found cracked, worn, or have a spot of grease on the belt surface, causing slow or erratic tape speed. Replace the rubber drive belt when the belt is cracked or stretched and will not stay on the motor pulley. Remove the motor drive belt and inspect for worn or torn areas. The motor belt is too loose when the motor capstan/flywheel is held by hand and the motor pulley turns inside the belt area. Clean off the normal belt with alcohol and cloth for slow speed problems.

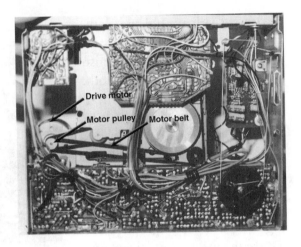

FIGURE 9-3. Here the cassette drive motor has two different flat belts.

Check the capstan/flywheel drive area for loose or stuck-on particles from the belt area. Replace the motor belt if a black area is found on the capstan or motor pulley, indicating that the belt is slipping and can cause slow and erratic speed problems. If in doubt, replace the motor drive belt, as it is easy to replace and costs very little.

PROBLEMS WITH THE AUTO CASSETTE PLAYER

No Tape Motion!

Suspect that the motor drive belt is either broken, or is lying in the bottom of the cassette player, if there is no tape motion. Notice if the cassette motor is rotating without any tape movement. You can hear the motor spin, with your ear close to the cassette opening. Replace the drive motor if it keeps coming off of the motor pulley.

A dead or defective drive motor can cause loss of tape rotation. Often the motor will run fast; you can hear it if the belt has slipped off the motor pulley. Suspect open or worn brushes inside the drive motor when it turns erratically. Take a continuity measurement with the ohmmeter, rotate the motor pulley by hand, and watch the meter change resistance. No continuity or resistance indicates an open motor winding. Check for a defective speed circuit if the motor is running too fast or too slow.

Rotate the capstan flywheel by hand to see if it has a dry or gummed-up capstan bearing. If so, remove the bottom metal plate holding the capstan in place and remove the capstan drive shaft from the bronze bearing. Clean up the old grease from the bearing area and capstan shaft. Apply light grease to the capstan bearing before replacing the capstan/flywheel. Clean up excess grease and dirt with alcohol and cleaning stick.

Running All the Time

When the cassette was inserted into a RCA 12R200 auto cassette player, the player would sometimes stop when the cassette was removed, and at other times it would keep on rotating. After removing the metal covers, a small micro-type switch seemed to stick. All switch contacts were cleaned with alcohol and cloth. Cleaning solution was sprayed down inside the switch area and excess fluid was wiped up. This cured the on/off switch-shutoff problem.

No Tape Action

For no tape movement, inspect the cassette leaf on/off switch. Press the contacts together and notice if the motor starts up. Clean up the copper spring-type contacts with cleaning fluid and cleaning stick. Really dirty silver contacts can be cleaned up with emery cloth or sandpaper. An emery board does a good job. Suspect a defective motor or motor circuit when the voltage source and switch are normal. A badly soldered connection on the speed control can cause loss of tape motion. Do not overlook a broken or loose motor belt.

Check for a broken solenoid lock arm when the tape stops in play mode. Check for a shorted diode to common ground when the tape locks up at the time it is inserted. A frozen capstan assembly with built-up oxide and tape residue can cause a lack of tape action. A washer missing on the supply reel assembly can also cause a loss of tape deck operation.

Slow Speeds

Slow speed can result from oil on the motor drive belt or from a worn drive belt. Clean up the motor drive belt, motor pulley, and flywheel belt area with alcohol and cloth. Check all idlers and pulleys for dry or worn areas. A dry capstan bearing can cause slow speed. Intermittent or erratic speed is often caused by a dirty drive belt or cassette motor.

The defective motor can cause slow or fast speeds while rotating (**Fig. 9-4**). Dry motor bearings or worn brushes in the motor can cause slow tape movement. Sometimes, tapping the end of the motor with a screwdriver handle can cause the defective motor to speed up or slow down. A defective motor speed circuit can cause a fast or slow speed symptom. Some motors have a hole in the rear of the motor for speed adjustment.

Slow speeds can be caused by a packed tape head or excess tape wrapped around the capstan drive area. Dirt or oil on the capstan drive can produce slow speeds. Check both the drive shaft of the capstan and the pinch roller for excess tape oxide down inside the bearing area. Clean all oxide residue off of the pinch rubber roller assembly. Notice if the pinch roller is worn or out of round. The pinch roller can cause improper speed and can pull out excess tape that the take-up reel does not pick up. Clean the rotating surfaces before starting to replace any moving parts.

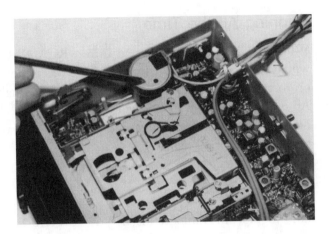

FIGURE 9-4. Adjust the speed control inside the hole on the back side of the drive motor.

Fast Speeds

Excessive speed can be caused by a defective drive motor or by a motor belt riding on the rim of the motor pulley. When the motor is running at a faster speed, tap the end of the motor and notice if it slows down. Replace the defective motor if the motor runs slowly or erratically. A higher-than-normal speed can result from excess tape wrapped around the capstan drive area. Remove the excess tape and clean it up with alcohol and cleaning stick.

Suspect a defective speed motor circuit if the motor is normal. First try to adjust the speed control before tearing into the motor speed circuits. Check the voltage applied to the motor terminals with the DMM. If the speed and motor circuits appear to be normal, replace the defective motor.

Replace a bad pinch roller if the cassette tape is rotating too fast. Normal speed in one direction and fast speed in reverse can result from excessive tape wrapped around the capstan with two different capstan drive assemblies.

Check the motor drive speed with a 3-kHz cassette tape and frequency counter. Connect the frequency counter across the speaker terminals. If the frequency counter registers above the 3 kHz, the tape is going too fast. When the tape runs too slowly, the frequency counter will read under the frequency of the test tape. Try to adjust the motor speed control where the frequency counter is at 3 kHz or a little ahead. This adjustment should be made after all repairs are finished and after a thorough cleaning of the tape head and moving components (**Fig. 9-5**).

Running Too Fast—Pioneer

In a Pioneer KP-4000 audio cassette player, the speed of the tape was moving too fast. Cleaning up the belts and idlers did not help. No excess tape was found around the capstan. A quick check on the motor schematic showed no speed circuits. The only part

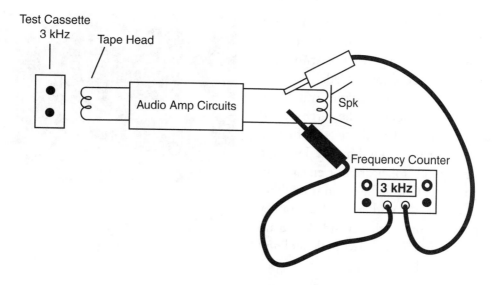

FIGURE 9-5. Tape speed can be checked with a 3 kHz test cassette and frequency counter.

left that could have caused the fast speed was the drive motor. Replacing the motor solved the fast-speed symptom.

No Fast-Forward (FF) or Rewind (RWD)

When the radio circuits are functioning and there is no tape action, notice if the tape motor is rotating. Check for a bad drive belt when the motor is running but there is no tape movement. Suspect dirty tape or radio switch contacts with no tape rotation. Measure the voltage applied to the motor terminals. If the tape is moving and there is no sound in the speakers, suspect a defective preamp or dirty tape-head switching assembly. Check for a bad idler or pulley when the tape plays normal with no fast-forward (FF) or rewind (RWD). A bad clutch assembly can cause poor rewind or fast-forward modes.

In fast-forward operation, the fast-forward roller or idler is pushed against the take-up reel at a faster rate of speed. Sometimes in larger cassette players, a different voltage is applied to the cassette motor to make it run at a faster speed. Suspect slippage on the idler drive wheel or pulley when the tape moves slowly or erratically in fast-forward mode. Clean up the gear teeth for cracked, broken areas or jammed gears for improper speed. Usually the tape head and pressure roller assembly are disengaged in fast-forward and reverse motion (**Fig. 9-6**).

You might find that plastic gears are switched in fast-forward mode in some tape players. Check the gear teeth for broken, cracked, and jammed gears. Notice if one of the gears or pulleys is missing and loose in the bottom of the radio-cassette player. Usually, gear-driven play and fast-forward modes very seldom lose speed.

Fast-forward and rewind run faster than the play modes. In the early rewind operations, the idler wheel is shifted to the supply reel in rewind mode. The tape head and pinch roller are not engaged in rewind mode. Check for oil on the rubber drive surfaces, worn idlers, and dry pulley components for erratic rewind. Clean up all idlers and drive areas with alcohol and a cleaning stick.

FIGURE 9-6.   A set of belt idler wheels rotates the take-up and supply reels.

No Fast-Forward—Panasonic

In this example, the service symptom in a Pioneer RS-2465 auto cassette player was that the tape did not go into fast-forward, and would sometimes pull out tape. Something appeared loose and made a rattling noise in the bottom of the radio cassette player. After removing the bottom cover, a pulley wheel and cam assembly were found in the bottom of the player. A "C" washer had come off and had let the wheel fall off.

All heads and tape moving parts were cleaned up and replaced. The pulley wheel was cleaned and replaced. A drop of light oil was applied to the bearing area. A good clean-up and replacing of the pulley solved the loss of fast-forward motion.

Keeps Reversing Directions

When the tape reaches the end of the cassette, some auto cassette players have an automatic reverse circuit that mechanically plays the tape in reverse direction. The early car cassette player had a magnetic switch and magnet that would reverse the direction of the motor. A magnet was mounted on the bottom area of a rotating turntable or reel, and underneath was a magnetic switch. When the end of the cassette was reached and the tape stopped, the magnet turned on the magnetic switch and reversed the direction of the tape and drive motor.

In older auto cassette players, an automatic reversing process included a commutator ring with spring-like tongs that rode upon the commutator rings. When the cassette reached the end of rotation, the commutator quit rotating and the electronic reversing circuit changed the polarity of the drive motor. Dirty contacts, dust, and dirt collected on the commutator rings can cause the auto cassette to automatically keep reversing directions. No reverse action can result from a bad zener diode in the reverse circuits.

Go directly to the auto reverse circuits when the auto cassette keeps reversing back and forth until the player is shut off. Clean up the commutator rings and tongs. Poor or bent prong contacts can result in rapid reverse procedures. Clean up the tongs and rings with alcohol and cleaning stick. Make sure the prongs are riding on each commutator ring (**Fig. 9-7**).

Do not overlook a loose belt or defective cassette when the player will not change directions. Sluggish reverse action can result

FIGURE 9-7. Small wire-like tongs lay against the rotating commutator rings with dual flywheels.

from cigarette and cigar ashes dropped down into the cassette player. Replace the main drive belt when the speed changes in both directions.

Audio Erase Head

The audio erase head removes any previous recordings from the cassette tape during the recording process. The erase head is always mounted ahead of the playback/record tape head. The erase head is smaller than the play/record head. Distorted and jumbled music can be heard when the erase head is not functioning. A packed erase head with tape oxide dust can cause poor recordings and distorted music.

The early and lower-priced cassette players with a DC voltage erase head were switched into the circuit for recording modes. Later on, oscillator bias circuits were designed to completely erase the previous recordings and provide high fidelity to the tape recordings. The bias oscillator waveform can be taken with the scope off of the tape-head winding terminal. Always clean up the erase head when cleaning up the play/record heads and rotating parts within the tape path.

Tab Erase Protection

Remove the small plastic tab at the rear of the cassette if you want to save the new recording. Notice that in all music recordings bought at electronic stores, no tab is found so that the cassette can be kept permanently. When the auto cassette is pressed for a recording, a lever pushes up into the knocked-out area and will not erase the new recording. If the tab is still intact, the lever is held back and the new cassette can now receive another recording.

If you want to preserve the recording, remove the tab from the back of the cassette. When the recording is no longer required, a piece of plastic tape can be placed over the opening and the old recording will be erased as the new recording is applied to the tape heads. Mechanically, the small lever is applied against the tape and lets the cassette to be erased. When the cassette player will not record, take a peek at the back of the cassette and notice if the tab has been removed.

No Automatic Shutoff

When the tape stops and comes to the end of rotation, the excessive tension of the tape engages and triggers a small ejection lever, which releases the play/record assembly and shuts off the tape operation. The detection lever pushes against the tape and trips a leaver to shut off the cassette player.

In most models with the auto cassette player, the tape will keep playing over and over again until the cassette is ejected. The car receiver with auto reverse circuits will not shut off, but will change direction of the tape and play over again. The auto reverse and stop circuit can fail because of a bad electrolytic in the main stop circuits.

No Ejection of Cassette

Some auto cassette players have an eject button that applies DC voltage to a solenoid plunger, which can eject the cassette part way out of the slotted area. The eject switch makes a DC voltage contact switch to a transistor and applies a DC voltage across the relay, energizing the solenoid winding and kicking out the cassette.

A shorted diode across the solenoid winding can defeat the eject system. Suspect dirty eject switch contacts when the cassette will eject sometimes and not other times. Check for a DC source applied to the solenoid when the eject button is pressed. An open solenoid winding or broken wire can cause lack of ejection of the cassette. Often this solenoid or plunger is tied directly to the auto battery supply source (**Fig. 9-8**).

Capstan Motor Problems

A defective drive motor can cause slow and erratic tape speed. Replace the motor and drive belt. A defective drive motor can also cause excessive speed. Loss of motor rotation can result from an open isolation resistor or series diode in the voltage supply source. Poor tape rotation can be caused by a low-voltage supply source or a dead motor. An open 0.5-ohm resistor in one leg of the drive motor can cause a dead motor operation.

Replace the drive motor for fast speeds or when the speed varies, causing a wow and flutter condition. A bad motor regulator transistor can cause really fast speed. When the tape runs slow and sometimes stops, suspect a defective regulator transistor in the motor

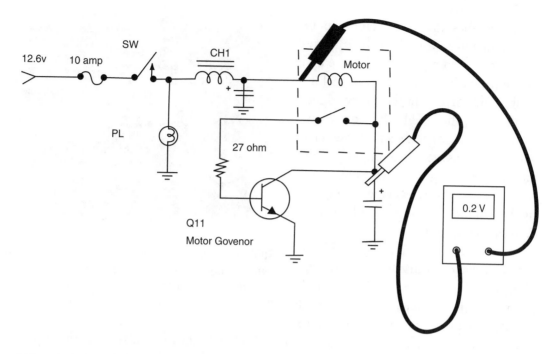

FIGURE 9-8. Check the voltage across the motor terminals to determine if the motor is defective.

circuits. An open motor control transistor can cause slow or fast speeds. If the tape runs too fast, check and adjust the tape speed control for correct speed. A bad IC regulator in the tape circuits can cause a wow condition. When the motor runs slow, suspect 10 volts or less at the motor terminals.

Check the voltage tied to the motor terminals with lack of tape motion. A bad leaf or tape switch can prevent voltage from the motor circuits. Suspect a bad motor speed control IC or transistor circuits when a wow or a change in speed varies. A low voltage applied to the motor terminals will cause the motor to run slow or stop under load. Replace the defective motor if improper voltage is applied at the motor terminals (**Fig. 9-8**). If the motor runs backward, replace the polarity protection diode.

Tape Being Eaten

The auto cassette player can eat or spill out tape when the take-up reel assembly is not taking up the excess tape. If the take-up reel stops or slows down, the excess tape might spill out and wrap around the capstan and the pinch roller assembly. Check for a spill-out of tape when the tape music slows down or stops. Eject the cassette and notice if a lot of tape is still inside and connected to the cassette.

Sometimes the excess tape can be unwound and rewound on the cassette-supply reel with a pencil. Stick the end of the pencil inside the take-up reel assembly and rotate it in the correct direction to wind the tape back into the cassette. A stuck cassette and loose tape within the cassette area might be removed by rotating the flywheel by hand. In most cases, the excess tape is wrapped tightly around the capstan and ends up damaged (**Fig. 9-9**).

Excessive tape wrapped around the capstan drive can cause a slow and warbling sound when the tape is played. A piece of tape wrapped tightly around the capstan drive can cause the tape to

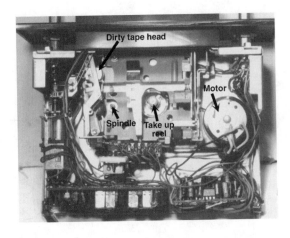

FIGURE 9-9. The different components that can cause excessive tape to pull out and wrap around the capstan and pinch roller.

rotate really fast. Pulling of tape can result when the hubcap is missing at the top of the take-up reel. Check for a worn or uneven pinch roller pulling out tape. If a sticky substance is found on the capstan or pinch roller, excessive tape can be pulled out that is not picked up by the take-up reel. Replace the worn rubber pinch roller for pulling out tape.

A sluggish take-up reel assembly or dirty belt can pull out tape from the cassette. Suspect a bad idler pulley rotation when the tape is pulled from the cassette. Clean up the rubber drive surface on the take-up reel, idler, and pulleys. Inspect the gear-driven idler or take-up reel for broken teeth or dry bearings.

A defective cassette can cause the tape to spill out; try another cassette. Clean up all around moving components for speed and the eating of tape. A missing washer, bad pulley wheel, and cam lying in the bottom of the cassette player can cause a pulling of tape and the loss of fast-forward motion.

Pulling of Tape—Realistic

In this example, a Realistic model 12-1942 AM/FM stereo cassette player came in with no rotation of tape. The cassette could not be removed from the auto tape player. After removing the top and bottom covers, the cassette player had two capstan flywheel assemblies. The tape was cut loose, as no flywheel could be turned to free the excess tape. Some of the tape was wrapped around the capstan drive shaft.

After removing all excess tape, the cassette could not be saved because several feet were damaged. The tape heads, capstan, pinch rollers, and reels were cleaned up with alcohol and cloth. Again, within a short time, excess tape started to run out between pinch roller and take-up reel. The take-up reel was found riding high and tape spilling out. Replacing the polyslider cut-washer at the top of the take-up reel cured the cassette of pulling and eating up tape.

Stuck Tape

Suspect that excess tape is wrapped around the capstan when the cassette will not eject or cannot be pulled out of the player. The radio cassette player must be removed from the automobile and both covers removed. Slowly rotate the capstan/flywheel backward with several fingers. Notice if some of the tape is starting to come loose, so the cassette can be pulled out.

The tape may have to be cut in several places before it can all be removed. Of course, the tape is now ruined and damaged. After the cassette has been removed, pull out all loose tape wrapped around the capstan and pinch roller assembly. Make sure that no pieces of tape are wrapped around the pinch roller bearings. If not, the pinch roller will rotate slowly and the music will have a wow condition. Clean up the pinch roller and capstan drive shaft with alcohol and cleaning stick before trying to play another cassette.

Replace the main gear assembly when the player stops in play mode and the cassette cannot be removed. The main gear might have to be removed in stop mode before the cassette can be removed from the player.

Flywheel Problems

Check the alignment of the capstan/flywheel assembly if it makes a noise while rotating. The flywheel might be rubbing against the bottom metal retainer. Try to replace the defective flywheel or try to repair it. Most flywheel shafts work down on the drive shaft as they are pressed together. Remove the flywheel and see if the flywheel cannot be returned to the correct spot on the drive shaft. Sometimes a missing washer under the flywheel assembly can cause the flywheel rotating noise.

If the flywheel has worked its way down the drive shaft and makes a noise as it rotates, slightly tap the top of the drive shaft bearing on a piece of wood until the flywheel rests in the original spot. Temporarily replace the capstan/flywheel to see if the motor belt now lines up with the drive motor. Once again, remove the flywheel and apply epoxy cement to the bottom shaft area. Drop the flywheel in a hole in the board so that the flywheel is level and the epoxy does not run off. Let the epoxy set overnight and replace the flywheel the next morning.

The drive surface of the capstan/flywheel can become very shiny-looking after several years of operation and can cause slow speed problems. Remove the flywheel and rough up the shiny drive area with PC board etching solution. Apply several coats on the shiny drive area and let it stand with the drive shaft in a downward position so that it will not run down into the bearing area. Let each coat dry before applying another. Wipe up the excess with a dry cloth. Of course, this is only a temporary drive shaft repair when the capstan/flywheel is not available.

Playback Problems

The tape will not move if improper voltage is applied to the tape drive motor. Check for an open or increased resistance in the power supply circuits. A bad leaf switch can cause loss of playback modes. Take critical voltage measurements on the voltage supply source for a possible defective transistor or IC regulator circuit. The leaky IC regulator can also cause intermittent stop in the playback mode. A defective regulator can cause the speed to vary in playback.

Replace the motor drive belt when the cassette begins to play and then stops. Bad contacts on the play switch can cause low and distorted playback sound. Suspect a leaky 4.7-uF electrolytic with weak and no audio in playback. Audio flutter in playback can result from a dry supply reel. Poor alignment of gears between the take-up pulley cam assembly and the take-up reel can cause a grating noise in playback.

Lubricate the idler gear pulley for a grating noise during playback. Rotate the supply reel by hand to notice if it makes a scraping noise during playback. A chattering noise during playback can be caused by foreign material between the flywheel assembly and metal support. Check for a bad solenoid when the tape will not rotate in playback mode. Replace the torsion spring if the tape stops rotating in playback.

Wow and Flutter

Wow and flutter conditions in the radio cassette player might occur during playback operations. The short change of sound might be called flutter, while the longer duration of sound might be called a wow condition. Some technicians might not hear the wow and flutter sounds, while those with better hearing can hear these changes in sound.

The wow and flutter within the tape deck can be located with a test tape and a wow-and-flutter test instrument. The frequency counter and test tape can be used to check wow and flutter problems. Insert a 3-kHz test tape and connect the frequency counter to the speaker output terminals. Simply notice the deviation in the frequency counter or meter as the tone test tape is played in the auto cassette player. A greater deviation of frequency (3 to 10 kHz) of the test tape indicates a wow or flutter condition.

A dirty or worn drive belt can cause a wow or flutter sound in the speakers. The dirty or binding capstan/flywheel, defective cassette, bad motor, and defective speed control circuits can result in wow and flutter. Suspect a nylon bearing working on the capstan that can cause the cassette to run slow with a wow symptom. The capstan might run slow and drag when particles of rubber from a bad belt are found on wheels or pulleys.

A dry or bent out-of-line flywheel can cause wow and flutter. Look for a missing or out-of-place idler spring that can cause a noisy wow and flutter symptom. Replace the idler spring for a knocking noise with wow and flutter. Suspect a dry supply reel table for wow and flutter in the playback modes. A worn pinch roller or a pinch roller spring needing adjustment can cause high wow and flutter conditions. The motor pulley too far out on the shaft, making the belt slip, can cause wow and flutter in playback.

THE VARIOUS CIRCUITS

The cassette player has many different circuits that must perform as the cassette is played. The tape-head circuits pick up the music from the tape, and feed it to the preamp circuits. A preamp circuit might consist of a directly coupled transistor or IC components. The tape heads might be switched into the preamp circuits with the output music switched to the audio output circuits.

Automatic reverse, automatic shutoff, and auto-stop circuits are found in the most deluxe auto radio cassette players. The automatic speed circuits are not found in the early auto cassette players. The Dolby and equalizer circuits are found in the latest cassette players. Last but not least, the audio line output circuits are found in the radio cassette players and can be connected to high-powered watt amplifier systems.

Tape-Head Circuits

The magnetic tape head picks up the music from the revolving tape and couples the music to the preamp circuits. Because the tape-head signal is quite weak, it must be amplified before it is applied to the audio AF or driver circuits. A mono tape head has only two terminals on the backside of the tape head, while the stereo tape head has at least four terminals. The magnetic tape head is wound around a metal framework with a slotted area at the front of the tape head (**Fig. 9-10**).

A typical tape-head winding can be checked with the ohmmeter for a continuity test. The tape-head resistance might vary between 200 and 800 ohms on the DMM. Check both tape-head windings and compare them, for they should be quite close to the same resistance.

Often, only one channel appears dead, intermittent, or weak. The tape head should be cleaned up before tearing into the tape-head circuits to search for the cause of weak or distorted sound. A packed tape head with tape oxide can cause loss of sound out of one channel and distorted sound out of the other stereo channel.

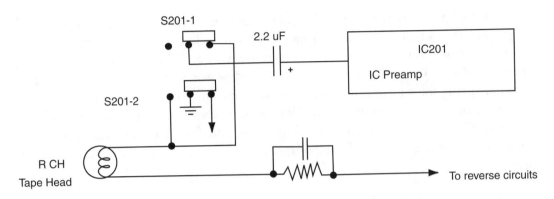

FIGURE 9-10. The tape head is switched into the tape head preamp circuits.

The defective tape head might have an open winding, bad terminal connections, and a worn front surface. Poor soldering on the tape-head connections or poor tape switching can cause intermittent music. Simply press the rubber end of a pencil on each head terminal connection to make it act up. Sometimes the internal winding soldered to the outside terminal can come loose and cause intermittent reception. After cleanup, if the distorted sound is still heard, check the worn gap area, and replace the tape head.

Check the tape head for poor high-frequency response. Notice if one mounting screw might be loose and if the tape is only sometimes engaging the tape head. A missing azimuth screw can cause lack of highs and a muffled sound. Poor cross talk can result from improper tape-head azimuth adjustment. Improper tape azimuth adjustment can cause a loss of high frequency sound. Try another cassette as this cassette might be defective.

An open tape head will cause loss of left channel with a normal right channel. An open tape head or disconnected tape head wire can create a loud rushing noise heard (but no other sound heard) with the volume turned up. Often, a loud rushing noise and a slight hum in the speaker indicate an open tape-head winding. Weak sound with a loud hum can result from one side of the tape head having a broken lead. A dirty tape head can cause low volume.

A bad tape-head connection can cause intermittent right channel. Most tape head cables broken off of the tape head are caused in the auto cassette players with reverse modes. Loss of left stereo channel can be caused by a poor adjustment of the tape-head height adjustment.

Suspect a broken ground wire when no sound is heard in reverse direction and audio in only one direction. Replace the tape head when the tape makes a squeaking sound. A bad radio/tape switch can cause loss of tape or radio sound. Check the preamp circuits for a distorted right channel and a weak left channel when the tape head circuits are normal.

Directly Coupled Preamp Circuits

The tape-head output signal is switched into a preamp circuit that might include two audio transistors that are directly coupled together. A tape-head signal is coupled to the base circuit through a 10-uF electrolytic. The collector terminal of TR4 is directly coupled to the base of TR5. Notice that the collector voltage of TR4 is very low, to prevent picking up noise in the input circuits. TR5 amplifies the picked-up head signal of TR4 and is capacity coupled to the audio output circuits (**Fig. 9-11**).

In the later preamp head circuits, the left stereo tape head signal is connected to a 4.7-uF electrolytic capacitor to terminal 1 of a dual-preamp IV. Likewise, the right tape-head signal is capacity coupled through C803 to input terminal 8 of IC701. The left output audio signal is amplified by the preamp IC701 and the output signal is taken from pin 3, while the left output signal is found at pin 6. The supply voltage of 3 volts is found on pin terminal 4 of IC701 (**Fig. 9-12**). Pin 5 is at ground potential.

Check the preamp circuits for loss of sound, for a dead left or right channel, for an intermittent or weak channel, or for one channel distorted and the other channel normal. When one preamp transistor is found open, the other transistor might appear normal or have a weak sound. You might find the second transistor open and the directly coupled transistor leaky.

Check all bias resistors when a transistor is found leaky or shorted. Remove one end of the transistor from the circuit for a good resistance test. Open or leaky coupling electrolytics can cause a dead, weak, or distorted sound symptoms. Improper supply voltage to the preamps can cause weak and distorted audio.

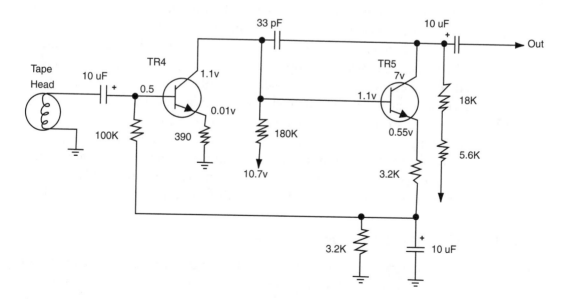

FIGURE 9-11. A directly coupled tape head circuit with direct-driven transistors.

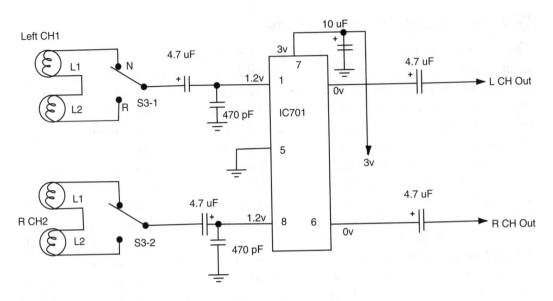

FIGURE 9-12. Usually, low supply voltage is applied to the first preamp transistor or IC to prevent a noisy pickup stage.

A slightly weak left channel can cause poor audio balance with the balance control. A dirty mechanical switch might cause one channel to squeak and the other channel to be normal. A bad jumper wire to the B+ supply source to the directly coupled preamp circuits can cause loss of sound output. Low and distorted audio can result in a leaky or shorted dual-preamp IC. No voltage found on the supply pin of the preamp IC might cause distortion in both right and left stereo channels.

The distorted left channel can result from a leaky preamp transistor. An increase in resistance to the preamp transistor and open electrolytic bypass capacitor can cause a very weak channel. Check for a shorted or leaky decoupling capacitor when distortion is found in one channel and the other channel is dead. Replace the leaky or open voltage regulator transistor for weak or no voltage to the preamp circuits.

The noisy preamp transistor can cause noise in one channel or hum in the other preamp circuit. Replace both preamp transistors if you have a noisy right channel after the cassette plays for several hours. The noisy left channel can be caused by a defective preamp IC. Noise found in both channels can result from a noisy dual-preamp IC.

Dead Right Channel—Loud Rushing Noise

In this example, a loud rushing noise was heard in the right channel of a Sanyo FT20 auto cassette player. The left channel appeared normal. After removing the top covers, a screwdriver blade was touched to both sides of the tape heads and only a hum was heard in the right channel. The left channel appeared normal.

When the external amp was connected to the right tape head, the signal was good in both channels. The right channel was signal-traced right up to pin 8 of the preamp IC701. No right channel sound could be heard at the right channel output terminal pin 6. The audio was normal in the left channel at pin 3. Replacing IC701 with a LA3161 dual-preamp IC resolved the dead right channel.

Auto-Reverse Circuits

The auto cassette player might have dual tape heads and dual capstan/flywheels found in cassette players with automatic reverse features. A solenoid energizes and pulls in another set of tape heads and capstan/flywheels, while switching the direction and polarity of the drive motor. The AMSS reverse switch circuits are found in some cassette players.

In the early Pace auto cassette player, the auto reverse circuit consisted of two transistors reversing with a rotating commutator. When the tape stops at the end of rotation, the commutator stops and reverses the signal to the reversing circuits (**Fig. 9-13**). The auto cassette will keep going in the same direction until the commutator stops and reverses the direction in the auto reverse circuits.

The tape end control or reverse circuits in an early AIWA CTR-2050Y auto cassette player has a more complicated circuit, with five different transistors. A magnet is attached to the rotary turntable or reel, and when it stops, the magnet switch mounted

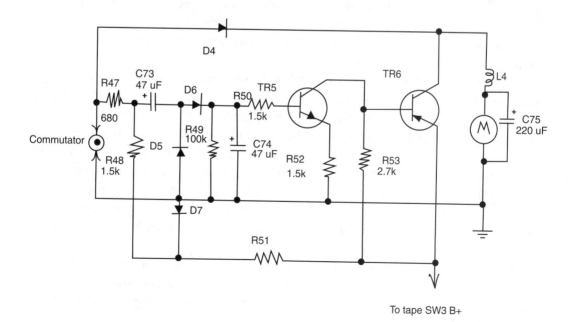

To tape SW3 B+

FIGURE 9-13. The auto reverse circuit in a early Pace auto radio cassette player.

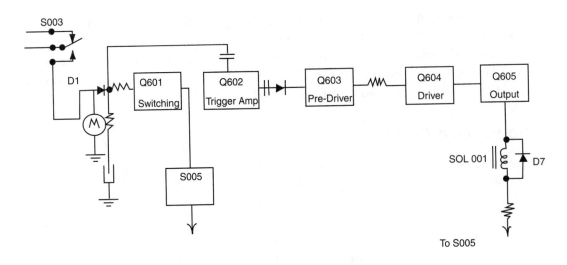

FIGURE 9-14. Block diagram of an end-control circuit found in a AWIA auto cassette player.

below turns on the end control or reverses the direction. The solenoid energizes and switches the power supply voltage to the tape motor.

A reversing S003 switching arrangement is switched and changes the direction of the motor by changing the polarity of voltage to the drive motor and audio preamp circuits. S003 is part of the solenoid switching and is triggered by transistor Q605. When the cassette comes to the end of rotation, a magnetic switch engages the input diodes of switching transistor (Q601). The output signal of Q601 is connected to another set of contacts of switching S005.

The output of S005 is coupled through a 10-uF coupling electrolytic to the input circuits of Q602 (**Fig. 9-14**). The trigger amp (IC602) is capacity coupled through another 10-uF capacitor to the pre-driver transistor (Q603) and driver transistor (Q604). Output transistor Q605 energizes the solenoid (SOL 001) and reverses the switching with S003.

The auto-reverse procedures can operate mechanically, like in the early cassette player, or electronically. In the reverse circuits, it can detect the no-signal segment of the stopped tape. After a few seconds, the tape is reversed by the end electronic control or reverse circuits. The tape-head reverse circuits can fail with a dirty commutator or defective magnetic switch assembly. Leaky or open transistors within reverse circuits can cause the auto reverse to not operate.

Excessive tape wrapped around the capstan drive can cause normal speed in one direction and fast speed in the reverse, with two different capstan/flywheel assemblies (**Fig. 9-15**). Sluggish reverse action can result from debris falling into the cassette mechanism.

Dirty commutator rings or tongs can cause the tape player to keep reversing directions automatically. Clean up the reverse disc for normal sound playing in the reverse mode. A dirty head switch can cause loss of reverse sound in the right channel. Clean up the dirty reverse pilot light switch for loss of reverse LED or pilot light.

Dolby Circuits

Dolby B and Dolby C circuits are found in some auto cassette players. A Dolby system is an electronic method or circuit for improving audio quality of the cassette player. For low-level sounds,

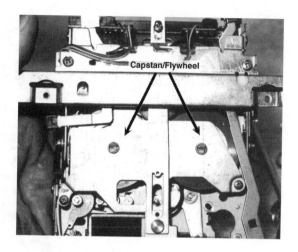

FIGURE 9-15. Two different capstan/ flywheels are found in the auto reverse cassette players.

the gain of the amplifier is increased during recording modes, and the low-level sounds are reduced in playback modes. The Dolby A system might have four ranges, while Dolby B has only one band-reducing noise in the amplifying circuits.

In some cassette players, Dolby noise-reduction circuits are included to reduce the level background noise normally found during recording modes. Dolby B noise reduction occurs only at the middle and upper portions of the audio spectrum. With the Dolby circuits, noise (NR) has been reduced to a minimum in portions of the recording program. Dolby reduction does not change the frequency of the audio signal. At high levels, the noise is suppressed. At low input-levels, the signal-to-noise (S/N) decreases and the noise is heard. The amount of boost depends upon the level and frequency of the signal.

In the early cassette players, several transistors were found in the Dolby audio circuits. These Dolby circuits are located between the preamp and audio output stages.

Audio Radio Equalizer Units

Equalization is the use of a circuit or device to make the frequency response of a line, amplifier, or other device to be uniform over a given frequency range. The equalizer circuit might compensate attenuations for equalization. The equalizer circuit might be built into the car radio or a separate unit between radio and amplifiers (**Fig. 9-16**).

For instance, a Pioneer DEQ-7600 Digital equalizer gives you 15 bands of EQ for precise sound control, and can apply them to just the speakers, just to the rear, or all of them. The Sony XE-90MK11 has a nine-band equalizer and features two sets of full-range outputs with a set of subwoofer outputs (with selectable crossover and volume), nine frequency band controls for precise equalization, and a listening position selector for adjusting the stereo image. An Alpine ERE-G180 equalizer can be mounted in the auto dash or trunk area, and it has crossover, 11 EQ bands, and 12 dB maximum boost, with a 4-volt maximum output level.

FIGURE 9-16. The early equalizer unit is usually mounted after the radio and between the audio output amplifier.

Audio Line Output

The auto radio cassette player might have a set of left and right audio output jacks that can play into a CD player or a high-powered amplifier. The auto CD player can heave from one to three sets of preamp outputs with 1.8 to 5 volts. Like the CD player, the radio cassette line outputs might include a mute circuit. The mute circuits might consist of a transistor in each left and right output jack. The mute transistors are controlled by a system control or microcomputer IC.

No noise or sound should be heard in startup and during switching operations with a normal mute system. Determine if the audio is dead at the line output amp IC, at the mute transistor, or at the output line jacks. Check the audio out of each line output jack with an external audio amplifier. Remove the collector terminal of the remote from the PCB and notice if the sound returns. Repair the defective mute circuit if the sound is now heard.

Replace a leaky mute transistor for loss of audio line output sound. Check for a bad mute-switch mechanism when the audio becomes intermittent or when it cuts out. Suspect a defective control system IC when the mute circuits do not function (**Fig. 9-17**).

Speed Control Circuits

In the early motor drive circuits, the drive motor was adjusted for speed with a screwdriver slot at the end of the motor. Some of the early speed circuits used only a single transistor, while others had two or more transistors within the drive motor speed circuits (**Fig. 9-18**). Usually, a speed control can adjust the speed of the motor within the speed circuits. Make correct speed adjustments with a 1-kHz or 3-kHz audio test cassette and frequency counter for slow and fast speed symptoms.

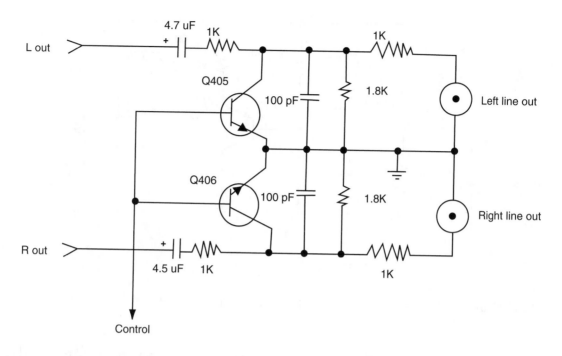

FIGURE 9-17. The audio line output jacks with muting transistors.

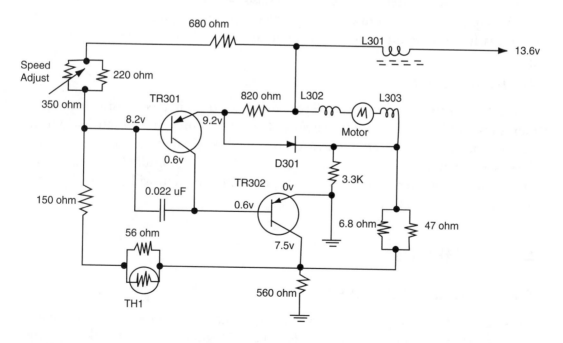

FIGURE 9-18. An early speed circuit that controls the drive motor with a speed adjust control.

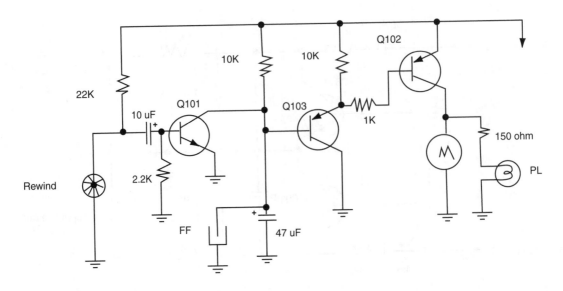

FIGURE 9-19. The early auto stop circuit with a pilot light indicator.

A defective speed control transistor can cause excessive or slow speed, or no speed at all, in the tape-drive motor. Check all transistors with the diode test of the DMM or with a transistor tester. Small diodes can also be tested with the diode test of the DMM. Open choke coils can cause slow or no tape movement. Take a quick voltage test on each transistor and across the drive motor terminals for no tape motor rotation.

Auto-Stop Circuits

In the early 8-track and cassette players the auto-stop circuits consisted of several transistors within the drive motor circuits. The auto-stop circuit operated within the play, rewind, and fast-forward modes. A revolving disk was mounted on the bottom of the tape reels and when it stopped rotating, the dive motor stopped. Here, three directly coupled transistors were found controlling the motor rotation with a pilot light tape end indicator (**Fig. 9-19**).

ADJUSTMENTS

Take-up Torque Adjustment

Measure the take-up torque in play mode, using a cassette-type torque meter (120 grams/cm max). Most measurements are between 30 and 50 gr/cm. Adjust the take-up torque by turning the friction star of the take-up spindle or reel assembly. Turn the friction star clockwise to decrease the tension, and turn counterclockwise to increase the torque (**Fig. 9-20**).

Pinch Roller Adjustment

Set the cassette player in the play mode, and adjust the clearance between the pinch roller arm and the solid base to more than 0.2 mm by bending the roller arm. Hook a spring balancer (measuring capacity 500 gr) to the pinch roller bracket, and pull the

FIGURE 9-20. For greater torque, move the star washer adjustment found mounted on one of the reel assemblies.

idler away from the capstan. The pinch roller pressure should be between 250 and 359 gr. Most pinch roller springs can be placed in another hole for a stronger pinch roller adjustment. Check the manufacturer's service literature for correct pinch roller adjustment.

Head Azimuth Adjustment

Connect a VTVM or FET-VOM to the left channel speaker terminals and play back an 8- or 10-kHz test tape. Turn the adjustment screw to the position when the meter swings to a maximum position. Likewise, connect the meter to the right channel and adjust the azimuth screw for a maximum reading. Split the difference between the two measurements and leave the azimuth screw set at that position. Drop or dab a spot of paint or glue on the azimuth screw so that it will not move (**Fig. 9-21**).

Tape Speed Adjustment

Connect a frequency counter to either the left or right channel speaker terminals with an 8-ohm load resistor across the speaker terminals. Play a 3-kHz test tape and if the cassette player has a speed adjustment, turn it until the counter reads 3 kHz (**Fig. 9-22**).

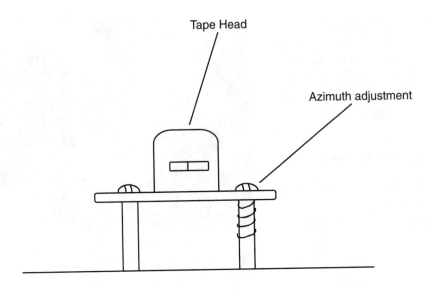

FIGURE 9-21. Adjust the azimuth screw with a test cassette and counter for maximum measurement upon the meter.

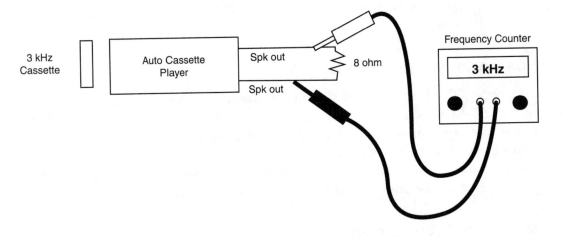

FIGURE 9-22. Check the tape speed adjustment with a 3-kHz test cassette and frequency counter.

CHAPTER 10

Servicing the Auto CD Player

The auto CD player has many circuits different from those of the auto radio or cassette player. The CD mechanism might consist of a disc sensor, chuck switch, in switch, limit switch, loading motor, SLED motor, spindle motor, and an optical pickup assembly. The optical pickup might include a laser diode, photo detector, focus, and tracking coils. The loading motor might have a transistor or driver IC. Both the SLED and disc or spindle motor might be driven from a servo IC (**Fig. 10-1**).

The optical pickup photo detector signal is fed to a servo signal processor and RF amp IC. Some CD players might have transistors within the RF amp circuits that receive the optical pickup signal. The RF signal eye pattern can be taken at the output of the RF amp, and indicates that the pickup assembly and RF circuits are functioning. The RF amp circuits must perform before the servo driver IC, digital processor, and microprocessor can operate.

A microprocessor IC operates the disc sensor, clutch switch, in switch, limit switch, loading motor and driver IC, laser on/off transistors, and digital processor. The microprocessor IC might have an input of a 5-volt backup and a 5-volt main voltage source, ACC, reset, data, CD in and out, power control, and eject circuits. The microprocessor IC might also control the servo signal processor, which in turn drives the

FIGURE 10-1. A factory-installed AM/FM/MPX auto receiver and cassette player with a CD player in an Oldsmobile 1998 model.

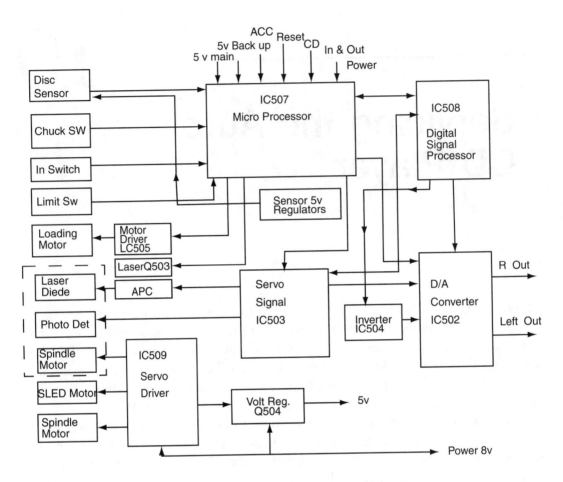

FIGURE 10-2. Block diagram of a simple auto CD player's circuits.

servo driver IC. A separate voltage regulator circuit can feed a voltage to the disc sensor and to the motor control 5-volt source (**Fig. 10-2**).

ELEMENTS OF THE CD PLAYER

A Typical Auto CD Power Supply

Several auto CD players have a conventional DC-DC converter power supply, while many others have a regular battery-voltage regulated system. The main car radio-CD player's 5-amp fuse and line choke might be tied to a common outlet board with a couple of 8- and 5-volt regulators. The 13.6-volt battery source is fed directly to the outlet board and to the transistor or IC voltage regulators.

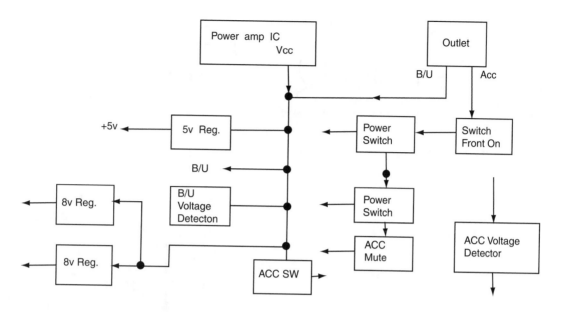

FIGURE 10-3. The auto battery-source power-supply circuits of a CD player.

A 5-volt regulator transistor is found in the back-up (B/U) Vdd power source feeding the microcomputer IC, 5-volt B/U source, Acc voltage detector, power amplifier, 8-volt voltage regulator, reset, and switch. The ACC 1-amp fuse protects the power line source that is fed to SW front-on, power switch, Acc, off mute, AF mute, Acc voltage detect, and reset IC. Both the Acc and B/U voltage sources are fed from the car battery source (**Fig. 10-3**).

Many different transistors are used in voltage regulators, such as Q723, within the battery source. The Vdd voltage regulator transistor provides voltage to the Vdd source with applied voltage from the B/U power source. The collector terminal of Q723 has a 4.7-ohm isolation resistor to the input B/U 12.7 supply voltage. A zener diode and the 47-uF electrolytic are found within the base circuit of Q723. The 5.3-volt regulated source form the emitter terminal, which is fed to IC701, Q601, Q604, IC702, and a Vdd voltage is fed to pins 3 and 4 of the microprocessor IC601. If voltage regulator Q723 fails, the whole control mechanism stops (**Fig. 10-4**).

The 8-volt regulator transistor (Q714) receives the power voltage from the Vcc source and is fed to the collector terminal. A zener diode and electrolytic capacitor are found within the base terminal, and an 8-volt regulated voltage is fed out of the emitter terminal to the 8-volt power circuits. The Vcc voltage source is fed to the power output IC and the input voltage from the back-up (B/U) battery source (**Fig. 10-5**). One 8-volt regulated source supplies voltage to the radio tuner and audio circuits, while the other 8-volt source feeds the LCD light display circuits.

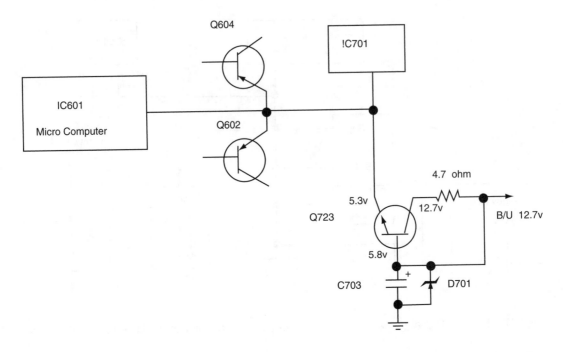

FIGURE 10-4. A 5-volt (Vdd) voltage regulator transistor (Q723) provides a 5-volt source to the microcomputer IC601.

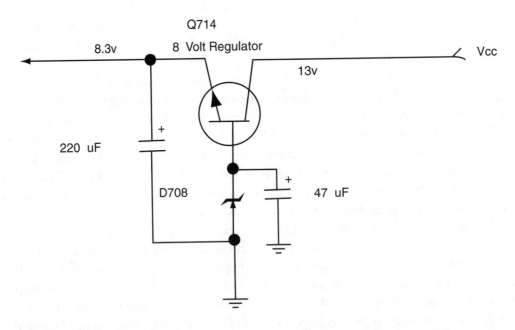

FIGURE 10-5. A simple 8-volt transistor regulator circuit providing operating voltage to the tuner, audio circuits, and LCD display.

Go directly to the regulator transistors when a voltage source is missing from the IC working circuits. Most regulator transistors will appear open without any voltage or will be leaky, causing weak or intermittent operation. Take a quick voltage measurement on the emitter terminal for a regulated voltage source and on the collector terminal for an input voltage source.

When the microcomputer IC has no power or system control, go directly to the Vdd voltage terminal on the micro IC and the Vdd regulator 5-volt source. Suspect the 8-volt regulator circuit when the tuner or audio amp stages are defective. If the LCD display will not light up, suspect another 8-30-volt regulator transistor circuit for low or no regulated output voltage. You will find many different voltage regulator circuits within the auto radio-cassette and CD players.

Check the IC5 voltage regulator if the CD player stops in the middle of the program or if it gets really warm. Voltage regulator (IC301) can cause a loss of play mode. Replace the voltage regulator transistor if the CD spins for a few seconds and then shuts down. Check the 3-, 5-, or 8-volt regulator source when the tuner presets will not hold. For a loss of LCD display, check the 8-30-volt regulator source. Replace the main voltage regulator on the main power supply board when the pickup assembly skips after warm-up.

Measure the voltage on the 8- or 10-volt voltage regulator supply for a dead CD player. Check the 5- or 8-volt regulator transistor located in the main power supply for a dead or intermittent display. Suspect the main voltage regulator for erratic loading and intermittent tracking modes. Replace open resistors within the main voltage source for loss of audio. Check the main IC voltage regulator for loss of play, fast-forward, or rewind modes.

A shorted motor regulator transistor can cause the drive motor to run with the power off. Replace the main IC voltage regulator when the CD player intermittently starts and stops in the middle of the program, usually after warm-up. Replace the 5-volt regulator IC with an intermittent or stop mode. Check the 5-volt regulator transistor when the disc will not rotate and the lens assembly does not operate. Replace the voltage regulator IC if the disc stops after loading, if it will not accept the disc, or if it has intermittent and erratic tracking. Inspect and check the main regulator IC when the focus coil oscillates.

The Optical Pickup Assembly

The auto optical pickup assembly might consist of a laser diode, lens, photodetector, focus, and tracking coils. A diffused laser beam is emitted from the laser diode (LD), and then goes directly through the lens assembly. The beam is flashed on a row of pits on the disc surface and reflected back through an objective lens. The LD emits a straight polarized beam, which is changed to a circular beam, and shines onto the disc (**Fig. 10-6**).

The laser beam strikes the surface of the disc on a row of pits, and the signal is reflected back to a photo-diode, which reads the digital signals picked up by the disc. The pit information picked up by the optical lens assembly is fed into the RF amplifier. In the early auto CD players, transistors were found in the RF amplifier circuits. Today, a large servo signal processor RF amp IC amplifies and controls the focus and tracking coils. In some models the focus and tracking coils

FIGURE 10-6. The top-side loading plate and keeper must be removed to get at the pickup assembly in a low-priced auto radio-CD player.

are fed directly from the servo IC. The focus error (FE) bias and tracking error (TE) bias are fed from the signal processor and RF amplifier (**Fig. 10-7**).

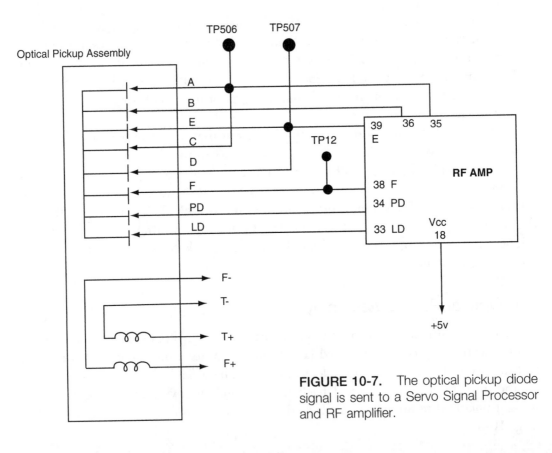

FIGURE 10-7. The optical pickup diode signal is sent to a Servo Signal Processor and RF amplifier.

The Optical RF Amplifier

The servo signal processor and RF amplifier provide an RF or EFM signal to the servo and signal processor IC. The RF signal at pin 31 of IC501 waveform is quite critical in the whole operation system of the CD player. The whole system is shut down if no EFM or RF signal is found at the output pin of the RF amplifier IC501.

Besides an RF output, the RF amp and servo IC provide a focus error (FEO) output signal at terminal 1. The focus error (FEI) input terminal is found at pin 2. A focus amp (FEM) inverted input is at terminal 7 and the focus drive signal (FEO) is at pin 6. The tracking error amplifies the inverted signal (TAM) at terminal 12 and the tracking drive output (TAO) signal at pin 13 of IC501. Besides the EFM, focus, and tracking signal, the servo preamp (IC501) provides a signal to the SLED motor driver IC.

Check the EFM/RF eye pattern at pin 31 of IC501 when the disc, SLED, focus, and tracking system do not operate. Notice if the focus and tracking coils are moving when the disc player is first turned on. You may have to remove some components to get at the optical assembly where the coils are located. If an eye-pattern waveform is found at TP505 or at pin terminal 31 of IC505, you can assume that the optical pickup assembly and RF amp circuits are normal (**Fig. 10-8**).

If no eye pattern is found at the RF output of the RF amplifier, take critical voltage measurements on the servo signal and RF amplifier. Look for the Vcc pin terminal that supplies B+ voltage from the main 5-volt power supply. If the voltage on pin 18 is low or has no voltage, suspect a defective RF amp IC, the 5-volt regulator transistor, or the IC that provides the 5-volt source.

Suspect a leaky RF amp IC when the supply voltage is low. Make sure the voltage on all other terminals is normal before trying to remove the servo signal processor and RF amp with 48 terminals or more. Remove the Vcc pin terminal with a solder wick and iron. Notice if the supply increases with pin 18 removed. The RF amp IC is leaky or pulling too much current if the supply voltage returns to a normal 5 volts (Vcc).

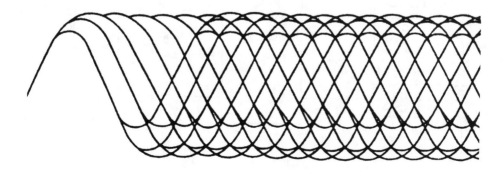

FIGURE 10-8. The EFM or eye-pattern waveform must be at the output of the RF amplifier before the CD player will operate.

Clean up the laser pickup lens assembly with alcohol and a cleaning stick when the disc spins slowly and sometimes will not play. Replace the laser scanner assembly if you find no CD operations or intermittent CD action. Clean up the laser assembly if the disc motor stops rotating. A bad laser assembly can cause the sound to fade with a static noise. Replace the optical assembly when static is heard while the disc is playing or if intermittent disc skipping is noticed. Check the optical pickup assembly when it will not read the table of contents. Remove a burr from the optical pickup assembly if there is audio distortion.

In addition, replace the optical pickup assembly if the SLED motor runs too fast or if there is intermittent play operation. Check the optical pickup assembly when the pickup will not track properly. If the CD player ejects after 5 to 15 seconds, replace the pickup head assembly.

Check the DC voltage supply source to the laser assembly and RF amplifier with a loss of eye pattern. Suspect a 5-volt regulator transistor or IC when the disc stops in the middle of the program after warm-up. Check the moving wires or cable to the laser assembly when the CD plays 30 to 60 seconds on the first track and then mistracks.

Focus/Tracking Coils

The signal servo processor operates the focus and tracking coils. A servo processor might be included within the RF stage. The focus and tracking error signals are fed to the servo processor, and RF amp is fed to a servo driver transistor or IC. Today, most servo driver circuits are controlled by a servo IC. One large servo IC might include focus, tracking, spindle, disc, and SLED motor circuits, while in some auto CD players, one servo driver IC might drive the spindle or disc and SLED motor operations (**Fig. 10-9**).

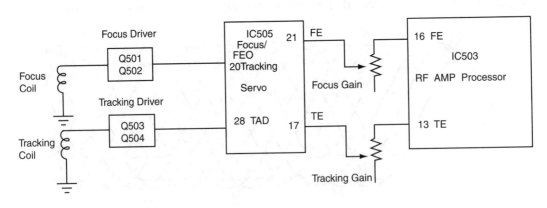

FIGURE 10-9. Block diagram of the focus and tracking signals to the focus and tracking coils, showing the gain controls.

The focus (FEO) and tracking (TRO) output signal from the signal processor and RF amplifier are fed to a servo driver IC. The focus positive (+) and negative (-) signal appears at the pins 26 and 27 of IC505 and are fed directly to the focus and tracking coils within the optical pickup assembly.

In the early auto CD player, transistors were found to drive the focus and tracking coils. The focus coil keeps the beam focused on the pit area at all times. The tracking signal keeps the optical assembly on the correct tracks of the rotating disc. Focus and tracking gain controls are found in the servo processor and RF amp circuits.

The focus and tracking coils can be checked with the low-ohm scale of the DMM. When checking the focus and tracking coils with the low-ohm scale of the DMM, the coils will start to move as the ohmmeter leads touch each coil, indicating the coils are normal. Remove the common ground lead of the focus and tracking coils and apply a 1.5-volt battery source across each of the coil terminals. The focus coil should move either up or down. Reverse the battery connections and the focus coil will move in the opposite direction. Test the tracking coil in the same manner.

Go directly to the servo focus and tracking IC when neither coil appears to be operating. Check the signal in and out of the servo IC. Measure the supply voltage terminal of the servo IC and compare it to the schematic. Usually a 5- or 8-volt regulated source operates the supply voltage to the servo driver IC. If the voltage supply (Vcc) is not present, check the voltage regulator circuits supplying the voltage to the tracking and focus servo IC (**Fig. 10-10**).

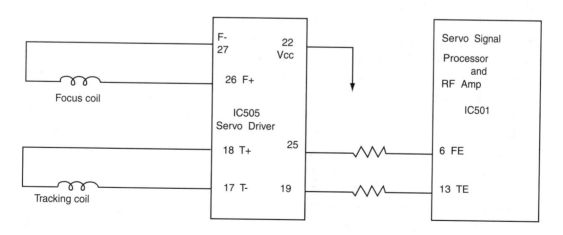

FIGURE 10-10. The focus and tracking coils within the pickup assembly are driven by a servo IC505.

CD PLAYER PROBLEMS

Poor Tracking And Skipping Problems

Make sure that the pickup lens is clean of dirt and dust. Check the laser power setting and auto focus gain, as they might be set too low when the pickup skips and jumps track. A defective pickup assembly can cause poor tracking. Check for correct auto backup, because the AC offset may be adjusted incorrectly. Incorrect turntable height can cause skipping. Replace the reset switch if the pickup skips at the first part of the tracking.

Mistracking and skipping can result from a bad plastic worm gear. Clean up the guide rails and lubricate for intermittent skipping. Clean off foreign matter on the slide mechanism when the pickup skips. Relubricate the slide shafts when intermittent track skipping occurs. It may be poor contact between clutch in the chucking arm assembly when the pickup skips or misses the complete tracks. Replace the loading belt when the pickup will intermittently not read the CD.

Replace the sticky slide motor when the pickup keeps skipping the track. A bad spindle motor can cause the disc to skip tracking. When the unit starts skipping on the first track, reposition the worm gear to mesh with the SLED gear, and loosen the LED holding screw to the mesh gear assembly.

Clean up the limit switch when the disc skips the first track. If the pickup skips at the end of a disc, check for a chattering of the SLED motor due to poor gear meshing between worm and first gear; shim up the motor assembly. A dead spot on the SLED motor can cause abnormal tracking; in this case, simply replace the motor. Readdress the SLED gears when the unit skips.

Reroute the flexible cable from the RF board when the disc skips. A worn receptacle can cause unstable clamping, also resulting in skipping. Replace the pickup assembly when the CD section skips tracks. Also, check for a bad spindle motor when the disc stops and skips the track. Replace the power switch when the pickup intermittently loses tracking after a few minutes of operation.

Replace the defective motor IC and microprocessors when the disc function is erratic and skips track. Check the system control IC when the CD player is in random mode and intermittently goes into search or skip mode. Mistracking can be caused by a defective optical pickup drive IC. Replace a leaky tracking transistor when the pickup mistracks toward the end of the disc. Replace the laser assembly and RF amplifier when the pickup intermittently skips around two or three discs. For erratic loading and intermittent tracking, replace the voltage regulator transistor or IC.

A low eye-pattern output of only 0.6 volts can cause intermittent skipping; readjust the control to at least 1.5 volts or until skipping stops. Resolder the tracking coil leads for loss of tracking action. A poorly soldered connection on the tracking coil can cause it to mistrack or either occasionally not read the disc. Check for a change in resistors in the open and close circuits for disc skipping. Replace the spindle motor if the pickup skips the tracks.

Readjust the focus-offset adjustment for loss of play or for disc skipping. When the music skips randomly, adjust the focus and tracking gain adjustments. Readjust the tracking error adjustment for erratic tracking. Perform the RF PLL frequency and focus gain adjustment when the pickup skips and jumps track.

Loading Motor Problems

The disc is pushed in a slotted area and the loading assembly pulls the disc inward, and then loads it on the disc spindle or platform. After the disc is loaded, stereo music is heard within a few seconds. If the loading mechanism does not operate, the loading motor will not load the disc inside the auto CD player.

The loading motor is operated from a signal of the main microcomputer, mechanism control, or master control IC. The loading motor driver transistors or IC provides a driving voltage to the loading motor with a positive and negative voltage. The negative voltage loads the disc, and when the eject button is pressed, the positive voltage to the loading motor unloads the disc or pushes it halfway out. The loading motor may drive the loading and unloading mechanism with a drive belt or with a meshed set of gears. Usually the loading motor is mounted toward the front of the CD player.

The loading and unloading signal is found on terminals 79 and 80 of the microcomputer or the mechanism control IC. A load and unload signal is sent to the motor driver IC and then to the loading motor. The loading negative voltage appears at pin 9 and the positive voltage at pin 7 of IC503 (**Fig. 10-11**). A 5- or 8-volt (Vcc) source

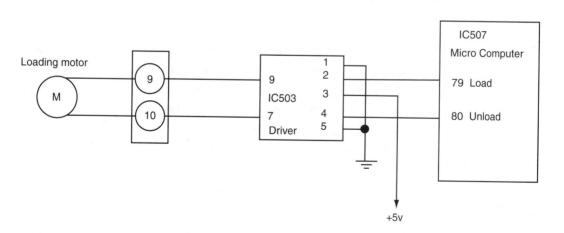

FIGURE 10-11. The loading motor signal is connected to a loading motor driver IC to load and unload the disc.

is fed to pin terminal 3 and common ground. Sometimes a connecting jack plug is found between the driver IC and the loading motor.

If the CD player does everything but load the disc, go directly to the loading motor and the motor driver IC. Check the 5-volt source applied to the loading motor and the motor driver IC. Measure the voltage across the loading motor terminals when the loading switch is closed. If the 5 volts is not found at the motor driver IC, then check the 5-volt supply and the transistor or IC regulator circuits in the power supply. A defective motor driver IC can prevent the disc from loading.

Check the drive voltage on the motor terminals when the auto CD player will not load or unload. Suspect a bad driver transistor or IC with no or low voltage terminal (Vcc) on the motor driver IC. Low or no voltage might result from a defective 5- or 8-volt regulated transistor or IC. A defective loading motor can cause loss of loading of the CD.

Check for a bad or improper voltage applied to the loading motor when it loads or unloads intermittently. A DC voltage source can be applied to the loading motor terminals to determine if the motor or circuits are defective. Remove one lead from the loading motor and clip the voltage source (5 volts) across the motor terminals to see if it rotates freely. A bad meshing of gears or slipping of the motor belt can cause erratic and poor loading.

If the disc does not load, then replace a cam, and search the track gear and the down-limit switch. Check for a missing loading voltage when the loading motor will not rotate. Measure the supply voltage applied to the loading motor IC when it stops loading. Replace a shorted motor driver IC, voltage regulator transistor, or IC when the CD will not load or take in the disc. Remove and replace the voltage regulator transistor for erratic and intermittent loading. Erratic disc loading can result from a intermittent motor driver IC.

Suspect a bent or broken leaf switch when the disc will not load. Readjust the loading switch when there is a loud clicking noise during the loading of the disc. Check the mechanism assembly for broken or missing teeth with a grinding noise and the disc will not load. A slipping loading motor belt can result in intermittent loading of the disc. Replace the loading belt if the disc loads and unloads rather quickly. Replace the roller assembly when the CD player will not load the disc. A badly soldered lead off of the loading motor prevents loading of the disc. When the disc will not load, replace the photo interrupter component.

Improper Ejection

If the player will not eject the disc, replace the roller shaft assembly with a loose gear, or also try replacing the hook assembly. Remove and replace the cam gear for intermittent ejection of the disc. Check the alignment and tightness of the plastic rack gear for a lack of ejection and a rattling sound.

Check for a loose gear or roller shaft when it will not eject the disc and hangs up before the eject cycle begins. Adjust the alignment of the phase lock control if the disc ejects after two to four seconds. Replace the pickup head assembly if the CD player ejects after five seconds.

OPERATIONS OF THE CD PLAYER

Disc or Spindle Motor Operations

The disc or spindle motor rotates the disc; the slide or SLED motor moves the optical lens assembly toward the outside edge of the disc. The spindle motor is usually directly coupled to the turntable, which the disc sits on. A disc motor can be controlled by a signal from the digital processor or servo IC. The servo driver IC applies voltage to the drive motor. The disc motor servo driver IC circuits can be driven by transistors or by an IC component (**Fig. 10-12**).

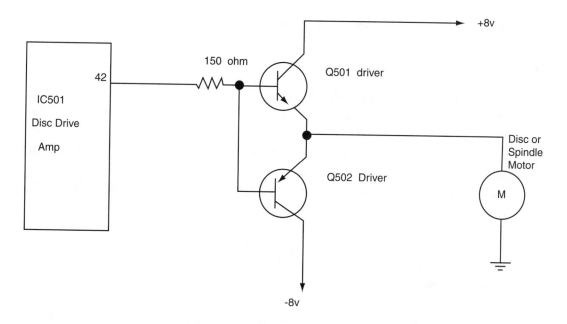

FIGURE 10-12. In the early auto CD players, transistor drivers were found to provide voltage to the disc motor.

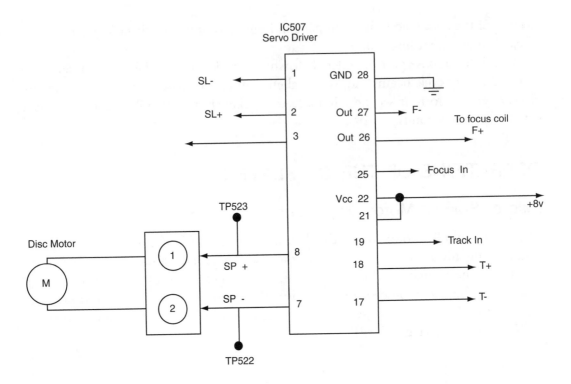

FIGURE 10-13. Check test points TP523 and TP522 for voltage to load and unload the CD.

In a Realistic 12-2151 auto dash CD player, the spindle motor is driven by the same servo driver (IC507) that operates the SLED, focus, and tracking coils. The SP+ terminal (8) is fed to a CN501 terminal block and onto the drive motor. An SP- out terminal (11) appears at pin 7 of the CN501 connection. TP523 and TP522 are the test points for voltage measurements on the disc motor terminals (**Fig. 10-13**).

Check the continuity of the disc motor winding with the low-ohm scale of the DMM. Measure the voltage applied to the disc motor from the servo IC. Suspect a leaky or open driver IC or transistor when there is low or no voltage applied to the disc motor. Check the voltage supply source of the servo driver IC if no run voltage is found at the spindle motor. A defective voltage regulator IC can cause a bad servo IC supply voltage (Vcc).

Replace the voltage regulator IC when the player will not accept the disc. Check for an open regulator transistor when the disc spins, the display is at zero, and yet there is no audio output. Replace the disc motor drive IC when the disc speeds up and the disc motor runs constantly.

Grease the slide mechanism for disc slippage. Check for a decrease in resistance when it will not play the disc. Replace the leaf switch when it will not rotate the disc spindle motor. Check for a cramped or grounded wire for intermittent rotation and when it will not read the disc. Replace the load in-switch when the disc will not move into place as the play button is pressed.

Replace the disc stopper in the carriage motor circuits if the disc passes beneath the disc stopper assembly. Lift up the spindle motor table from the bottom of the disc when it will not turn or when it rotates slowly with noisy operation. Clean up the laser lens assembly with Windex and cleaning stick if the disc will not play and tends to spin very slowly. A broken cam assembly prevents playing of the disc. Replace the flexible cable when the disc constantly rotates the disc after being loaded. Replace the disc assembly if it is found resting on the table assembly and the drive motor will not turn. Check for a bent holder and lever, which sits on top of the disc, if no disc rotation occurs.

Replace the focus and tracking IC if the optical assembly does not move outward and yet still accepts the disc. Replace the leaky servo driver IC or transistor when the disc motor slows down and then ejects the disc.

Readjust the focus offset and RF PLL adjustments if the disc motor will not rotate. Adjust the VCO frequency when the spindle motor rotates at a fast rate of speed, tries to eject the disc, and scratches the disc on top and bottom.

SLED Motor Operations

The slide, carriage, or SLED motor moves the optical lens assembly from the center toward the outside edge of the disc. The SLED motor starts out slowly and speeds up as the disc approaches the outside rim. The signal that operates the SLED motor is driven from a servo signal processor and RF amplifier. A SLED amp in the early SLED motor circuits provided voltage to a pair of transistors or to an IC driver circuit. The SLED motor might be driven from the same servo driver IC as the spindle or disc motor.

Notice if the SLED motor is driven from the same driver IC that drives the focus and tracking coils, spindle, or drive motors. The slide or SLED motor slides the pickup assembly down sliding rods to reach the outside of the disc. The SLED motor signal (SEO) is fed from pin 16 of the servo signal processor and RF amp IC to pin 3 of IC507. The negative voltage to drive the slide motor appears at terminal 1 and the positive terminal at pin 2 (**Fig. 10-14**).

Check the SLED motor winding with a continuity test of the DMM. Measure the voltage on the slide or SLED motor terminals when the disc is rotating. No or low voltage on the SLED motor might be caused by a bad motor or by improper drive voltage from the servo driver transistor or IC. When no voltage is found at the output terminals of the servo driver IC, suspect a leaky or open servo driver IC. Check the servo supply voltage source (Vcc) on the servo driver IC with no SLED motor movement.

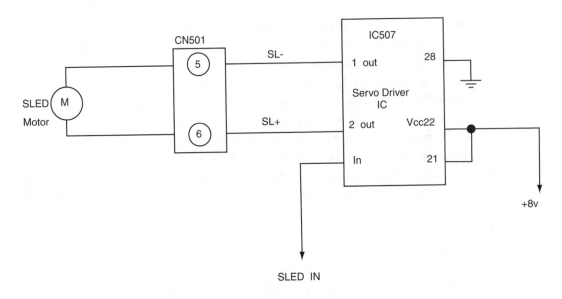

FIGURE 10-14. A servo driver IC507 provides voltage to the SLED motor.

Replace the D/A crystal with intermittent play speed and CLV drive. Suspect a damaged gear or replace the servo IC driver when the SLED motor continuously drives the pickup assembly outward. Replace the microprocessor or microcomputer when the SLED motor moves to the outer edge or when the disc will not rotate while it moves to the outer edge of disc. Check the SLED, spindle, focus, and tracking servo IC when the player accepts the disc and will not move the optical assembly outward (**Fig. 10-15**).

Suspect that the mechanism assembly gears are bent out of line and preventing the slide motor to back off of the pickup "in" switch. When this happens, the CD player will not play and the play arrow in the display will keep flashing. Replace the sticky slide motor when skipping of the track is noticed. Readdress the SLED gears when the pickup keeps skipping. For a chattering noise at the beginning of the disc, readjust the SLED limit switch.

Replace the optical pickup assembly if the SLED motor runs too fast. When the music begins

FIGURE 10-15. The servo driver IC might have regular IC or gull-wing IC terminals.

skipping randomly, lubricate the slide shaft; reposition a worm gear to mesh with the SLED gear by lowering the SLED screw to properly mesh the gears.

The Signal Circuits

The optical pickup assembly reads the pits on the disc and feeds the RF signal to a large preamp and servo processor IC at terminals 35 and 36 of IC501. The laser diode signal (PD) from the optical assembly is fed to terminal 34. A focus signal (F) is applied to pin terminal 38, while the error signal (E) attaches to terminal 39 (**Fig. 10-16**).

Here the preamp, tracking, automatic focus, error, and signal processing circuits are located in one large IC501. Because the signal from the photodetector diodes is very weak, it must be amplified before it can be used in the signal or data signal processor IC. In some CD circuits, the signal processor circuit might contain another IC component.

The most important EFM (8-to-14-bit modulation) is found at pin 31. The RF signal terminal might have a test point (TP505) for a quick test that shows that both the optical assembly and RF amp are functioning. The eye-pattern waveform can be checked on pin 31 of IC501 with the oscilloscope. Besides the EFM signal, the signal processor provides a tracking error (TE) offset output (TAO) signal found at pin 13 and a focus error (FE) output signal (FEO) at terminal 6. Both the TE and FE signals are sent to the analog control servo driver IC507.

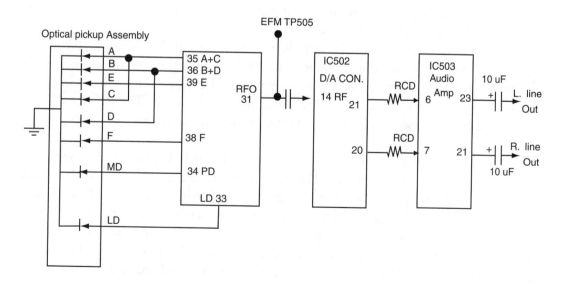

FIGURE 10-16. Block diagram of the optical pickup assembly through the D/A converter IC and line output amp IC503.

The block diagram is useful in determining how the different stages and circuits are connected together. Locate the possible trouble symptom on the block diagram and then proceed with waveforms and critical voltage measurements to locate the defective component. Check for correct test points on the chassis, if they are shown on the schematic, to test each circuit function. Make sure that all tests are made on a suspected LSI or a microprocessor IC, as they are difficult to remove and replace.

FIGURE 10-17. The EFM eye pattern found at the output terminal of the RF amplifier.

Check the RF eye-pattern waveform at pin 31 or, if a test point (TP505) is handy, check with the scope, to determine if the optical assembly and RF circuits are normal (**Fig. 10-17**). A defective optical assembly or RF IC might cause a lack of EFM signal. Improper supply voltage on terminal 18 of IC501 can cause a weak or absent eye-pattern waveform. Check the +5 volt supply source to a possible 5-volt regulator IC or transistor for possible leakage, shorted, or open conditions.

Data Signal Processor

The RF or EFM output signal is passed on to pin 14 of IC502. The signal processor or data signal processor is a large-scale integrated (LSI) or microprocessor IC with 64 or 80 terminals. A signal processor IC might have surface-mounted component terminals (gullwing) or regular IC terminals. Besides accepting the RF signal, the signal processor might be tied to other circuits.

The RF signal processor might include a clock generator, EFM modulator, error, timing control, oscillator, digital filter, data, error connection, servo, servo system control, and mute circuits. Also, the circuits tied to the signal microprocessor might include a VCO, crystal control, Ram, and system control circuits. The RF signal is fed into the signal processor IC, and is then sent to the D/A converter and servo control circuits (**Fig. 10-18**).

Remove and replace the signal processor IC when it is found defective with the scope and by voltage measurements. The larger chips are more difficult to replace than a regular IC. Check for terminal 1 of the signal IC and mark it on the PCB if it is not already stamped on the board. Make sure the processor part number is the same. These large LSI components should be replaced with the original part numbers. Remove the large chip by cutting the flat terminals close to the body of the IC with a sharp knife or Xacto cutting tool. Remove the old soldered terminals and solder from the board with soldering iron and solder wick.

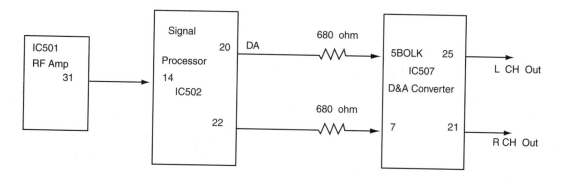

FIGURE 10-18. The RF or HF pattern is fed from the RF amplifier to the D/A converter IC507.

Make sure each terminal is straight and in line at the right position before soldering in place. Apply a minimum amount of flux on the terminal areas. Align the IC over each correct PC wiring. Do not smear flux on unwanted areas or components. Solder each corner terminal so that the IC cannot move when iron pressure is applied. Use thin solder to make a good soldered connection on each thin terminal. Make a good soldered joint but do not leave the iron on for too long. Check each terminal with a magnifying glass to be sure that no two terminals are soldered together.

Digital and Analog (D/A) Processor

The D/A converter IC (DAC) changes the digital signal in a voltage or analog audio signal. The input of the digital signal is fed into the input terminals of the D/A converter, and the stereo audio signal is taken out of pins 21 and 23. The left audio channel signal is fed at pin 21 and the right stereo channel out of pin 23. IC507 operates from a Vdd +5 volt source at pin 24. The D/A output audio connects to the line output jacks, or to another audio output circuit, or to both.

The D/A IC can cause many different problems within the audio output circuits. A leaky or open D/A can cause weak and distorted sound. Replace the D/A converter IC for lack of sound in each stereo channel. Distorted audio in one or both channels can result from a defective D/A IC. A defective D/A converter can cause noise in the background while the music is playing. Suspect a bad filter network feeding the input D/A signal for loss of audio output. Replace the D/A crystal for intermittent play of the SLED and CLV drive.

Signal-trace the audio signal in both left and right output terminals of the D/A IC with the scope or external audio amp (**Fig. 10-19**). The audio signal can be checked

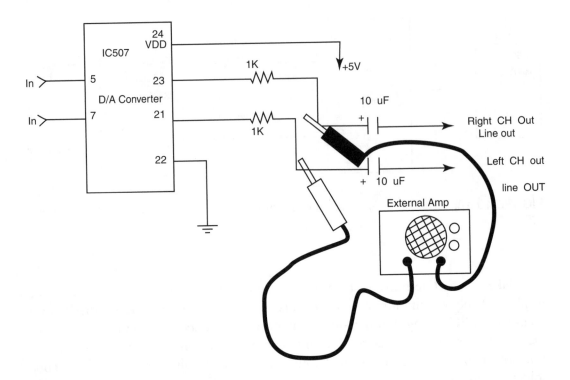

FIGURE 10-19. The audio stereo output left and right channels can be signal-traced with the external audio amp.

at the line output jacks. The external audio's distorted or weak channels can be compared with the external audio amplifier. If the left channel indicates low distortion, take a critical voltage measurement on the D/A IC. Check the left channel with the right channel to be sure of the same amount of audio with the external amp.

Check for correct (Vdd) voltage at pin 24 of IC507. Suspect a leaky D/A IC or improper voltage source. Check the +5-volt voltage regulator transistor or IC for improper supply voltage at pin 24.

Microprocessor or Mechanism Control

The microprocessor, microcomputer, or mechanism control IC provides many signals to the various mechanisms within the auto CD player. A microprocessor might control the disc sensors, chuck switch, in switch, limit switch, loading motor and motor driver IC, laser on/off transistor, APC, signals to the digital processor IC, and to the servo signal processor and RF amp IC (**Fig. 10-20**).

In the CD trunk changer, the microprocessor might control the analog switch, carousel driver and motor, buffer transistor, loading motor, loading switch, channel in

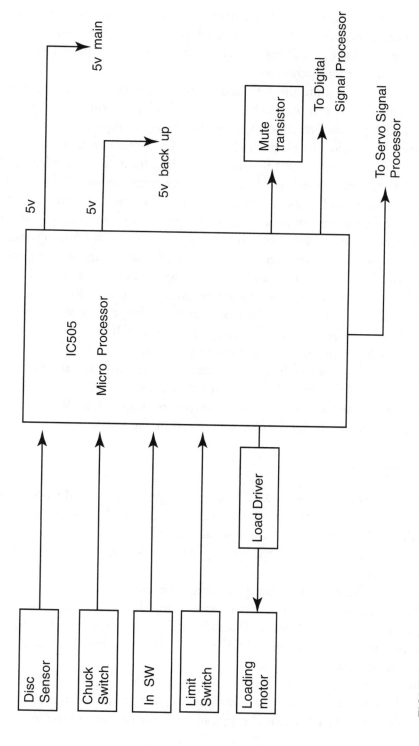

FIGURE 10-20. Block diagram of the microprocessor IC505 and connecting circuits.

and channel out, operation key, tracking, reset, eject, and data to the signal processor IC. The same microprocessor might be connected to a carousel stop-position transistor, carousel detection transistor, oscillator crystal, remote operation, data, CD in and out, power control, and muting operations.

The muting system within the CD audio output circuits are controlled by the microprocessor, microcomputer, or mechanism control IC. Most muting systems within the auto radio CD players are controlled by the main microprocessor with muting transistors within the audio line output jacks and audio power output speaker systems. In auto CD changers, audio circuits that have front left, front right, rear left, and rear right auto amplifier circuits, as well as muting transistors, might be included between each set of input audio signals to the audio output ICs and within the audio line output jacks.

Suspect a defective microprocessor or microcomputer IC if the loading motor is not rotating but the motor and driver circuits are normal. Poor or lack of muting operations can be controlled by a defective microprocessor. Check the voltage source (Vcc) of 5 or 8 volts provided by the backup and main voltage sources. Check for low or no voltage at the supply pin terminal that might be caused by a 5- or 8-volt transistor or IC regulator. Make sure that more than one circuit either does not operate or has incorrect supply voltage before removing and replacing this IC with many terminals. Always replace the microprocessor with the exact part number.

Replace the 8-MHz crystal if the microprocessor has no audio and the SLED motor moves to the outside of the disc. No sound in either the CD or AM/FM radio can be caused by a defective zener diode in the main power supply. Check for overheated resistors in the power supply for a noisy sound within the speakers.

Power Output ICs

The low-wattage powered radio and CD players might have an RMS power output from 15 to 50 watts and a peak power of 35 to 52 watts. The line output jacks can be included in the car CD player with a single-dual output IC or a larger IC that may drive many speakers. The left and right preamp output line jacks include one, two, three, or four sets of outputs.

The lower-priced audio output amplifiers might be connected directly to several left and right PM speakers without a higher-powered amplifier attached to the auto CD player. For instance, a Power Acoustik (Pacd-500) CD player receiver does have 4x20 watts of output power. Most low-priced CD players can easily drive front left and right speakers in each audio output channel.

The power output ICs might have two IC outputs in two separate outputs of both the right and the left channels. Muting transistors might be found within the input and output stereo circuits. The left channel might have a stereo output to the front left and

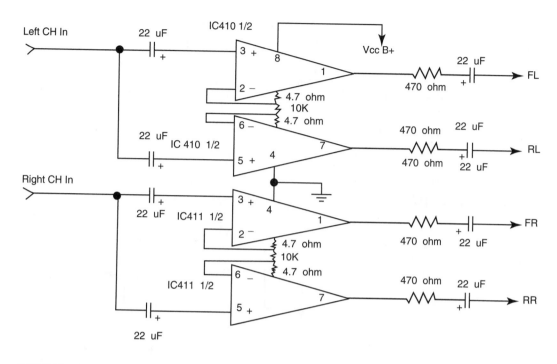

FIGURE 10-21. A dual-audio output IC is found in the front left and front rear of the left channel output circuits.

another IC for the rear left speaker. One half of each stereo output is found in another dual-output IC. The right front speaker and the right rear speaker have a different audio output audio IC (**Fig. 10-21**).

In the lower-priced auto CD player-receivers, a separate power output IC is found in both the left and right channel audio output circuits. Some auto output ICs might be found in each left and right tweeter and left and right woofer speakers.

The quad-four Radio Shack 12-1999 AM/FM stereo car radio player contains one large Power IC (TDA7370) that serves as a nominal 16-watt power output with 2-CH/4-CH at 4-ohms output impedance. The quad-four audio inputs are found on pins 4, 5, 11 and 12 with 1-uF electrolytic coupling capacitors. The audio output terminals are at pins 1, 2, 14, and 15. A separate speaker common terminal is coupled to ground through a 1000-uF electrolytic.

The quad-four power-output IC has a 4-ohm output impedance. Notice that a DC voltage (6.7 volts) is found on each quad output IC terminal, as the 1000-uF capacitors keep the speakers above common ground. Accurate voltage measurements on each IC power output terminals can indicate if the IC is leaky, shorted, or open (**Fig. 10-22**).

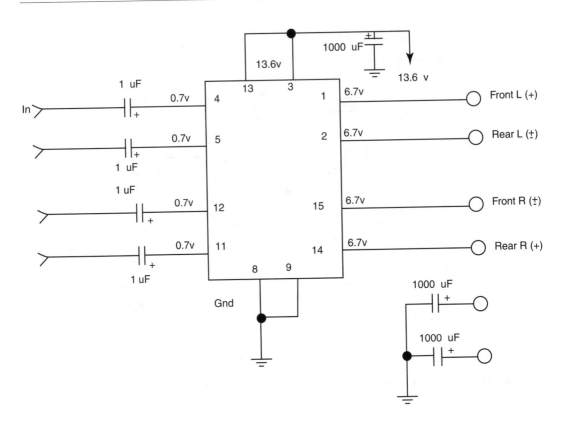

FIGURE 10-22. A quad-four power-output IC (TDA7370) with four different output speaker circuits in a Radio Shack 12-1999 AM/FM car radio.

CD OUTPUT PROBLEMS

Audio Output Power Problems

The left and right analog audio circuits are developed within the D/A converter amp stage. The right and left audio channels go directly to the line output jacks and feed into the audio output circuits. In some of the latest auto audio circuits with front left and front right, right left, and rear right audio circuits, they are amplified and muted before they appear at the large audio dual-output IC circuits. After the first IC amplifier, the right and left output jacks are found connected to the external high-powered audio amps. Transistor mute circuits are found in each set of front and rear amp circuits (**Fig. 10-23**).

A large, high-powered IC402 amplifies both the Front L and Front R, Rear R, and Rear L speaker outputs. All four output-speaker terminals are fed to a rear block in the

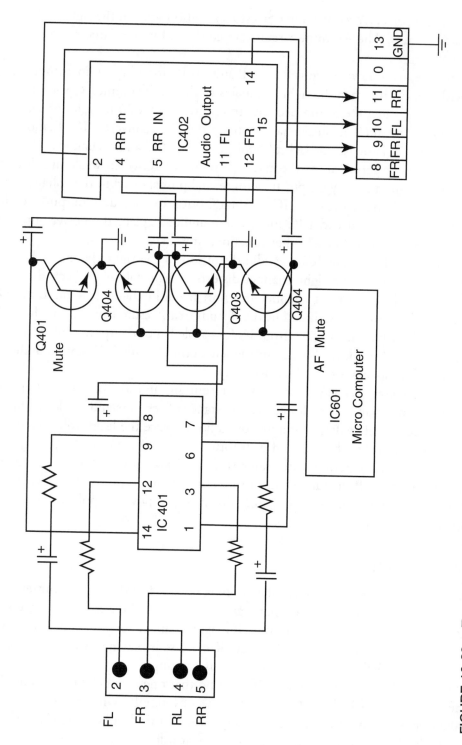

FIGURE 10-23. Transistor mute circuits are found in the left and right channels before the final dual-output IC402.

back of the CD player. A set of mute transistors are also found in the right and left line output jack circuits. The AF mute circuits are controlled with a large microcomputer IC601.

Replace the main power output IC or FL and FR IC for weak audio output. Check the driver FL or FR IC when the right channel is weak at half volume. Replace the main power output IC when the sound becomes intermittent after 20 minutes or so. If you find no audio from the left output jack, replace the D/A converter IC. Remove and replace the demodulator IC for lack of audio out of either audio channel.

Resolder the pin terminals of the audio amp output IC when weak and distorted sound occurs. Suspect a leaky output IC with several pins shorted to the voltage supply source pin (Vcc). Replace the main IC on the PCB for a weak audio output. Resolder all bad terminals on the decoder IC for a loss of audio. Replace the power output IC if there is a loss of audio but all other functions are okay.

Replace the audio output IC for audio distortion in the right channel. Check the digital control IC when the right channel distorts at half volume. Check the power output IC for a distorted right channel and no left audio channel. Replace the D/A converter IC for distorted sound. Distorted audio can result from a bad Ram IC. Replace both the driver and audio output IC for garbled audio output. Replace the low-pass filter network for a loss of audio in the left output channel. Distorted audio can result from a low laser-power pickup assembly.

A leaky IC voltage regulator can cause intermittent and loss of sound after 20 to 30 minutes of operation. Check for correct resistance of resistors and diodes within the voltage regulator supply voltage for a loss of audio. Replace the leaky voltage regulator transistor for loss of left channel and a distorted audio right channel. A leaky zener diode in the low-voltage power source can result in a distorted sound. Check for a leaky electrolytic within the B+ regulator transistor circuit in audio playback mode. A leaky decoupling electrolytic can cause a loading down of the power-supply voltage with a loss of audio in play mode.

Noisy Operations

Check for a missing reference voltage if a popping noise is heard with intermittent static in the sound. Replace the noisy output transistor or IC if there is noise in the audio output circuits. Check the audio output IC right channel for low and noisy audio. Replace the Ram IC with static in the audio if the disc is playing. A bad Ram IC can also cause a high-pitched noise when the disc is rotating.

Replace the dual-power output IC if you find that one channel has a hissing noise and the other channel is dead. Check the D/A converter IC with background noise while the disc is playing. Replace the PLL crystal if you find intermittent noise in the audio and also hear cracking and popping noises at a low-level sound setting.

Replace the timer display switch with a popping noise in the speakers. For a clicking noise, check for a skipping screw that is not removed or a maladjustment of the close switch. A turntable set too low can cause a grinding noise. Adjust the SLED limit switch for a chattering noise at the beginning of the disc's playing. Check all decoupling electrolytic capacitors with the ESR meter for noise in the speakers.

CD CIRCUITS AND WAVEFORMS

Line Output Circuits

The audio line output circuits might include an additional audio amp IC before the audio is fed out of the line output jacks. The left and right line output jacks might be tapped off of the audio circuits before the last power audio IC in the CD player. The set of line output jacks can be fed to a high-powered external amplifier system for greater power output and to a CD changer located in the trunk area (**Fig. 10-24**). Several sets of preamp line output jacks might have up to four sets of outputs.

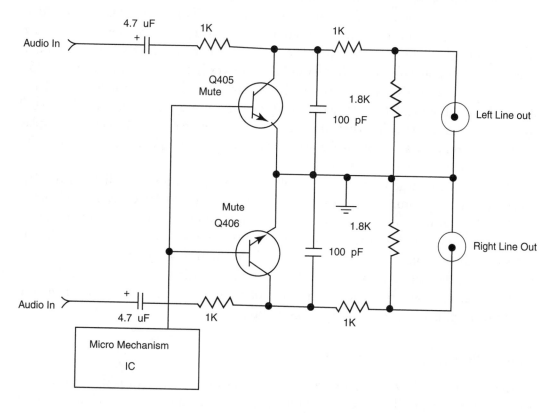

FIGURE 10-24. Audio mute transistors might be found in the audio line output jacks.

The muting transistor circuit suppresses noise that might be produced when the power is turned on. Muting in some players might be automatic when the disc stops, during accessing operations, and during pause mode. The muting transistors are found in most line output jack circuits. The muting transistors are controlled by the microprocessor, computer, or control mechanism IC.

Replace the main microprocessor for a loss of muting in the line output jacks or audio circuits. Check the left muting output transistor for distorted sound within the left muting circuits of the left channel. A dead right channel can result from a leaky right muting transistor. Replace both muting transistors with a loss of output from the front speakers.

Display Circuits

The LCD (liquid display) IC circuits include a common driver, clock generator, interface, control register, shift register segment driver and latch, key buffer, and key scan circuits. Many different momentary switches are connected to the LCD driver, which might include CD, LO/DX, loud, mode, scan, clock, seek, and band circuits. There are many segment signals for the output of the LCD from the LCD driver, data output, reset, key scan, oscillator, and common signal output for the LCD display. Many LED light indicators correspond to the momentary switches when pressed and selected. The supply voltage applied to the LCD driver is 5 volts or more (**Fig. 10-25**).

Install a new focus assembly when the pickup assembly will not focus and the display keeps blinking. Check for a defective coupling and bypass capacitors (0.1 to 0.2 uF) in the tracking circuits when the display shows no tracking. Replace the display assembly if you have a loss of display while everything else is operating. Check the Vcc and Vdd voltage source on the display driver IC.

Measure the voltage (Vss) on the display IC if there is no sign of light or display. Check all filter capacitors in the power supply circuits with the ESR meter for incorrect display at power turn-on. Replace a leaky or shorted diode in the power supply source with a low illumination of the display assembly. If there is no display or if the display fades out, there could be an intermittent diode in the display circuits. Loss of display functions can be caused by a shorted electrolytic in the display power source. Suspect a bad power regulator transistor with a loss of display. Resolder the power regulator transistor terminals for loss of display assembly.

The Different Motor Circuits

In the early motor CD changer circuits, transistors were used to provide drive voltage to the motor circuits. The tracking, spindle, and carriage motors were often driven by a large IC component. Likewise, the elevator, or up and down motor, plus the tray or

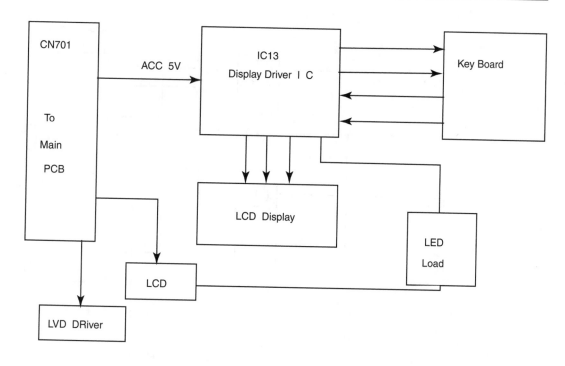

FIGURE 10-25. The LCD driver display IC controls data to the LCD display assembly.

magazine motor, often operated from another large IC component. In a Pioneer CDX-FM45 changer, the motor driver IC801 operates from a 10-volt transistor regulator source that controls the elevator and tray motors (**Fig. 10-26**). The tray or magazine motor and elevator, or up and down motors, are controlled by a signal from the system control processor to the motor driver circuits.

When a motor does not rotate, check the DC voltage across the motor terminals, with the motor circuits functioning. Suspect a motor driver IC with a loss of voltage applied to the motor. Check the regulated voltage source pin on the motor driver IC. If voltage is present on the IC supply pin (Vcc) and you find no voltage to the motor circuits, suspect a leaky or open motor driver IC. If voltage is present on the voltage supply pin terminal and you find no voltage at the motor circuits, suspect a leaky or open motor driver IC.

Check the motor continuity of the motor winding if voltage is present at the motor terminals but the motor does not rotate. Replace the defective motor with open windings. Remove the positive terminal wire of the motor and apply 1.5 to 10 volts to the motor terminals. Rotate the motor pulley or shaft and notice whether the motor takes off. Replace the defective motor if it stops in one spot or with an open winding.

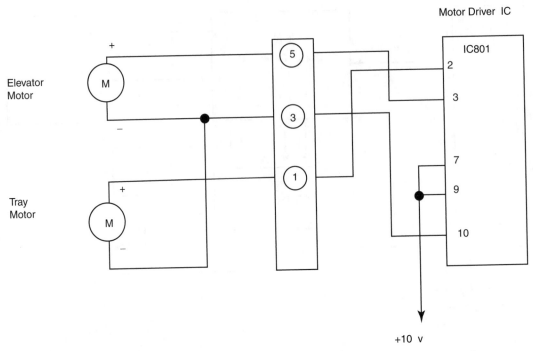

FIGURE 10-26. The elevator and tray motor in a Pioneer CDX-45 CD changer is controlled by a motor driver IC801.

Replace the open low-voltage regulator transistor with improper voltage applied to the motor control circuits. Suspect a leaky transistor regulator if the disc moves slowly and the disc possibly ejects. Replace the disc motor when the disc plays intermittently. When the disc motor is rotated manually and will not rotate, replace the motor. Replace the motor driver IC when no voltage is applied to the elevator or magazine motors. Intermittent operation might be caused by a bad motor terminal connection.

Replace the flat cable in the changer when the disc stops between the magazine and the MD mechanism. Add a cushion to the upper side of the chucking arm when the disc rotation is noisy. Check the waveform on the crystal tied to the servo or system control IC when the disc motor runs at power-on with no loading or with chucking operations. Replace the drive belt with no chucking action. Replace an open or leaky voltage regulator transistor for a slow tray or magazine movement and no display. Check the magazine assembly when the disc player inserts a disc in the wrong slot.

Critical Waveforms

The most critical waveforms are the RF, EFM, or eye pattern on the RF IC amplifier (Fig. 10-27). If the waveform is not present, very few CD operations are performed. The high frequency (HF) signal should be present and stable in the play mode with a

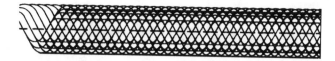

FIGURE 10-27. Scope the RF amp IC for an Eye EFM pattern to determine if the optical pickup and RF amp are functioning.

FIGURE 10-28. The focus coil waveform taken across the focus coil assembly.

FIGURE 10-29. The focus error (FE) waveform from the RF amp IC.

FIGURE 10-30. The tracking coil waveform taken across the tracking coil winding.

sharp diamond shape in the middle of the waveform. Critical waveforms taken on each IC can determine if these circuits are performing.

A focus coil waveform (FE) taken across the focus coil winding in play mode can indicate if the focus coil is functioning (**Fig. 10-28**). The FE signal out of the RF amp to the second driver IC indicates that the RF amp IC and focus circuits are working (**Fig. 10-29**). The tracking coil waveform (TE) taken across the tracking coil indicates that the tracking coil circuits are okay (**Fig. 10-30**). The focus (FEO) signal out of the RF amp is shown in **Fig. 10-31**.

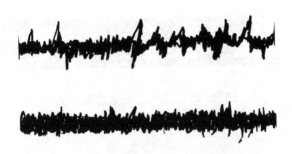

FIGURE 10-31. The FEO signal out of the RF amp connected to the servo driver IC.

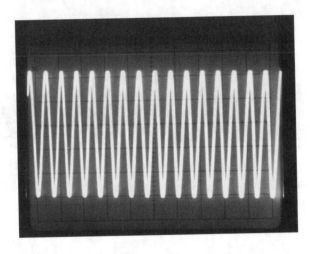

FIGURE 10-32. An 8-MHz crystal waveform taken on the microprocessor or control IC.

An 8-MHz crystal waveform taken from the microprocessor or controller IC indicates that the IC and circuits are good (**Fig. 10-32**). Check the manufacturer's waveforms found in the service manual of the CD player you are now servicing for correct CD operation.

CD CHANGERS AND ADJUSTMENTS

CD Auto Changers

Practically the same operational circuits found in the auto CD player are also found in the auto CD trunk changer unit mounted in the trunk area. The 8-track CD changer can load up to eight separate discs with in-dash control, with features of pause, repeat play, and random play. Disc scan provides a quick sample of every disc in the changer, while track scan lets one hear the beginning of each disc. Today, an MP3 changer is small enough to fit in the auto dash and glove compartment.

Besides more low-voltage regulator power sources, the CD changer might have a tray or magazine, elevator, or up and down motors. Most CD changers have a carriage, SLED and spindle, or disc motors plus the magazine or elevator motors. The tray or magazine motor provides loading of the correct disc, while the elevator motor positions the CD pickup assembly. Most of these motors operate from the system control IC circuits to a motor driven IC. The motor driver IC might provide operation for several different motors.

The CD changer should be installed where the temperature is not high, not in direct sunlight, nor where heavy vibration might occur. Do not install the CD changer over electrical wires or cords, pipes, near the spare tire, or where other components can damage the CD player. Most CD changers are mounted horizontally with metal brackets provided by the CD changer. You can mount the CD changer vertically, except that it must be mounted so the unit will not move or vibrate. Mount the changer to the metal deck, under the trunk carpet.

CD Player Adjustments

After replacing critical components within the auto CD changer, you should make correct adjustments. Checking and making the tracking error, diffraction grating, tracking error balance, tracking gain, and APC are the most important adjustments. The test disc (YEDS-7) or (Sony type 4) can be used in making most auto CD player adjustments. The oscilloscope, DMM, or FET-VOM and small screwdrivers are the required tools and test instruments for electronic adjustments. Always follow the manufacturer's adjustment procedures, when available.

CHAPTER 11

Troubleshooting High-Power Amps

The high-powered amplifier goes from 50 watts to 2000 watts of power output. Most high-powered amps employ transistors instead of IC components to acquire high outputs. The metal-oxide semiconductor field-effect transistors (MOSFETs) are used within the power supply and output for greater efficiencies and higher speed switching, and they generate much less heat than the ordinary output transistors (**Fig. 11-1**). Usually the auto AM/FM/MPX, cassette, and CD player provides only 14 to 50 watts of peak power output, using transistors and IC output components. Today, front-end tube preamp circuits can be found in the high-powered amplifier systems.

AMPLIFIERS

Key Amplifier Features

The key features in a good and dependable amplifier are stereo outputs; bridged mono operation; 2-, 3-, 4-way channel output; low-impedance high-powered MOSFET power supply; adjustable bass boast; and built-in selectable high and low pass crossover controls. Other features might include a crossover mode switch, input level control, low-level input (high impedance), high input (low-level impedance), protection LED indicator, and a power LED indicator, which are found in some of the high-powered output amplifiers.

The correct input level controls can handle low- or high-level inputs to avoid noise and distortion with a bad input level matching of impedance. A frequency response of 10-60,000 Hz with bridged mono output and stereo operation is ideal. The common ground or floating ground for 2/4-channel input and bridged input connections make for easy hookup of the head unit. Some amplifiers have insulated mounting feet to eliminate chassis ground noise.

The Infinity Kappa 202a amplifier has an RMS power of 200X2 at 14.4 volts and a peak power of 300X2 with bridged 600 RMS power output watts to one speaker load. The Kenwood KAC-818 amplifier has an RMS power of 200X1 at 14.4 volts with a peak power of 400X1 watts. An RF Punch 500a 2-amp has an RMS

FIGURE 11-1. An electronics technician checking the power output transistors in a 700-watt amplifier with the diode test of the DMM.

power of 250X2 watts with a bridged RMS power of 500X1 watts. The RMS at 2 ohms with the same amplifier is 250X2 watts of power. The RF Power 1000a 2-amp has an RMS power of 125X2 watts at 13.8 volts and a peak power of 210X2 watts with a bridged RMS power of 500X1 watts. The RMS at 2 ohms impedance with the same amplifier is 250X2 watts of power.

The Root-Means-Square (RMS) measures the power in watts that the amplifier can produce continually. A peak power rating is maximum power that the amp can deliver during a brief moment of peak power operation. RMS power at 2 ohms can tell you how much power the amplifier delivers into a 2-ohm load or speaker. Speakers should be wired in parallel to reach a 2-ohm load, or you can just use a 2-ohm speaker.

The measured output power depends upon the automobile's input battery voltage. The average voltage from a battery with the car operating can be around 14.4 volts, and around 12 to 13 volts when the engine is shut off. The battery voltage should never be under 12 volts to prevent damage to the high-powered amplifier. Likewise, the amplifier will have a higher wattage output at 14.4 volts than at 12.6 volts of input battery power. Most amplifier input voltage ratings are provided in 12.5, 13.8, and 14.4 volts. For instance, an MTX RT2600 amplifier has an RMS power of 200X2 at 12.5 volts input with a 375X2 of peak power, and 600X1 watts of bridged power output.

A bridged power amplifier is when, in mono mode, the left and right channel stereo outputs are tied together to power a single speaker (usually a 4-ohm speaker). The bridged single speaker is often a very large 10- to 15-inch sub woofer PM speaker (**Fig. 11-2**). Connect the positive output of the left

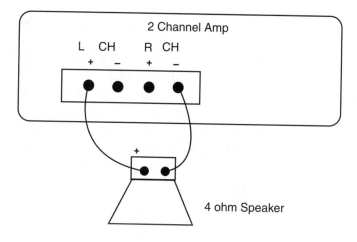

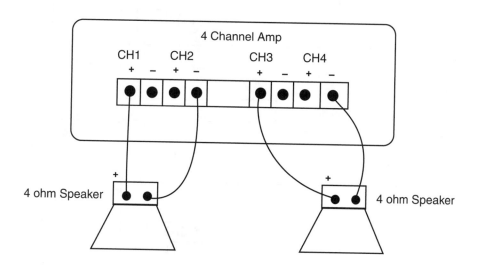

FIGURE 11-2. A two-channel and four-channel amplifier with bridged speaker connections.

channel to the negative terminal of the right channel to provide mono operation to a minimum 4-ohm speaker load. Two mono speakers can be connected to a four-channel amplifier with a minimum speaker impedance of 4 ohms on some amplifiers.

The signal to noise ratio (S/N) is usually between 90 and 110 dB, with the average around 100 dB. Most audio high-powered amps have a speaker level input and preamp output circuits. Some amplifiers have a built-in high/low-pass network, while a few have only a low-pass crossover filter network. Besides a built-in crossover network, some amplifiers have a bass boost circuit.

An input level control is used to match the output of the head unit to the amplifier. The head unit is often an auto receiver, cassette, and CD player or a TV LCD screen system. You can reduce the setting of the level control to prevent audio distortion. In low-level inputs (high impedance), the amplifier might have gold-plated RCA input jacks and RCA-RCA-type patch cords to connect these inputs to the RCA outputs from the head unit. The high-level input (low impedance) employs RCA output jacks for the automobile's stereo channels that connect to the speaker output from the head unit to the high-level input RCA jacks (**Fig. 11-3**).

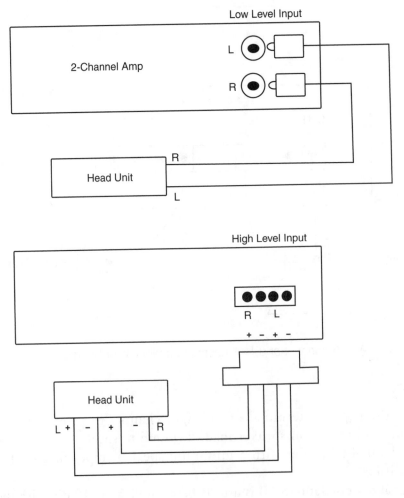

FIGURE 11-3. The low- and high-level input connections of the head unit (receiver-cassette and CD player).

The two-channel stereo output from the AM/FM/MPX, CD, and cassette player are connected from the head unit L and R outputs to the left and right low-level inputs with output patch cords. From the left and right speaker terminals of the head unit, the left positive terminal is connected to the left high-input level input, and the left negative output is tied to the left negative high-level input plug. Likewise, the right positive terminal from the head unit is connected to the right positive head input, and the right negative stereo channel is connected to the negative right high-level input terminal. The L and R audio output of the head unit connects to the low-level inputs of the amp, while the speaker terminals of the head unit are plugged into the high right and left input terminals of the amplifier.

Two- and Four-Channel Amplifiers

A two-channel high-powered amplifier has two left and two right input channels to feed the right and left output channels of the head unit. The left (L) output of the head unit plugs into CH3 and CH1 low-level inputs of the amplifier, and the right (R) output of the head unit plugs into the low-level inputs jacks of CH4 and CH2 of the high-powered amplifier. The four-channel head unit has two left and right (L&R) rear outputs, and two L/R front outputs that feed into channels 1, 2, 3, and 4 of the high-powered amplifier.

A floating ground system is used when connecting the four-channel head unit to the high-powered amplifier. The front left head unit output cables are connected to channel 1 and the front right channel connections are connected to the high-input terminals of channel 2 of the amplifier. The rear left output terminals of the head unit connects to channel 3, while the rear right output terminals go to the high input of channel 4 (**Fig. 11-4**).

The common ground connection from the head units are tied together in the front negative right and left ground terminals that tie the high-input terminals of channel 1 and 2 of the high-powered amp. The common ground (black) terminals of the rear left and right outputs are tied together and go to channels 3 and 4. The front positive outputs are connected to the positive terminals of channel 1 and 2, while the rear output terminals are connected to the high-level input of channels 3 and 4 (**Fig. 11-5**).

In a two-channel amplifier, the head unit (receiver) can be connected to the low-level inputs by inserting the left output of one channel into the low-level input jack with patch cords. The other left output can be inserted into the low-level input jacks. Then, patch the right output source into the left and right low-level input terminals as shown in **Figure 11-6**.

To connect the head unit with a four-channel output to the high-powered amplifier with four output cables, connect the left and right rear outputs to the low-input jacks.

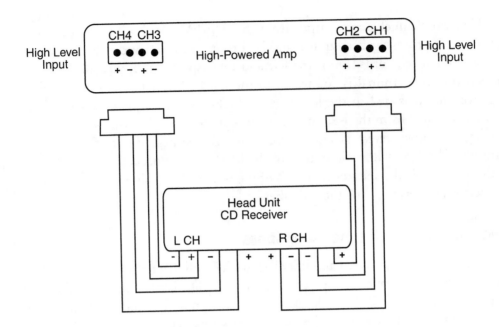

FIGURE 11-4. The floating ground connection on the high-level input of head unit.

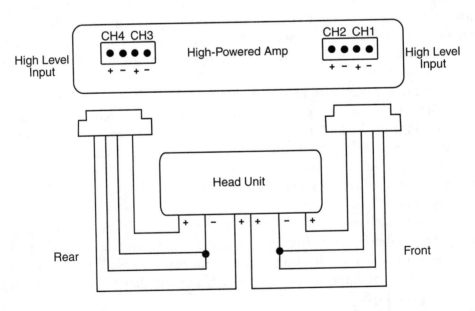

FIGURE 11-5. A common ground connection on the high-level input of the head unit to the high-powered amplifier.

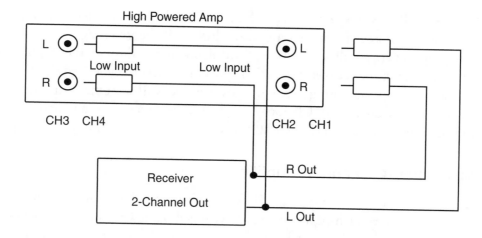

FIGURE 11-6. Connect a two-channel connection of the head unit to the low-level input with patch cords.

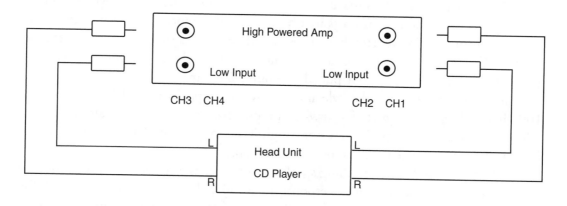

FIGURE 11-7. The left and right rear outputs of the head unit are plugged into the low-level inputs of the amplifier.

Patch cord the left and right front outputs to the other set of input level jacks of the high-powered amplifier. In other words, the L/R outputs are inserted into the low inputs of CH3 and CH4, and the L/R front outputs of the head unit are plugged into the low-level inputs of channel CH1 and CH2 of the high-powered amplifier (**Fig. 11-7**).

Some high-powered amplifiers have a bass boost feature that enables you to hear and sometimes feel the boosted bass from the amplifier. Actually, the bass can be adjusted in some models. For instance, a JVC KS-AX6300 amplifier has a 0-12 dB bass boost, while a Kenwood KAC-628 has no bass boost circuits. Likewise, a Profile 400

SX has a 0-12 dB bass boost and an RF Punch 200a4 has no bass boost circuits. The range of the bass boost circuits has a 0-12 and 0-18 dB boost rating.

A preamp-outlet jack allows a no-amplified signal to pass through the amplifier and into another amp in the speaker system. The preamp outlets lets you feed signals to the other amplifiers without using separate crossover networks. Some amps have this feature and others do not. A Clarion APX 400-2 amplifier does not have a set of preamp outputs, while an RF Punch 400a4 does have a set of preamp output jacks.

A Typical 250-4 Channel Amplifier

A Pyle PLA540 amplifier is a typical 250-watt high-powered amp at a 4-ohm speaker load. Many of the new high-powered amps have metal-oxide semiconductor field-effect transistors that have a higher switching speed than regular bipolar transistors and they generate less heat. With the field-effect transistors used in the various power circuits of the amplifier, the results are a faster response and high-efficient amplifier (**Fig. 11-8**).

The two-channel Pyle PLA540 amplifier provides power to a 4-ohm load speaker at 2X85 watts and 2X125 watts in a 2-ohm load. The maximum power output is 2X350 watts or a total of 700 watts in a bridged speaker mode. To produce this kind of power output, the power supply voltage should be around 14.4 volts with a negative ground. The minimum voltage is at 10.5 volts and the highest operating voltage is at 16 VDC; the maximum current pulled is around 20 amps.

The input impedance of the Pyle amp at low-level input is around 10-K ohms and 100 ohms at high-level signal. The low-level inputs have a 250 mV, and at high levels, 2.5 input volts. This amplifier has both high-level and low-level inputs. The power LED is illuminated when the remote is turned on.

In the stereo mode, the matching speaker impedance is 2-4 ohms, and in a bridged mode, 4-8 ohms. The frequency response is 10-30 kHz. The amplifier has a fully adjustable electronic crossover network with variable bass boost in a positive or negative

FIGURE 11-8. A typical Pyle 250X4 channel amplifier to be connected to the receiver and CD player.

18 dB circuit. The high-pass control permits you to adjust the crossover frequencies from 35 Hz to 400 Hz, to suit the subwoofers. In a full-range system, the crossover mode switch is set full, and if the amp is being used to power a crossover system, set it to either HPF or LPF as needed.

The amplifier's Protection LED Indicator protects the circuits in the amp and will disable the amp if it senses an overload, speaker short circuit, or thermal overload condition. Should this occur, the protection LED will light up. At that time, it is important that you check to determine what has caused the protection circuit to become activated. If the amp shuts off because of a thermal overload, allow the amp to cool down before attempting to turn it on again. If the shutdown occurred because of an overload, or because of a speaker short circuit, be sure to correct these conditions before attempting to turn on the amplifier. To restart the amplifier, turn the remote power off and then on again.

Mount the amplifier under a seat or within the trunk area. When placed under the seat, the battery leads are shorter. When mounted in the trunk, the power cables are longer, but the amplifier is closer to the rear speakers and subwoofer speakers. Make sure that there is plenty of ventilation so the amplifier does not overheat.

Low- and High-Powered Amplifiers

The high-powered auto receiver that includes the AM/FM/MPX, cassette, and CD player might have a total output of 14 to 40 watts or 2X20 watts. The left channel contains 20 watts of power and so does the right channel. The AM/FM/MPX receiver with a CD player might have 14 to 50 watts times 4 outputs of RMS power watts, while the peak power output might be from 35X4 to 50X4 watts of power. The auto receiver audio output components are usually power output transistors or a dual IC component.

The audio components within the high-powered amplifier circuit consist of many transistors and IC input components. A high-powered amplifier might operate with an output power of 50 to 2000 watts. Most of these high-powered amplifiers contain MOSFETs and high-powered output transistors within the audio circuits and power supply.

The average high-powered auto amps might have a 50X2 to 400X2 RMS at a 2- or 4-ohm speaker load. The two-channel amplifier might start at 50X2; some are found at 300X2, at 14.4 input voltage. A MTX Thunder 1000D has a 500X1 RMS power in watts with a 12.5-input voltage and a peak power of 1500X1 or a 1000X1 RMS at 2-ohm impedance rating.

A four-channel high-powered amplifier output average might be 50X4 channel outputs up to 100X4 RMS watts at 13.8 volts, with a 400X2 RMS bridged power output, and a peak power at 150X4 watts, with 200X4 RMS watts in a 2-ohm load. The high-powered amp might have a built-in low-pass (LP) or a high- and low-pass crossover network. The signal-to-noise ratio might be from 90 to 110 dB with the average noise ratio around 110 dB.

The Root-Means-Square (RMS) measures the power in watts that the auto amplifier can produce continually. The measured output power of an amplifier depends on the automobile's input voltage to the amplifier. The average voltage from a battery, with the auto operating, can be around 14.4 volts and 12 to 13 volts when the engine is shut off. Likewise, the amplifier will have a higher wattage output at 14.54 volts than at 12.6 volts of input battery power.

High-Powered Amplifiers

The high-powered amplifier might appear in 50 to over 2000 watts of power to amplify the music. The lower wattage amplifier might have power output ICs, while the high-wattage amplifier contains many transistors in directly coupled power output circuits. A typical 150- to 200-watt amplifier might have two ICs in the preamp circuits, a single transistor in the muting system, 10 power amp output transistors, and a single transistor in the overload and shutdown circuits, within each stereo channel.

The 170- to 250-watt amplifier might consist of two directly coupled AF transistors, four driver transistors, and four output transistors in a push-pull directly coupled power output circuit. Q109 and Q110 are NPN power-output transistors connected to the + 35 volt source. Q111 and Q112 are PNP output transistors connected to the -35 volt source. Q109 is a 2SC3421 or 2SC600 transistor, while Q110 is a 2SC3907 or 2SC3519 NPN transistor. Q111 is a 2SA1358 or 2SB631, while Q112 is a 2SA1516 or 2SA1386 PNP transistor (**Fig. 11-9**). Replace all power output transistors with original part numbers, when available.

A typical Pyle high-powered PLA540 2-channel amplifier that has a power RMS of 2X125 watts in a 2-ohm load and a maximum power output of 2X350 watts, will have many small signal transistors, power output transistors, and preamp ICs. In the bridged mode to a single speaker, the 2-channel amplifier has 1X700 watts of power. The input transistors of each channel consist of nine transistors and two MOSFET power output transistors. The whole semiconductor line consists of five IC components and 22 transistors within the input and driver audio circuits. A total of 32 transistors are found throughout the power supply and audio circuits (**Fig. 11-10**).

HOOKUPS AND FUSE BLOCKS

Battery Cable Hookups

The battery cable should never be less than 10-gauge wire; 4- and 8-gauge wire are better choices. Always use a lower gauge wire for quality sound and for very little

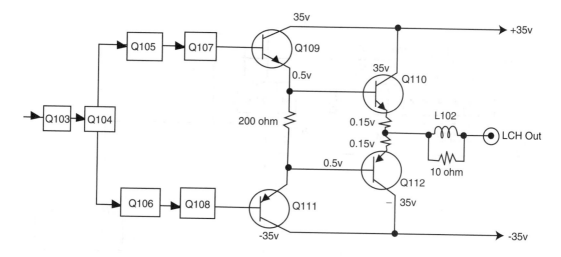

FIGURE 11-9. Suspect a leaky or shorted polarity diode when the main fuse to the amplifier blows.

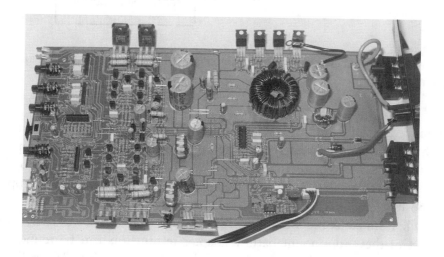

FIGURE 11-10. Checking low signal transistors within the high-powered amp with the diode test of the DMM.

voltage loss. An easy amplifier cable hookup to the battery can be made with a side-mount post adapter. The side-post adapter helps to attach the battery cable to the battery post. The automobile positive power cable attaches to the adapter post, and the amplifier power cable eyelet goes under the top-mount adapter post. The top-mount terminal attaches to the battery cable. A single battery adapter-post of solid brass eliminates the need for a factory harness cable (**Fig. 11-11**).

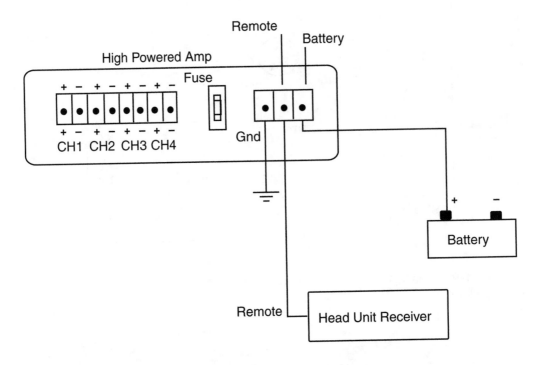

FIGURE 11-11. The battery and remote connections of the head unit to the high-powered amplifier.

To remove the auto's battery cable, replace it with a top-mount terminator, and then attach the power amp's battery cable and the car's power cable to the top-mount adapter post. A multi-feed battery connector is a connector that can hold two 2-gauge and two 4-gauge cable wires. Two multifeed gold battery terminals can be applied to both the positive and negative battery post connectors. A machined battery post terminal made from stock brass can hold two 8-gauge auxiliary power cables with set screws. Most of these top-mount battery adapter terminators can be found at auto supply stores, electronics outlets, and automobile audio system installation firms.

Battery and alternator damage can be caused by the amplifier system pulling too much current from the output of the battery and the alternator. If the battery goes dead after the car sits for 10 to 24 hours, suspect that the amplifier equipment is pulling excessive power and current. The alternator cannot keep up with the battery charging if the drain of current is too large. Suspect a defective component within the amplifier, battery cable, or alternator. Check the current drain of the amplifier with a good current meter.

If voltage from the battery and alternator is too high, the amplifier circuits can be damaged. A higher input battery voltage can cause the transistors to overheat and become damaged. Improper voltage can cause the CD player to skip. If the CD player skips during driving, suspect that either the vehicle hit a bump or that the player is not mounted properly. Secure the unit in the front dash. The input voltage should not be over 16.5 volts for normal amplifier operation.

The ground cable should be at least a 6-gauge cable wire. The ground cable should connect to a terminal block so that all components can be connected to the same terminal block. A gold-plated brass power distribution block is very important because it can be mounted in places where space is at a minimum. These distribution blocks can accept up to 8-gauge power cables. Gold terminal connectors are best with a fused distribution block assembly.

Distribution and Fuse Blocks

You can connect many different types of distribution blocks and fuse holders to the power amplifier system. There are several 24-karat gold distribution blocks with 2 or 3 inputs up to 4-gauge cable, with three outputs up to 8-gauge wire or cable. Another 24-karat gold distribution block might have one input that can handle a 4-gauge cable, and four outlets that can handle up to 8-gauge wire.

A 24-karat gold pulse plate fuse contact can hold an SE in-line fuse holder that is machined from solid-brass stock, in which you can see through the glass if the fuse is blown or open. Gold finish fuse blocks with plastic covers can hold 4- to 8-gauge cables or wire with a fuse block of 150A and 225A current fuse holders. Usually these fuse holders are rated at a maximum of 32 volts DC and from 75, 150, 225, and 300 amp current fuses.

Speaker and Amplifier Hookups

Today, a high-powered amplifier might produce 2000 watts bridged into a 4-ohm impedance load. High-powered speaker hookups from the amplifier to the speaker may be rather simple or may be quite complex in the high-powered amplifier systems. A simple speaker hookup in a 2-channel left and right system (L&R) might consist of just connecting a large subwoofer speaker on the left and right speaker terminals.

The speakers should be greater in wattage than that powered by the amplifier. If the power amplifier is a 2X50 watt job, then the speakers should be able to handle at least 25 to 50 watts with each channel. When smaller wattage speakers are used in a system that is higher in wattage, the voice coils can be blown out or damaged when the volume is raised up to peak power. (**Fig. 11-12.**)

Simply connect the left speaker to the left side of the amplifier terminals. Connect the positive terminal of the amplifier output to the positive terminal on the left speaker. The negative terminal of the amplifier is connected to the negative terminal of the

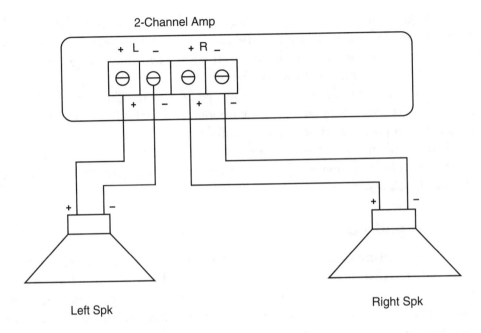

2-Channel Amp

Left Spk

Right Spk

FIGURE 11-12. The basic speaker connections to a stereo two-channel high-powered amp.

speaker. Use 12-gauge wire for a two-speaker hookup and 14- or 16-gauge for high-powered systems with several different speakers.

Sometimes the speaker wire will have a flat white line or contain a different color so that color polarity can be made at the speaker and amplifier. Use speaker wire connectors on both ends to ensure a better connection. Multi-stranded pure copper wire for maximum signal flow can be ordered in packages of 5-, 10-, 20-, 30-, and 60-foot lengths in 16-, 14-, or 12-gauge flat or twisted speaker wiring.

A basic speaker hookup for a four-channel amplifier can contain a separate speaker for each channel. Two of the channels might go to the front deck or the front doors, while the other two channels are connected to two separate speakers in the rear deck. Again, make sure that each speaker is polarized by the amplifier outputs. Channel 1 and channel 2 can serve the front speakers, and channels 3 and 4 to the rear speakers. Usually the basic four-channel system contains speakers with a 4-ohm output impedance (**Fig. 11-13**).

To bridge a two-channel high-powered amplifier, the positive terminal of the left channel is connected to the positive terminal of the speaker. The negative terminal of the right channel is connected to the negative terminal of the mono or single PM speaker. For example, a two-channel of 2X85 watts at 4 ohms with a maximum output of 2X350 watts at 2 ohms can be bridged in a mono operation at 1X700 watts. The

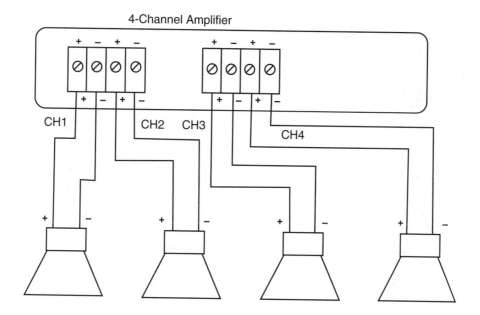

FIGURE 11-13. The basic four-channel amplifier hookup to four different subwoofer speakers.

single-bridged speaker should be able to handle at least 700 to 1000 watts of power. Many of the high-powered amps recommend a minimum of 4-ohm impedance when bridging a single speaker to a high-wattage amplifier.

In a four-channel high-powered amplifier (2X500 watts) of power, channels 1 and 2 are connected to a single (mono) PM speaker. The positive terminal of channel 1 is tied to the positive terminal of the bridged mono speaker, while the negative terminal of channel 2 is tied to the negative terminal of the same mono speaker. Likewise, the positive terminal of channel 3 is tied to the positive terminal of another PM speaker, and the negative terminal of channel 4 is connected to the negative speaker terminal. Again, the minimum speaker impedance should be not less than 4 ohms (**Fig. 11-14**).

In a four-channel amplifier (2X700 watts), the left and right speaker terminals are connected to channels CH1 (left) and CH2 (right). Again, each speaker has the correct positive and negative terminals of the amplifier tied to the same polarity on the left and right speakers. With channels 3 and 4, the positive terminal of channel 3 is connected to the positive terminal of a mono-bridged high-power subwoofer, and the negative terminal of the amplifier is connected to the negative terminal of channel 4. The minimum speaker impedance of 4 ohms is recommended for this type of hookup (**Fig. 11-15**). Simply follow the manufacturers recommendations for correct speaker connections.

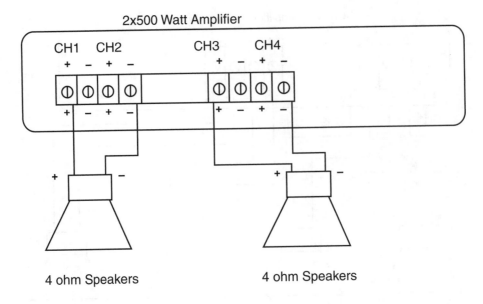

FIGURE 11-14. The bridged speaker system tied into a 2X500 watt high-powered amp, and into a 4-ohm speaker.

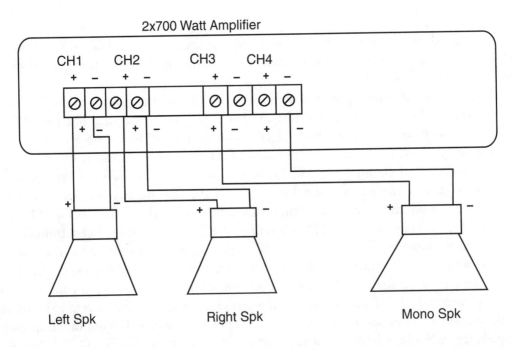

FIGURE 11-15. A 2X700 watt amplifier speaker connection to a 4-ohm impedance left and right and a mono sub woofer.

Choose the correct speaker size and wattage to be able to handle a total of 1400 watts in a bridged connection. The left and right speakers should be able to handle 350 watts each (2X350), and the single subwoofer should handle 700 watts of power. In the stereo mode, the left and right speakers can have an impedance of 2-4 ohms, while in a bridged mode, the mono speaker should have an impedance of 4-8 ohms.

A single high-powered speaker hookup connection to a single amplifier can operate several different sizes of speakers. The head unit (receiver) may be connected to a CD player or CD changer. The full-range signal from the head unit is connected to a 500-watt amplifier and high-level input terminals. A high-pass speaker level output of the 500-watt amplifier is tied into a crossover that controls the midrange and the high-range speakers in the left channel. Another high-pass signal level from the 500-watt amplifier is tied to another crossover network that controls another midrange speaker and tweeter speakers in the right channel.

A low-pass signal from the 500-watt amplifier is fed directly to mono subwoofer that has high wattage capabilities. The subwoofer speaker is mounted in the center of the back area of the auto or truck for the bass frequencies, and each side of the car has a midrange speaker and tweeter speakers to pass the high frequency sounds. The small single 500-watt amplifier is capable of handling several different speakers (**Fig. 11-16**).

Another high-powered speaker hookup might consist of two different sizes of amplifiers that control six different speakers. The head unit (receiver-CD changer) output feeds into a control-equalizer unit and this output is fed to a 2X500 high-powered amp and a 2X75-watt amplifier. The 2X500 high-powered amplifier drives two 10-inch subwoofers that can be connected in a series or in a parallel speaker hookup. Here, each speaker should be able to handle 500-1000 watts of power.

The second high-powered amplifier fed from the control unit is a high-pass network to a 2X75-watt separate amp. A high-pass crossover network feeds midrange and tweeter speakers. Each of the tweeter and midrange speakers are fed from a separate crossover network, and each speaker should be able to handle 75-100 watts of power. One amplifier provides the bass frequencies (2X500), and the second amplifier (2X75 watts) furnishes the higher frequencies to the two tweeters and two midrange speakers (**Fig. 11-17**).

In more complex high-powered speaker connections, two separate amplifiers provide high, low, and midrange frequencies to drive 10 separate speakers. A CD changer is connected to the head unit (AM/FM/MPX receiver) and a full range of control is fed to a control unit that connects to a 2X450-watt and a 2X500-watt high-powered amplifier. Some head units might have a small TV system tied to it. The TV monitor with a color LCD screen can be converted to the automobile high-powered audio systems.

A low-pass line is fed from the 2X500-watt amplifier to a bass cube that ties into the input of the 2X500-watt amplifier. The 2X500 amplifier with a low-pass line drives

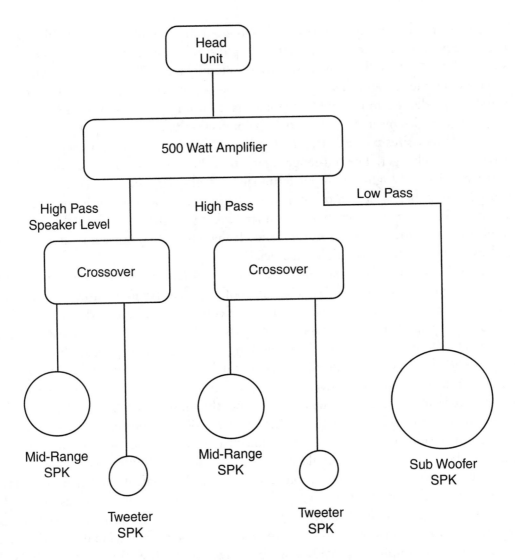

FIGURE 11-16. Here, one 500-watt amplifier is operating five different speakers including the tweeter, midrange, and low-pass subwoofer.

two subwoofer speakers directly, and a speaker relay controls two very large subwoofer speakers. A total of four subwoofers are driven by the 2X500-watt amplifier.

The 2X500-watt amplifier provides a left and right channel speakers to a midrange line and high-pass line of speakers. The two midrange line speakers have a separate crossover network for each midrange speaker. The 2X450-watt amplifier also drives four high-pass PM speakers with corresponding networks. A small

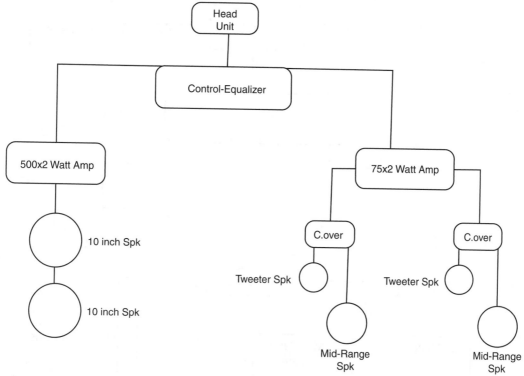

FIGURE 11-17. The head unit drives two different wattage amplifiers that feed the low frequency to two different wattage amplifiers, which then feed the low frequency to two 10-inch sub woofers, midrange, and tweeters.

speaker, such as a tweeter or horn, might be connected to the high-pass frequency network. Here, the 2X450 watt amplifier is connected to six different PM speakers (**Fig.11-18**).

DISMANTLING THE AMPLIFIER

Most high-powered amplifiers have a heavy heat sink to dissipate heat from the MOSFET power supply and audio power output transistors. Some of these amps have outside metal covers or cabinets, such as the molded heat sink. The inside electronics components are mounted on a PC-board chassis that slides in a groove to hold it up from the metal bottom section. The rest of the electronic components are found mounted on the metal end pieces that can easily be removed (**Fig. 11-19**).

Here is how to remove the typical auto amplifier from its cover:

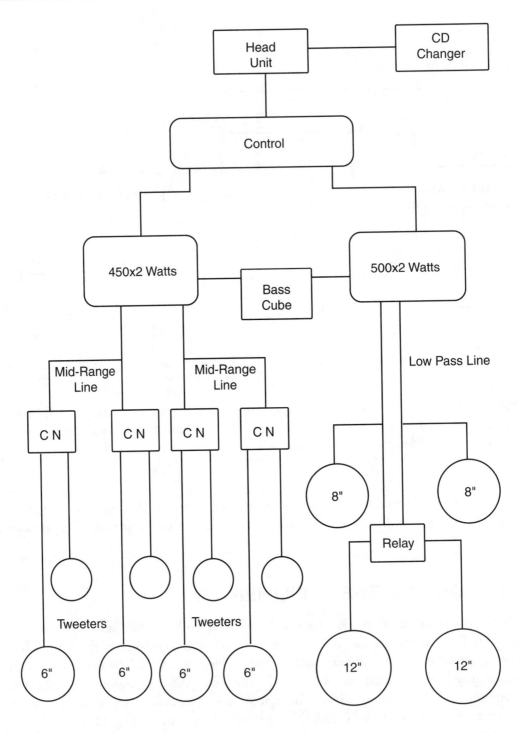

FIGURE 11-18. A really high-powered system: 10 different speakers with 450X2 and 500X2 amplifiers.

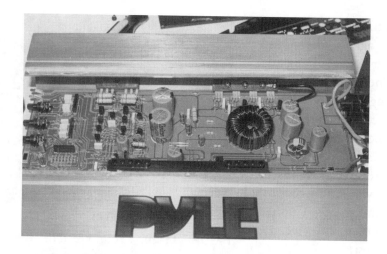

FIGURE 11-19. Remove the plastic top cover that is glued down to reach the components; then remove the metal screws on the top side of the high-powered amp.

1. Pull off the top plastic cover to inspect the electronic components and the transistor mounting plates and screws. The clear plastic plate is glued in place after all repairs are made with a few drops of regular cement.

2 Remove 5 screws in each end of the amplifier to remove the metal cover that covers the controls and contains the speaker terminals and fuse container. Most of these components are bonded or riveted to the black metal end-piece.

3. Remove two metal screws that hold a pair of power output transistors to the outside metal cabinet or heat sink. A long Phillips screwdriver is necessary to remove the mounting screws at a slight angle.

4. Slightly and carefully pull up the transistors from their mounting insulation and mounting position so that the chassis can be removed.

5. Slide the chassis out toward the end piece that is attached to the speaker terminal connections. Make sure each power output transistor slides out easily. Remove the insulator that is mounted between each transistor bonded to the outside metal heat sink (**Fig. 11-20**).

Some of these metal screws might be difficult to remove as they were probably fastened down with a power screwdriver. If necessary, choose a large-handle screwdriver that can provide greater leverage to remove those difficult screws. You might run into separate ground wires with spade-like terminals connected to the metal outside heat sink before the PCB can be removed.

After all repairs have been made, replace each mica or cloth-like insulator for each transistor. Apply a coat of silicone grease to each side of the insulator and place the insulator behind each transistor that mounts against the outside heat sink. Make sure the insulators are in the same spot so that the mounting screws can be replaced with ease. Tighten down the metal mounting plate over the transistors for a good transfer of heat from each transistor. Replace the two end pieces with the same metal screws. Connect up the amplifier and give it a good bench test.

To remove the amplifier board from a high-powered amp that is completely covered with a metal heat sink, remove the 4 to 6 metal screws that cover the bottom area of the amplifier (**Fig. 11-21**). To remove the front panel that holds level input controls and jacks, remove the two or three metal screws. Remove the rear panel from the unit by removing 2-6 metal screws. In some models, the metal strips holding the power transistors to the outside metal heat sink must be removed to release the whole PCB. Now the components can easily be checked for leaky or shorted conditions.

FIGURE 11-20. You must remove the screws that are holding a metal plate over the transistors before the chassis can slide out one end of the amplifier.

FIGURE 11-21. The outside metal cover on a high-powered amplifier serves as a heat shield for the high-powered transistors.

THE POWER SUPPLY

The 50X2-watt amplifier might have a power supply with DC-DC converter circuits with two oscillator transistors and two MOSFETs operating in a transformer rectifying circuit. If an overload or short occurs in either the left or right audio output channels, the overload circuits will shut down the DC-DC power supply. Two DC detector transistors protect the high-powered output transistors that provide power to each right and left speaker output terminals.

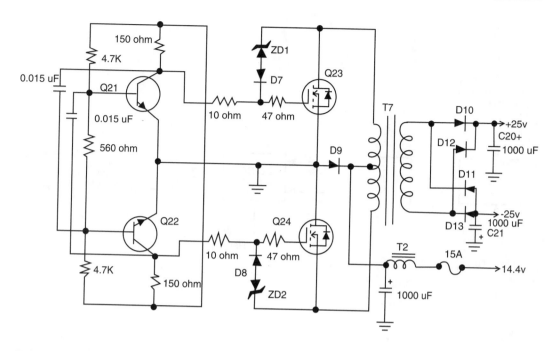

FIGURE 11-22. A DC-DC converter power supply found in 50X2 amplifier circuits.

The 14.4 volts from the auto's battery is applied through a 15-amp fuse in the 50-watt 2-channel amplifier and onto a filter choke. A polarity diode with the cathode tied to ground protects the transistor circuits if the battery terminals are reversed. Otherwise, the battery might be charged up backward. A large 1000-uF 35-volt electrolytic provides filtering action at the center-tap of the transformer (T1). Q21 and Q22 provide an oscillating circuit that is fed into the MOSFET output transistors with the output of each transistor tied to the primary winding of T1. An AC voltage is developed in the secondary winding of T1 to the bridged rectifier circuits.

Diodes D10 and D12 are wired with the cathode terminals to a 1000-uF 35-volt electrolytic capacitor, and the positive (+) 25 volts is applied to the collector terminals of the NPN transistors in the high-powered amp circuits (**Fig. 11-22**). Diodes D11 and D13 rectify the AC voltage in the negative 25-volt source and feed the negative voltage to the PNP power output transistors within the high-powered amplifier. A positive and negative 25-volt source is fed from the DC-DC converter and rectifying circuits to each of the left and right power output circuits. Electrolytic capacitors C20 and C21 provide filtering action in the output power supply circuits with the center-tap secondary transformer winding at common ground.

In the larger high-powered amplifiers of 150 watts or more, a pulse width modulator (PWM) IC drives a DC-DC converter circuit containing 8 transistors with the last six as MOSFETs. The bridged rectifiers and filter circuits consist of four silicon diodes

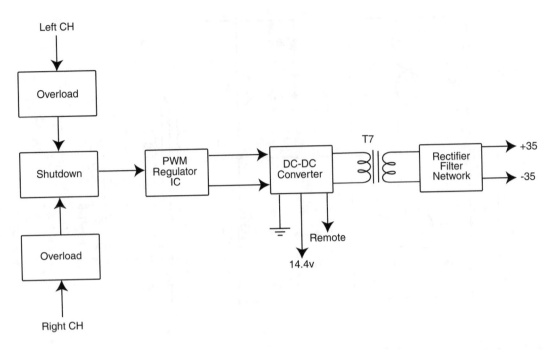

FIGURE 11-23. The overload transistor from each channel output shuts down the shutdown transistor, and in turn, the DC-DC power supply.

and two 2200-uF electrolytic capacitors. A shutdown circuit connects to the PWM IC, shutting down the power supply with a short or overload in the power output transistors circuits. The shutdown transistor is controlled by several overload transistors within the overload circuits that tie to each of the left and right power output transistors (**Fig. 11-23**).

The pulse-width modulated IC100 provides a drive signal to transistor Q3 and Q5, which drives six MOSFETs in a push-pull transformer circuit. Q8 and Q10 provide a series-parallel operation in the top half of the primary winding of power transformer (T7). MOSFETs Q5 through Q7 drive the other leg of the power transformer (T7) with a center tap on the primary side that is connected to the 14.4-volt battery source. The DC-DC converter is designed to provide a higher DC voltage to the power output transistors instead of only 14.4 volts from the car battery.

The primary winding of the transformer (T7) battery voltage source contains a filter network of two 2200-uF electrolytic capacitors with a choke transformer and capacitor input filter network. Diode D1 provides a correct polarity battery voltage to the DC-DC converter circuits. Another 1000-uF electrolytic is found in the input side of the filter network after the 30-amp fuse. A transistor-regulated 13.7 volts is fed to the PWM IC off of the 14.4-volt battery source (**Fig. 11-24**).

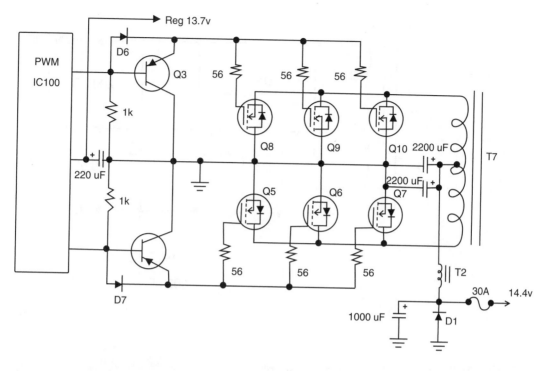

FIGURE 11-24. The PWM and DC-DC converter power supply found in the high-voltage amplifier circuit.

The AC voltage developed in the secondary winding of the transformer is center-tapped to common ground. Each secondary winding contains two silicon diodes that rectify the positive and negative 34.9 volts to the power output transistors. D2 and D3 provide a negative 34.9-volt source to the PNP transistors in the output circuits, while D4 and D5 are connected to the cathode terminals of the diodes, providing a 34.9-volt source to the NPN power output transistors. Both 34.9-volt sources have 2200-uF 50-volt electrolytics (C13 and C14) in each power source for filtering action. When a shorted or leaky component occurs in the high-power transistor output circuits, the shutdown circuits cause IC100 to shut down the power supply (**Fig. 11-25**).

Power Supply Damage

A shorted or leaky transistor within the power supply can cause damage to the driver or MOSFET power transistors. Since all six MOSFETs are directly connected, check all of them connected together, because one that becomes shorted can destroy or change the voltage on the output transistors. Check all diodes in the power supply with the diode test of the DMM. Discharge all electrolytic capacitors in the power supply before

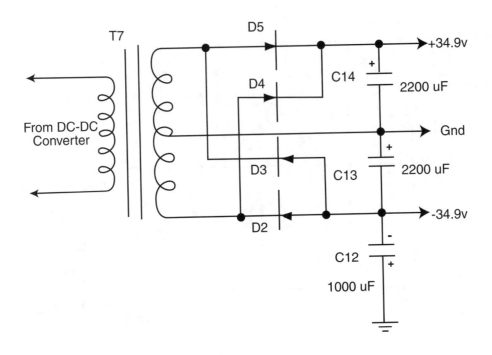

FIGURE 11-25. The DC-DC power supply with a bridged rectifier circuit that provides a 34.9-volt positive and negative voltage for the output transistors.

taking leakage or ESR meter tests. First check each electrolytic with the ohmmeter scale of the DMM for being open or having leakage, and then take a test across the electrolytic terminals with the ESR meter.

Critical voltage measurements on the PWM IC100 and on all transistors can determine a leaky or shorted component. Disconnect the negative and positive 34.9 volts from the power supply that feeds to the high-power amplifier. Some amplifiers have a terminal board for easy removal. Sometimes a shorted transistor within the amplifier can shut down the power supply and also lower the applied DC-DC converter voltage. Isolate the transistor shutdown circuits from a possible leaky component in the power supply.

An overloaded circuit in the secondary winding of transformer T7 can defeat the PWM system. Double-check each silicon diode in the power supply output voltage source for leaky or shorted diodes. Leaky or shorted filter capacitors can lower the voltage or prevent the PWM circuits from functioning. If the 30-amp fuse keeps blowing, suspect a shorted MOSFET, diodes, or electrolytic capacitors in the DC-DC power supply. Do not overlook a shorted polarity diode (D1) when the main fuse keeps blowing open.

CIRCUIT SYSTEMS

Low-and High-Level Amplifier Input Circuits

Most commercial audio amps have both high- and low-level impedance input circuits. The high-level jacks can be used with the power speaker output of an auto receiver with a cassette player, while the low-level input might match the stereo CD player line output jacks. Usually the high- and low-level input might be found in the auto stereo equalizer/booster and in high-powered amplifiers. A low-level input might connect the preamp output circuits to the low-level input of the high-powered amplifier.

The high-level radio and cassette player outputs might be switched into the IC preamp circuits of an equalizer-booster circuit or a low-level power audio amplifier. The high and low-level inputs in a resistance network are switched by S-1 into the input of IC2. Both the left and right stereo channels are switched with a dual S-1 into the preamp circuits (**Fig. 11-26**).

The high-level input might match the input impedance with transformer coupling. A transformer input is found in both left and right audio channels. Some audio amplifiers have speaker level inputs from the radio receiver, CD, and cassette player. The multichannel amplifiers might have both speaker level input and preamp jacks.

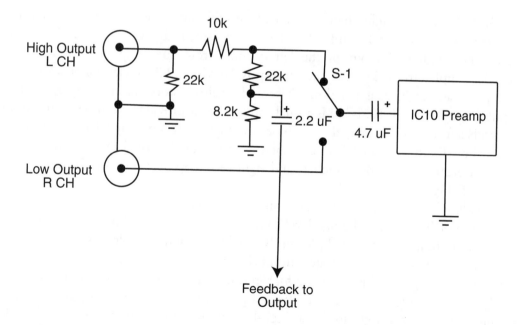

FIGURE 11-26. The high- and low-level input circuits are switched in from the radio and CD player to the preamp IC10.

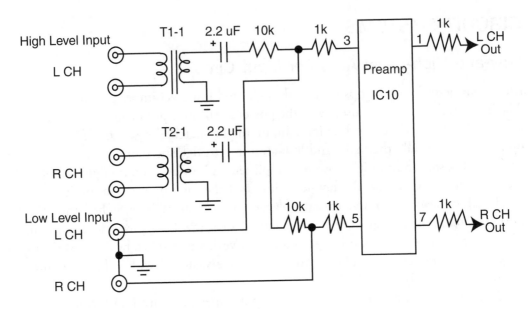

FIGURE 11-27. The high and low-level inputs are coupled by transformer, and the low-level feeds directly into the preamp IC10 circuits.

Both the left and right channels are tied into the high-level input terminals of the transformer-coupling of T1 and T2. A 10-ohm resistor is found across each high-level input. The high-level input of the left channel from the coupling transformer is capacity-coupled to the input of the preamp IC10 with a 2.2-uF electrolytic capacitor. The input audio signal is fed through several small resistors before being applied to input terminal 3 of IC10. IC10 is a dual input IC that provides preamp amplification for both left and right channels. Likewise, the high-level input signal is transformer-coupled by T2 to another 2.2-uF electrolytic, and through several resistors to pin 5 of the dual preamp IC10. Both the left and right secondary windings of T1 and T2 are grounded (**Fig. 11-27**).

The low-level left RCA input jacks are tied directly to a 1K-ohm resistor and to terminal 3 of IC10. Also, the low-level right channel input signal is tied to a 1K-ohm resistor to terminal 5 of the dual preamp IC10. The amplifier low-level signal is fed out of terminal 1, and the right channel output signal appears at terminal 7 of IC10. Preset gain controls might be found in the high-powered amps at the low-level inputs.

In the larger high-powered amplifiers, a preset gain control is found in the input circuits of the left and right low- and high-level IC input circuits. Separate or dual IC components might be found within the low- and high-level input circuits. The high level of both the left and right transformer-coupled audio signal is fed through a

100K-ohm resistor to the left and right gain controls. The low-level input of the left and right RCA jacks are tied directly to the preset gain controls of the amplifier. The adjustment of VR1 and VR2 will determine the amount of audio that is fed into the preamp Of IC10. Sometimes if the preset gain controls are set too high with certain receivers and head units, some distortion or clipping might occur.

Directly Coupled Audio Circuits

Two preamp or audio AF and power output transistors might be directly coupled to one another and can provide more audio gain and power with fewer components in the audio circuits. A directly coupled amplifier occurs when the output of the first transistor is wired directly to the input of the following stage. The collector terminal is connected directly to the base terminal of the next audio transistor. You might find in some high-powered audio circuits that the emitter terminal of one transistor might be tied to the base of the next power output transistor. Sometimes a small ohm resistor might be connected between two high-powered transistors.

The directly coupled amplifier circuits have a wide frequency response and can handle either AC or DC signals. There are no coupling capacitors or resistors between the collector of the first audio amp and the second transistor. You might find the directly coupled transistors are NPN types or that in other audio circuits, one transistor is a PNP and the other an NPN type power transistor.

In **Figure 11-28** the audio directly coupled circuits are identical to both left and right AF or preamp audio circuits. The audio signal is fed through a 1-uF electrolytic and a 1000-ohm resistor before being applied to the base terminal of Q10. A 470-ohm resistor within the emitter circuit of Q10 provides forward bias to the AF amp transistor. The collector terminal of Q10 is fed directly to the base of Q11. A DC voltage is fed through a 560-ohm and 1.5 K ohm resistor to the collector of Q10 and the base of Q11. The amplified signal out of Q11 is coupled to the next audio transistor through capacity-coupling of a 4.7-uF electrolytic. These directly coupled transistor circuits are not only found in the high-powered amplifiers but also located in tape, cassette, or radio circuits.

The defective preamp or AF transistors might become leaky, open, and intermittent. If the first or the second transistor becomes leaky or open in a directly coupled amp circuit, the voltage will also change on the other transistors. When the first directly coupled transistor becomes leaky, the base and emitter terminals might go to zero or to very little voltage. The collector voltage will be lower than normal, and this means that the second transistor base voltage should be lower, and should equal the collector terminal voltage of the first transistor. The voltage on both directly coupled transistors will change from those shown on the schematic.

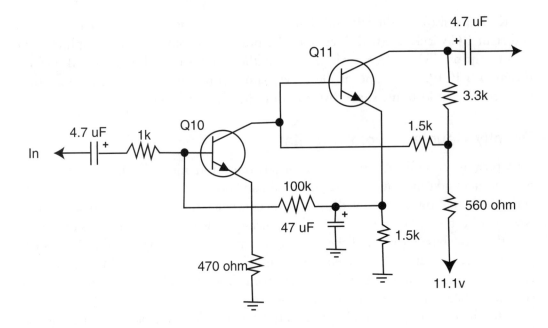

FIGURE 11-28. Directly coupled audio transistors are found in the input and output circuits of the high-wattage amps.

When the first directly coupled transistor goes open, the measured voltage on the emitter and base terminals will decrease to about one half of the original normal voltage. Because the transistor opens up, the collector voltage will increase in voltage equal to the supply voltage for the directly coupled transistors. This also increases the DC voltage to the second directly coupled transistor. Sometimes when the meter probe touches the base or emitter terminals, the intermittent transistor might return to normal operation. At other times, the suspected transistor might test good out of the circuit and test open under load. Replace both suspected transistors.

Directly coupled transistors are found throughout the high-powered amplifier. Sometimes the collector of one transistor is tied directly to the base of the next one. In other directly coupled circuits, the emitter terminal of one transistor might be connected to the base of the following transistor. Always remember, when one directly coupled transistor is defective, the other transistor is also affected with voltage change and possible damage.

Tone Controls

In some high-powered amplifiers, the bass and treble tone controls are found after the first preamp IC or transistor, and are fed to another preamp circuit (**Fig. 11-29**). The bass and treble controls are found between two audio dual-IC preamp circuits. Notice

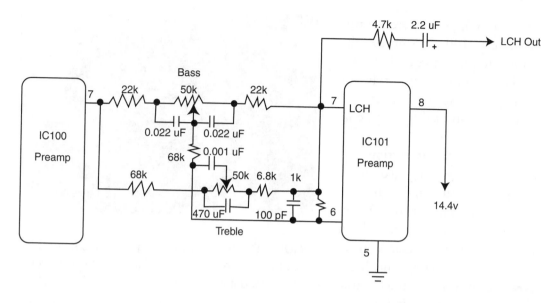

FIGURE 11-29. The bass and treble controls might be located between two preamp IC components.

that larger bypass capacitors are found in the bass control circuits compared to the small pF capacitors within the treble control circuits. You will find a bass and treble control in both the left and right channels of the stereo amplifier.

The defective bass or treble controls can change the volume or cause intermittent audio. You can solve a noisy bass or treble control by spraying cleaning solution down inside the control, or if it's really bad, replacement is the answer. The open or damaged bass or treble control can affect the gain of the amp in a series-parallel circuitry. Simply check the resistance across the controls and then vary the center tap with the ohmmeter connected to one side of the control and wiping blade. Notice if there is a loss of, or a drop in, resistance as the control is rotated.

By shunting a fixed 50K-ohm resistor across either control, if the volume is low after going through the controls, the volume should increase, indicating a possible open control. Since these preset controls are not rotated back and forth like a volume control, they should not become worn or cause problems, unless the control is damaged (**Fig. 11-30**). Replace the damaged tone controls with the same part number, or if not available, universal PCB replacement controls are still available from the electronic parts firms or depots. Replace the tone controls with a linear taper.

Muting Circuits

The muting circuits within the high-powered amplifier might be located after the preamp stages. A muting type of transistor is found in each preamp stereo circuit. In the audio high-powered amps from 50 to 100 watts or less, the mute circuits might be found before the power output IC or transistors. Usually the mute circuits are controlled by a system-control IC.

Besides relay-protection circuits, the large integrated stereo component system and high-powered amplifiers might contain several mute circuits. The sound can be muted after the gain control or the preamp stages. Some high-powered amp circuits might mute the power output before the

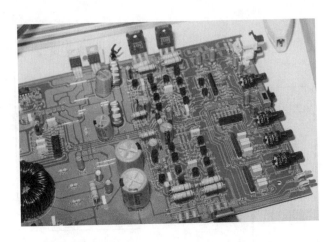

FIGURE 11-30. The preset gain controls are mounted at one end of the high-wattage amplifier and are mounted directly on the PCB.

speaker-relay switching. Each right and left channel does have a separate muting circuit.

The system-control IC or microcomputer controls each muted circuit. The amplifier input circuits are muted so as not to hear the operation of power or function-switch noise. The output circuits of the CD player might be muted for operations of motors and changes in loading. Q500 and Q501 mute the incoming signal from entering the high-powered amp IC500. The muting circuits tie the audio output to common ground. A defective muting system can cause loss of sound in either the right or left stereo channels, or both.

The muting circuits can be signal-traced with the external audio amp or scope. Monitor the audio signal at pins 5 and 9 when the system-control IC is functioning. The base voltage on Q500 and Q501 are at -3.5 volts in normal operation (**Fig. 11-31**). Both transistors act as a switch when the system control IC is in the muting mode. The audio on both left and right channels is grounded between the two 4.7-uF electrolytics with transistors Q500 and Q501. Q500 and Q501 switch the audio signal to common ground. When the muting takes place, no sound should be heard at pins 5 or 9 of the power amplifier IC500. Most audio signal-tracing methods and control voltage tests can locate the defective component in the muting circuits.

Overload Output Protection

The overload protection output circuit is designed to shut down the power supply when the output circuits become unbalanced. The directly coupled output circuit to the left and right channels is balanced at zero volts. No DC voltage is found on the

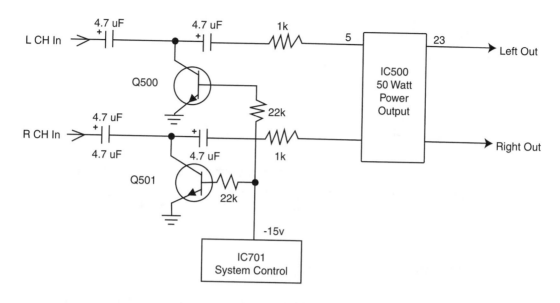

FIGURE 11-31. System-control IC701 controls the input muting system before the power output IC in a 30-50 watt audio amplifier.

voice coil of a subwoofer-connected speaker when one terminal is wired to common ground. If DC voltage is found on a normal speaker that has a large electrolytic placed between the speaker winding and ground, then the speaker is working in a DC voltage circuit that measures voltage on each side of the speaker terminals.

When a driver or a directly coupled power-output transistor becomes leaky or open, the output voltage of Q111 and Q112 change, resulting in a DC voltage at the speaker output terminals. Anytime that a change occurs in either output-transistor voltage or condition, a DC voltage will appear across the speaker terminals (**Fig. 11-32**).

If a DC voltage is left too long on the voice coil of a speaker, the winding of the voice coil will become warm and will drag. The voice coil might result in a frozen or burned coil on the speaker magnet pole piece. Not only do you have an amplifier that needs repair, the damaged speaker must also be replaced. The overload output circuit is supposed to protect the subwoofer speaker and to shut down the low voltage or DC-DC power supply.

The DC voltage on the emitter terminals of Q111 and Q112 is zero. This same voltage is applied to the base and emitter circuits of the overload transistor Q113. D111 isolates the DC-DC power supply from the overload transistor. When the power audio output circuits fail, a DC voltage is found at the base and emitter of Q113, upsetting the balanced output circuits. The overload transistor conducts, applying a signal voltage to the protection circuits in the DC power supply (**Fig. 11-33**).

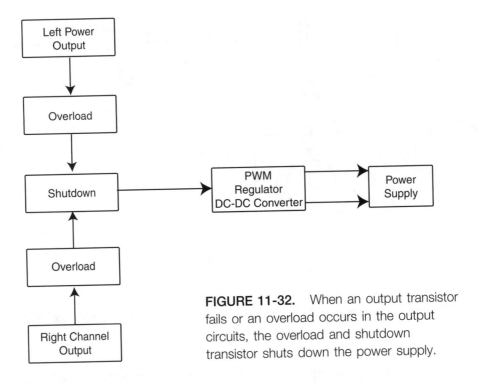

FIGURE 11-32. When an output transistor fails or an overload occurs in the output circuits, the overload and shutdown transistor shuts down the power supply.

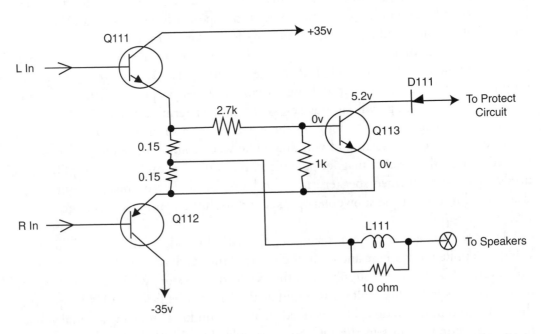

FIGURE 11-33. Overload transistor Q113 is connected to the emitter circuits of the high-powered amplifier output transistors.

The protection circuit shuts down the power supply, disconnecting the positive and negative 35 volts to the output transistors. Each stereo channel has the same type of overload protection circuit and shutdown transistor.

The defective audio output circuits might come up and then shut down at once. Sometimes the audio output transistor or IC might become leaky and blow the main power fuse. The DC voltage on the speaker fuse might cause the fuse to open. When a blown fuse is found or the audio chassis fires up and shuts down, check for a DC voltage at the speaker output terminals.

Do not attach a test speaker to the high-powered amplifier that has a DC voltage at the speaker terminals. Connect the high-powered output speaker terminals with a 10-ohm 50-watt load resistor instead. Then take a low-voltage measurement at the left and right output speaker terminals. This only applies to high-powered audio output transistors working in a balanced output circuit, when one side of the speaker is grounded.

SERVICING AND TROUBLESHOOTING

Precautions

A person should take many precautions before attempting to make holes and mounting the different speakers within the automobile or truck. Before you drill any holes, check the automobile's layout very carefully. Be very careful when working near the auto's gas tank, fuel lines, hydraulic lines, and electrical wiring. Make a careful inspection of where the speaker's hole is to be mounted, and that it is clear of moving operations, especially in car doors and where windows are involved. Then recheck that mounting space.

When mounting the high-powered amplifier, be careful that wire connections aren't unprotected or getting pinched, or that they aren't likely to be damaged by the movement of seats and nearby components. Select a location to mount the amplifier inside the auto with proper ventilation. Poor ventilation and excessive use of driven high-power output can damage components within the amplifier.

Do not operate the high-powered amplifier without a speaker or a load across the amplifier's speaker terminals. Make sure that a speaker is connected or an 8- to 10-ohm 10-watt resistor, and is placed across the speaker output terminals while working on the amplifier on the service bench. Always keep the volume turned down on the amplifier while repairing it. Just turn up enough volume to do the repair.

Attach all system components securely within the auto or truck to prevent damage, especially in case of an accident. Loose amplifier and connecting components should be tied down, so to speak. Replace speaker and amplifier wire cables under the carpet or where they cannot be tripped over or worn excessively. Keep all wiring cables hidden away from view.

Before making or breaking different connections in the audio system, disconnect the automobile battery cable. Do not try to make audio connections on the speakers or the amplifier while the amp is operating. Shut the amplifier down before attempting any type of repair of damaged or worn cables. Do not turn the amplifier on after hearing a popping or cracking noise in the speaker. Try to locate the shorted cable or wires or install a new set of wires. Make sure the amplifier is turned off when trying to repair the hookup jacks and speaker terminals.

Replace the exact size of power fuse identical to the one that is supplied with the amplifier. Using a fuse of a different size or rating can result in damage to the amplifier, which may not be covered by the manufacturer's warranty. Do not temporarily replace the fuse with tin foil wrapped around the fuse so that the amplifier will operate. Sometimes, if you forget to remove the tin foil, the amplifier can be destroyed internally. This might result in a much more expensive repair than replacing a low-priced fuse.

Typical Troubleshooting Symptoms

For a loss of sound within the auto tape player or radio, determine if the tape player or car radio (head unit) is defective. Check for a blown fuse with a dead amplifier system. Inspect for a disconnected cable between the battery source and amplifier. Check for a disconnected or bad contact of the input or output connections on the head unit and amplifier. Suspect a faulty connection between the metal frame of the vehicle and ground wire. Check for a faulty connection of the remote wire or cable.

The typical no-sound symptom in the amplifier might be a faulty connection of the remote wire. If the pilot light is out, suspect a blown fuse. Check for an open circuit or a cold solder joint around coils, choke transformers, and a DC-DC transformer. A defective MOSFET power transistor can cause loss of DC-DC converter output voltage. Suspect drive transistors, diodes, and a PWM IC for loss of or low DC-DC converter output voltages.

When the pilot light is on and there is no sound from the amplifier, suspect poorly soldered connections or open transformer windings of the high-level input circuits. Check for open or bad solder joints on the input jacks where the patch cords are attached. Check for defective transistors and IC components within the front-end audio circuits for loss of sound. Make sure that the correct DC voltage source is found on each transistor or IC from a zener diode regulated source.

For intermittent sound, suspect any transistor in the front-end or high-powered output circuits. Check for an open circuit or cold soldered joints on the emitter terminals of the final audio output transistors in both audio output channels. Just about any intermittent component within the amplifier can produce an intermittent symptom.

Servicing the High-Powered Amplifier

The high-powered amplifier might have a dead left channel, weak right channel, distorted sound, intermittent audio, hum in the around, and a noisy left channel. The dead channel is the easiest service problem to solve in any amplifier. A dead channel might be caused by a defective transistor or IC component, open coupling capacitors, shorted or leaky capacitors, loss of voltage applied, or really low voltage from the power supply circuits.

The dead channel can be located by taking critical voltage and resistance measurements on the transistors and IC components. Check the dead channel by injecting a 1-kHz sine or square wave signal into the input, and scope the various stages until the signal stops. Locate the dead channel by checking each audio stage with another external audio amp.

Nothing usually operates in a dead channel. Most dead channel problems originate in the low-voltage power supply and power output circuits of the audio amplifier. One channel in a 2- or 4-channel amplifier might be dead while the other channels are normal. You can compare the different dead circuits to the normal audio or stereo channels to help locate the defective stage or circuit. When a schematic is not available, use the voltages on the normal channels to check against those in the defective or dead audio circuits.

The dead audio amplifier symptoms might be caused by loss of sound in the speaker of either channel while no pilot lights are on. Check for a blown fuse or a bad battery lead connection. Test each silicon diode after replacing the line fuse. The dead symptom might result from an open regulator transistor or from IC and power resistors. A leaky power output transistor or IC can cause a dead symptom. The dead chassis might be caused by a bad switch or by the remote turn-on system. A dead speaker relay or speaker might cause a dead chassis symptom with loss of sound in the speakers.

Both speakers might be dead with a defective speaker relay, relay circuit, or protection component. Check to see if there are fuses protecting the speakers. The amplifier can be dead with a blown power fuse that was caused by a leaky and shorted output transistor. A dead symptom with arcing noises might result in a defective power-on push switch.

A loss of sound with the pilot lights on might be caused by a leaky zener diode, open regulator transistor, or a filter capacitor in the power supply circuits. Do not overlook a blown fuse caused by a leaky or shorted polarity diode on the battery source line. This diode is found in the power cable of the amplifier so that correct battery power polarity is applied to the amplifier (**Fig. 11-34**).

The dead symptom within the high-powered amplifier might result from a blown fuse and the DC-DC converter MOSFETs. Locate the MOSFETs in the power-supply circuit and check them for leakage or shorted conditions. Remove the power-output transistors from the DC-DC power output, which includes both positive and negative voltage to the amplifier circuits. Determine if the dead amplifier components are within the power supply or the amplifier circuits. Do not overlook a defective voltage protection circuit causing a dead chassis.

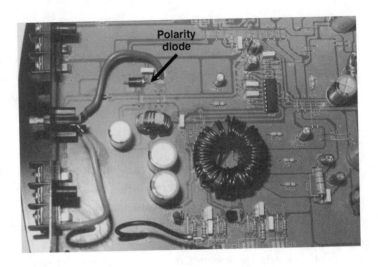

FIGURE 11-34. Suspect a leaky or shorted polarity diode when the main fuse to the amplifier blows.

Noisy Conditions

In the early days of the car radio, the spark plugs, distributor, and generator caused motor noise. By placing a noise resistor adapter in the center of the distributor, the high-voltage tension noise was eliminated. One would simply cut the high voltage center wire in two and screw in the eliminator. A 0.5-uF bypass capacitor placed at the input battery terminal of the radio helped to eliminate noise coming in the 6- or 12-volt line. Another 0.5-uF bypass capacitor was placed on the generator bell area and connected to the positive terminal of the generator to help prevent noise provided by the noisy brushes of the generator.

Today there are many different types of noise eliminators to help erase the noise picked up by the high-powered amplifier system. Popping, whistle, and cracking noises can be caused by different components within the automobile and high-powered amp system. A speaker wire grounding to the metal car body or a defective audio channel might cause a popping and cracking noise in the speakers. Install a new speaker wire after disconnecting it from the high-powered amp that produces the noise, instead of trying to find where the worn spot or shorting out of the speaker cable lies. Poor grounding of the audio patch cords can cause the high-pitched squealing noise.

Placing a 0.5-uF bypass capacitor across the brake or power switches can eliminate a popping sound in the speakers that occurs when the brakes are applied or when the lights are turned on. Noisy suppressors can be placed on noisy fans, motors, and light and heater switches. Place a filter or choke transformer in the battery line to eliminate noise on the battery cable. Heavy-duty 10-35 amp noise line suppressors can protect the stereo system from engine noise in the radio or amp power lead. Automotive noise filters for use with radios, equalizers, tape players, and CD players help to filter out the engine noise with a 10, 15, and 20 amp rating and can handle up to 150-350 watt amplifiers.

A radiated alternator can cause a high-pitched noise that varies with the automobile. Radiation noise can be picked up by a magnetic field of coiled wire or cable, voice coil winding in the speaker, and a crossover unit. Simply try to move the speaker or relocate the speaker to help eliminate the noise picked up by the speaker, or place a metal shield between the noisy source or components. Intermittent alternator whining noise can result by a passive crossover or coils of wire in the speaker hookup wiring. A radiated ground loop can cause noise in the amplifier. Connect all amplifier components to a center grounding box or distribution block.

Poor wiring or bad electrical grounding of amplifier components can cause a faint whistle when the level of the motor is running and when the tape player is operating. A hissing or background noise can be caused when the CD player is playing; it might actually be in the player, or caused by poor grounding. Transistor and IC components within the radio, CD, or cassette player can cause a hissing noise in the background.

Disconnect each component to determine which unit is causing the hissing noise, and then repair it or take it to the audio service shop. A crossover or speaker voice coil might pick up a ground loop (noise) and appear in the sound system. Remove each component, one at a time, until the noisy component is located. A good common ground connection block tied to all units can prevent radiated noises.

Fuse That Keeps Blowing

When the auto radio system shuts down with blown fuses, determine if the high-powered amp fuse is blown or if the head unit fuse is open. The left channel in the AM/FM/MPX head unit might become dead from an open AF transistor, driver, or audio output transistor. Sometimes you cannot hear even a slight hum in the speakers. Check the audio output transistor or IC component when the fuse keeps blowing. A dead head unit might be caused from a defective voltage regulator in the power supply.

When the main power-line fuse of the high-powered amp keeps blowing, a short or leaky component is likely down in the DC-DC converter power supply, a polarity diode, or electrolytic capacitors in the fuse input circuits. A blown fuse might be caused by a DC voltage placed on the speaker terminals or by too much volume applied to the speaker; the speaker protection circuits may have shut down the amplifier circuits and power supply. Sometimes you cannot see if a fuse is open, so test the continuity of the fuse with the low range of the ohmmeter. Some high-voltage types of fuses are enclosed and you cannot see the glass area like you can in the ordinary type of auto-radio fuse. After repairing the defective component, test-run the amplifier for several hours to see if the fuse opens.

Suspect a leaky or shorted component within the low-voltage power supply or DC-DC converter supply. Check for a shorted or leaky silicon diode. Check each diode in the DC-DC converter circuits for leaky or shorted conditions. Test each MOSFET within the DC-DC converter system. Check the continuity windings of the secondary and primary of the transformer for a dead DC-DC power supply.

Critical voltage measurements on each transistor or component should determine the defective component. Test each electrolytic capacitor within the bridged voltage output and the 14.4 volt input circuits. A leaky or shorted transistor within the remote system might cause the main fuse to open.

Suspect a shorted or leaky power output transistors within the high-wattage amplifier. Quickly test each transistor with the diode test of the DMM. You might find one power output transistor open, the other leaky, and the driver transistor defective within the directly driven circuits. Very often, a shorted high-power output transistor can cause the amplifier to blow a fuse or to shut down the amplifier. Check each bias resistor within the emitter circuits for possible burned or open conditions. These emitter resistors are very low in resistance and might vary from 0.15 to 1 ohm resistance.

Defective Transistors

Each transistor within the high-powered output amplifier can be checked in minutes with the diode test of the DMM. There are many different transistor testers on the market that can test each transistor within the high-powered amplifier. A component analyzer or MOSFET analyzer can check MOSFETs as well as LEDs, diodes, and diode networks. Defective transistors, diodes, and electrolytic capacitors cause the bulk of the troubles in the high-powered amplifier.

The diode test of the DMM can not only locate a leaky or shorted transistor, but also can find an open transistor and locate the base terminal. The common terminal of an NPN transistor is the base terminal. The base terminal is common to both the collector and emitter terminals with the diode test of the DMM. The base-to-collector terminal of a small signal transistor is around .695 ohms. The base-to-emitter terminal is around .715 ohms.

Keep the red or positive probe on the base terminal when testing an NPN transistor. Likewise, reverse the probe terminals with the negative (black) probe at the base terminal with PNP transistors. (**Fig. 11-35**). Compare these transistor measurements with the good channel, as the resistance should be quite similar.

Distorted Audio

Distorted music may be noted in one or more channels if the gain of the amp is rotated too high and produces clipping within the amplifier. The extreme distortion symptoms

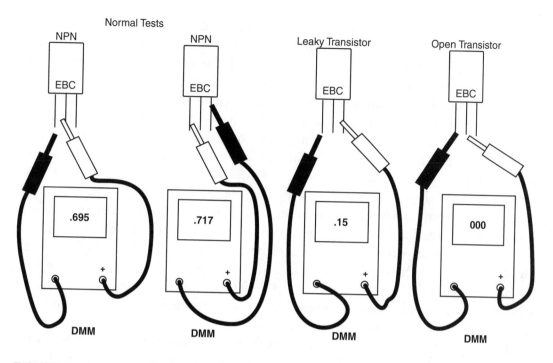

FIGURE 11-35. Checking low-signal transistors within the high-powered amp with the diode test of the DMM.

are generally located in the power output circuits of the high-powered amplifier or the head unit. Defective transistors, diodes, coupling capacitors, and bias resistors cause most of the distorted symptoms.

Sometimes a readjustment of the bias controls within the head unit (receiver) can help to solve the distorted left or right channel. When one channel is distorted, check the signal of each stage and compare it to the normal channel where the distortion might occur. Locate the distorted component with audio signal tests, using the scope or external amplifier as indicator.

Clip a 1-kHz audio signal from the audio signal generator or audio source to both stereo channels. In a 4-channel amp, clip the generator to only the distorted channel and one other channel with the speakers connected. Place a speaker load (10-50 watt resistor) on the other two channels within the four-channel amplifier. Scope the distorted signal with the oscilloscope until the distorted stage is located. Compare the same waveforms to those in the normal channel. After locating the distorted stage, take critical voltage and resistance measurements to locate the defective component.

If a scope is not handy, check each stage in the same manner with the external audio amplifier. Make sure the bench power supply is plugged into an isolation transformer.

Keep the volume of the external as low as possible. Proceed from the front toward the rear of the distorted channel until the distorted audio is found. Now compare the same distorted stage with one of the normal circuits. Keep lowering the volume on the external amp as you proceed through the amplifier circuits.

When two or more audio channels are distorted in the high-powered amplifier, look for a defective component that deals with both channels. A likely culprit is a dual-preamp IC. Sometimes the same coupling capacitor in the same circuit of both channels can become leaky and cause distortion in both channels. A dual-output IC within the receiver or head unit can cause distortion in both stereo channels. Eliminate the head unit from the high-powered amp by removing the different patch cords. When one patch cord appears to have distorted music, switch the two cords to determine if the distortion is in the head unit or amplifier and what channel the distortion is in.

Remember that extreme distortion in the audio channels of the high-powered amplifier can result from leaky transistors and IC components. Check for a change in resistance or a burned bias resistor for distortion. Usually the emitter resistor is damaged with a shorted output transistor. These bias resistors are tied to the emitter and bass terminals of the power output transistors.

Do not overlook an improper voltage source from the low-voltage power supply for causing distortion. Often both channels are distorted with improper voltage applied to the left and right channels. Check those small 1- to 2.2-uF electrolytic coupling capacitors for being open or having a loss in capacity. Check each electrolytic with the ESR meter.

TESTS

Critical Voltage Tests

Take a quick voltage measurement across the large filter capacitor to determine if the supply circuits are normal without a schematic. First take a voltage test across the input filter capacitor that forms the DC voltage supply at the choke coil or transformer. If the voltage at this point is 13.5 to 14.5 volts, proceed to the large filter capacitor within the voltage-output circuits of the DC-DC converter power supply. Really low voltage at the output diodes and electrolytic capacitor might indicate a leaky or open filter capacitor. Make sure each silicon diode is okay. Check each filter capacitor with the ESR meter (**Fig. 11-36**).

The voltage on the capacitor should be a few volts lower than the working voltage marked on the filter capacitor. Without a schematic to indicate the various voltages, check the measured voltage compared to those found on the body of the capacitor. If the working voltage rating on the largest filter capacitor is 50 volts, the actual voltage measured should be from 27 to 35 volts across the capacitor. Likewise, for input electrolytics and coupling capacitors that have a lower rating of 25 volts, the actual voltage measured should be around 13.5 to 20 volts across the capacitor terminals.

FIGURE 11-36. Checking filter capacitors within the high-powered amplifier with the ESR meter.

Although these capacitor voltage ratings might not be true in every case, you might find one with a lower working voltage. Remember, the actual voltage measured should be under the voltage marked on the electrolytic capacitor. By quickly checking the voltage across the various electrolytics, you can determine whether the voltage sources are normal or improper without a schematic.

A quick voltage test on a transistor will indicate if the transistor is an NPN or PNP type. The collector terminal of an NPN transistor always has the highest positive voltage. The positive voltage on a high-powered output NPN transistor might measure +35 volts, the base at 0.5 volts, and the emitter at zero volts. Likewise, the collector terminal of the directly driven transistors is the same voltage as the base terminal on the output transistor.

If the directly coupled transistor connected to the NPN output transistor is a PNP transistor, the emitter of the driver transistor is directly connected to the base of the power output transistor. The directly connected transistor emitter terminal would be 0.5 volts, a minus -1.1 volts on the base terminal, and -35 volts on the collector terminal of a PNP transistor (**Fig. 11-37**). A PNP transistor has the highest negative voltage or the lowest positive voltage on the collector terminal with a positive or negative voltage on the emitter, and a very low negative voltage on the base terminal.

The collector voltage is the highest positive voltage on the NPN transistor, while the emitter terminal is the most positive voltage on the PNP transistor circuit. Just compare the suspected transistor terminal voltage to the same transistor in the normal stereo channel without a schematic. When the voltages are incorrect on a NPN or PNP power output transistor, the emitter and collector voltages on the directly connected transistor will be found to be the same.

Try to locate the unmarked transistor terminals C, B, and E stamped on the PCB wiring when no schematic is available. Take transistor tests with the transistor or diode-junction test of the DMM when the terminals are not marked on the PCB, or when

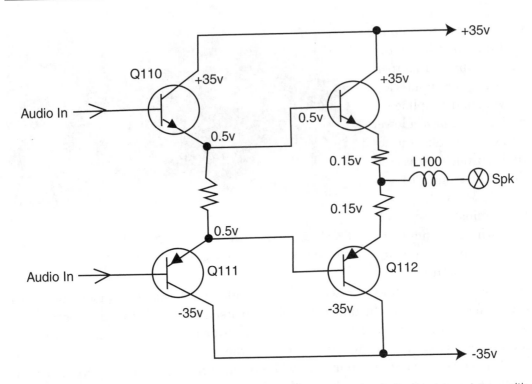

FIGURE 11-37. The different voltages found on the high-powered output transistors with a negative (-35) and positive (+35) volt source.

a schematic is not available. Locate the common base terminal with the same common measurements on the collector and emitter terminals. There will be only a small resistance between the collector and emitter terminals.

The highest voltage measurement, either positive (NPN) or negative (PNP), will indicate a collector terminal. Check from chassis to the emitter terminal with a low resistance measurement to common ground, since most audio emitter resistors are connected to common ground; the terminal remaining is the base terminal.

You can compare the voltages on the suspected transistor in one channel with the normal audio channel transistor. For instance, if the left channel is dead or weak, and really low voltages are found on the AF or driver transistor, then compare these voltages to the normal right stereo channels AF or driver transistors. Check the audio output transistors in the same manner.

A quick in-circuit voltage test on a suspected transistor can indicate if the transistor is open or leaky. A higher-than-normal voltage on the collector terminal and a loss of voltage on the emitter terminal might indicate that the transistor is open. Make sure the emitter transistor is normal and not open. A very low voltage on the collector, or on all terminals that are about the same voltage, indicates a leaky transistor.

Critical Resistance Measurements

A quick resistance junction test across any two transistor elements can indicate if the transistor is open or leaky. The diode-test of a DMM across fixed diodes can indicate if the diode is leaky, shorted, or open. A normal junction-diode test across the transistor or diode will show a normal reading in one direction, and with reversed probes, will show no measurement at all. The leaky diode or transistor will have a low resistance measurement in both directions or with the test leads reversed. The leaky transistor often has a low resistance measurement between the collector and emitter terminals.

Resistance tests within the audio circuits might have an improper measurement if a diode or coil is found in the base or collector terminal. If the resistance measurement is quite low from base to emitter, remove the base terminal from the PC wiring with solder wick and iron. Now take another measurement between the base and emitter terminals. Replace the leaky transistor if a low resistance measurement is found with the base terminal removed from the circuit.

A quick resistance measurement across a capacitor, diode, and resistor can indicate if the component is leaky or open. Remove one end of the diode or resistor from the circuit for a correct resistance measurement. Sometimes another component is shunted across the part to be measured and this arrangement results in a poor or improper measurement. When locating a leaky or shorted transistor in the audio circuits, always check the emitter bias resistors for correct resistance. Sometimes the resistor can become overheated or burned and have a change in resistance. A normal silicon diode should have a resistance measurement in only one direction.

A low resistance measurement between a fixed capacitor terminal indicates a leaky capacitor. The normal capacitor should have an infinite resistance measurement. A leaky capacitor will show a low measurement in both directions. The electrolytic capacitor should always charge up and down with a normal filter capacitor on the 20 K-ohm scale of the ohmmeter. Remove one end of the capacitor from the circuit for an accurate resistance measurement. Remove both terminals of a radial mounted electrolytic for charging tests. Check the suspected capacitor with the ESR meter, if handy.

Comparison Channel Tests

When the left channel keeps blowing the fuse or applies a DC voltage on the speaker voice coil, repair the chassis before another speaker is connected or it might also be damaged. Instead of a PM speaker, connect a 10-ohm 50-watt resistor across the left channel speaker terminals. Very often, one of the audio output transistors is shorted or leaky, and the directly coupled transistor might also become damaged at the unbalanced amp speaker terminal. Keep the amplifier gain as low as possible so as not to damage a connected speaker. Even after replacing the two output transistors, the left channel still produces a DC voltage on the left speaker load resistors.

To prevent damage to a new output transistor, take critical resistance measurements. Replace the bad output transistor and do not fire up the amplifier or apply a 14.4-volt supply to the amp chassis. Start at the speaker output transistors and take resistance measurements from each transistor terminal to common ground. Likewise, compare the same resistance measurements to the normal channel. Keep going backward in the audio circuits until the same measurement compares to the normal chassis. For instance, if you find the same or very close resistance measurement on the base terminal of the AF transistor in both the left and right channel, then take the resistance to ground measurement on the collector and emitter terminals.

When the resistance at this point is very different on the collector terminal of the AF transistor when compared to the same AF transistor in the right channel, suspect that a defective component is close at hand. Check the AF transistor and replace it, if in doubt. Take another resistance measurement. If the resistance is still way off from the normal channel, check all resistors and diodes within that stage. Do not overlook a bare wire as the jumper wire tying two circuits together. A poorly soldered connection on a wire jumper can cause many service problems. Check the jumper wire from one side of the foil to the other side or circuit.

Although resistance checks and measurements take a lot of service time, sometimes you can locate a very difficult component or connections with resistance tests. This is especially true if the fuse keeps blowing; the applied voltage can still damage the power output replacement. Always check the bad channel with comparison tests of the normal channel to help locate a defective component within either the left or right stereo amp channel.

REMOVING AND REPLACING

Replacing High-Powered Transistors

After several tests, if a transistor is found defective, remove it from the PCB with solder wick and iron. Be very careful not to raise pieces of the foil from the board chassis. The high-powered output transistor must be replaced in the same spot as the original one. If needed, cut the transistor terminals the same length as the old one. The replacement must be installed in the same spot, as it is mounted against the outside metal heat shield (**Fig. 11-38**).

Usually the power output transistors are mounted tight against the metal outside shield, with a metal plate and screw holding the transistor tight against a mica or cloth insulator to provide a transfer of heat to the metal shield. Place a dab of silicone grease on both sides of the insulator. Sometimes the old insulator will stick to the metal shield when the chassis is being removed from the metal chassis. The transistor terminals should not touch the metal mounting plate or the heat shield. Check each terminal of the transistor with the low-ohm scale of the DMM between metal shield and transistor terminals. Double-check the condition of each transistor after it has been mounted to ensure the amp will perform with the new replacement.

Locating Defective Audio ICs

FIGURE 11-38. Replace the high-wattage output transistor with the original part numbers.

The quickest method to locate a defective preamp or power-output IC is with audio signal in and out tests. Locate the input terminal by tracing the audio signal from the preamp input jack or gain control to the IC component when a schematic is not available. Often a resistor or coupling capacitor is found tied to the IC input terminal. The output terminal is generally capacity coupled or has a direct connection to the speaker terminal. Check for a 2.2- to a 4.7-uF electrolytic capacitor in the chassis close to the power IC, in order to locate the output terminal.

Another method of locating the various IC preamp or power-output terminals is in the semiconductor manual. Take the part number stamped on the top of the IC and look up the universal part number. Also, locate the preamp IC terminals on another schematic that has the same part number. Many of the manufacturers use the same component within other chassis. Now check the input, output, and supply terminals on the universal IC drawing. Besides finding the correct input and output terminals, you might find the required operating voltage on the supply terminal (Vcc).

Check the supply voltage applied to the IC component. This voltage is always the highest, and it is called the supply voltage. Suspect a leaky or shorted IC if the supply voltage is quite low. Feel the body of the IC and determine if it is running too warm. Remove the supply pin from the PC wiring with solder wick and iron; only the supply pin is required for this test. Now take another voltage measurement on the IC terminal pad and common ground.

Suspect a leaky IC when the supply rises higher than the normal supply voltage. If the voltage increases greatly, suspect a leaky IC. Double-check by taking a resistance measurement between the supply pin and common ground. Replace the leaky IC if the resistance is below 100 ohms.

Signaling in and out tests with the audio signal generator can locate defective transistors, ICs, and coupling capacitors. If one stereo channel is weak or dead, signal-trace the audio circuits with a 1000-Hz signal from a signal or function generator. Check the signal in and out on the suspected IC. Suspect a defective transistor when the audio signal is applied to the base and there is no signal at the collector terminal.

Check the suspected coupling capacitor by injecting audio into one side and then the other. When one side of the capacitor will not respond in the speaker, check the coupling capacitor for open conditions. Sometimes the intermittent capacitor might pop on with a loud signal tone in the speaker. Simply replace the intermittent capacitor. Check the electrolytic coupling capacitor with the ESR meter.

Removing And Replacing IC Components

After determining if the preamp IC components within the high-powered amplifier are open or leaky, remove the amplifier from the PCB with solder wick and iron. Hold the solder wick on a row of terminals and, as the terminals are heated, slide the solder wick down each row of terminals with the iron, to help to remove the excess solder. A couple of attempts should be made to remove the solder around the IC terminal. Then go to the outside of a row of terminals with the solder wick and iron to pick up any excess solder left on the row of terminals. Flick each terminal with a small screwdriver blade or a pocketknife to make sure the terminal is free. Apply the iron and solder wick on any terminal that is not free from the PC board wiring.

If the IC terminal 1 is not marked on the PCB with the "U" end or an indented dot on the body of the IC, then mark terminal 1 on the board area. Remove the defective IC and take a peek at the solder pads where the old IC was mounted. Run the solder wick and iron over the PC wiring to remove any sharp points and to clean up the soldering pads. Replace the new IC into the same holes and make sure that terminal 1 is at the right spot. Resolder each terminal with a low-wattage soldering iron to prevent too much heat and possible damage to the new replacement.

Double-check each terminal connection. Make sure that no two terminals are soldered together. Use a magnifying glass to check out each soldered terminal. If two different terminals look like they are soldered together, recheck with the low-ohm scale of the DMM and with a test between the two terminals. At the same time, take a low-resistance measurement between the soldered terminal and the PC wiring, to ensure that the soldered pad is not broken between the IC terminal and PC wiring. Sometimes these breaks cannot be seen, especially with excess rosin on the soldered terminal and PC wiring.

Parts Replacement

Exact replacement components are easy to replace because they are designed for a certain chassis and circuit. Always replace with the original replacement components whenever possible. Replace high-powered transistors and IC components with the exact replacements, and replace these high-powered output transistors found in the high-wattage amplifiers with the originals. Obtain the original part number from the manufacturer or replacement depot, and from electronic mail-order firms.

Low-signal and general-purpose transistors and ICs can be replaced with universal parts. Simply cross-reference the defective semiconductor to the universal semiconductor manual. Most resistors and capacitors can be replaced with regular replacements. You can replace silicon and zener diodes with parts found at any electronic distributor.

The power and special resistors should be obtained from the manufacturer. The dual-volume controls should be ordered from the parts depot. Special types of coils, chokes, and transformers should be ordered from the manufacturer. The special function and power switches should be replaced with original parts.

Filter capacitors can be placed in series or parallel when the original is not available. Most bypass and coupling capacitors can be obtained from local electronics distributors. Make sure the capacity and working voltages are the same or higher in replacing electrolytic capacitors. Test all new components before replacing them in the audio circuits. Here is a list of mail-order and electronics firms that might have the correct replacement:

All Electronic Corp
905 S. Vermont Avenue
Los Angeles, CA 90006
1-800-826-5422

Antique Electronic Supply
6221 S. Maple Avenue
Tempe, AZ 85283

B&D Enterprises
Main and Liberty Street
Russell, PA 16345
1-888-815-0506

Contact East Electronics
335 Willow Street
North Andover, MA 0185-5995
1-800-225-5334

Digi-Key Corporation
701 Brooks Avenue South
Thief River Falls, MN 56701-0677
1-800-Digi-Key

Electrotex Corp
2300 Richmond Avenue
Houston, TX 77098
1-723-526-3456

Fox Electronics International
23600 Aurora Road
Bedford Heights, OH 44146
1-800-445-7991

Hosfet Electronics, Inc
2700 Sunset Blvd
Steubenville, OH 43952-1158
1-800-524-6464

Jameco Electronic Components
1335 Shoreway Road
Belmont, CA 94002-4100
1-800-831-4242

JDR Microdevices
1850 South 10th Street
San Jose, CA 95112-4108

Kelvin Electronics
10 Hub Drive
Melville, New York 11747
1-800-756-1025

Meci
340 East First Street
Dayton, OH 45402
1-800-344-4465

Mouser Electronics
958 N. Main Street
Mansfield, TX 76063-4827
1-800-346-6873

Parts Express
340 E. First Street
Dayton, OH 45402-1257

Radio Shack (Local stores)
R&D Electronics
5363 Broadway Street
Cleveland, OH 44127

Tritronics Inc
1306 Continental Avenue
Abington, MD 21009-2334
1-800-638-3328

Union Electronics Distributors
311 East Corning Road
Beecher, IL 60401
1-800-648-6657

Wholesale Electronics
123 West First Avenue
Mitchell, SD 57301
1-605-996-2233

CHAPTER 12

Repairing the Antique Car Radio

THE ANTIQUE CAR RADIO

The early car radio operated within the AM band and consisted of tubes; it operated from a 6-volt source with a three- or four-prong vibrator. The 6-volt power supply provided 200 to 250 volts DC that were applied to the radio circuits. The IF frequency of the early car radio was 262 kHz later on, the standard 455 kHz frequency was adopted. The output transformer drove a four- or five-inch speaker with 3.2-ohm impedance.

The Different Tubes

The triode tube has three elements: a cathode, plate, and grid. A tetrode tube has a cathode, grid, screen grid, suppressor grid, and a plate. You might find a triode and tetrode or pentode tube in one envelope with different pin connections. Often the RF and IF tubes are tetrode or pentode types, with the AF amplifier as a triode and the pentode in the power output audio circuits. Besides these tube elements, a heater or filament lights up the cathode so that electrons will start to flow.

The rectifier tube might consist of a 6- or 12-volt tube with two diode plates and a cathode terminal. A 6- or 12-volt heater element is found inside the cathode element of each tube. A beam-powered pentode tube may be found as the audio output tube (**Fig. 12-1**).

When the heater element is lit, the tube heater lights up, and the cathode element begins to emit electrons (negative charge) toward the plate element. A wire grid is inserted between the cathode and plate elements to control the amount of electrons that are attracted to the positive plate element. The plate has a really high DC voltage applied, which helps to pull the electrons from the cathode, and thus the flow of electronics occurs inside the vacuum tube.

The second grid, known as the screen grid, is inserted to overcome the high-frequency response of the triode section, and it operates at a lower voltage level than the plate terminal. A third grid element, called the suppressor grid, is placed close to the plate terminal to leak off low-velocity elec-

FIGURE 12-1. Check all tubes of the car radio on a tube tester if one is handy.

trons that are bombarded against the plate terminal and that are attracted to the suppressor grid, which is tied to the cathode terminal inside the tube. A beam output pentode tube directs two electron beams from the cathode to the beam plates, which greatly increases the electron density.

After a few years, the 12-volt battery was used to supply the automobile, and the 12-volt vibrator power supply was designed to power the AM car radio. The early three-prong vibrator supplied from 200 to 265 volts to the audio power output stages. The standard 12-volt tube car radio operated with a 12BA6 as RF amp, 12BE6 converter, 12BA6 IF amp, 12AV6 as Detector-AVC-Amp, and a 12AQ5 output tube (**Fig. 12-2**). The broadcast band frequency was 540-1605 kHz and the IF frequency was 455 kHz.

Later on, a set of 12-volt tubes was designed to operate the AM car radio directly from the 12-volt battery. The large and cumbersome vibrator power supply was eliminated from the auto radio, replacing the physical size of the large firewall-mounted car radio. A miniature set of 12-volt tubes found in the 12-volt system consisted of a 12BL6, 12AD6, 12BL6, 12AE6A, and a 12AB5 or 12AQ5 output tube. With the 12-volt power supply, large filter capacitors (1000-500 uF) with a choke and resistor filter network were required to provide sufficient hum elimination from the power-supply circuits.

After a few years of operation and with the design of transistors, the hybrid or transistor output stage was born. The solid-state device was first found in the transistor driver and audio output circuits. Most of these output transistors were PNP types with a higher voltage found on the emitter and base terminals, and with very low voltage on the collector terminals of the output transistors.

An interstate transformer was found between the AF or driver transistor and the base terminal of the output transistor. The output transistor collector terminal connected to a tapped autotransformer to common ground. The 3.2-ohm voice coil of the

PM speaker was connected to the tap on the output transformer and ground (**Fig. 12-3**). You will find many different tubes in the early car radio lineup and transistors within the hybrid car receiver.

Early Integrated or Hybrid Auto Radios

The early antique hybrid auto radio has several different tubes in the RF, converter, IF, and detector circuits with a driver and power output transistor within the audio output circuits. The miniature 12-volt tube and integrated transistor output circuits might include a 12AD6, 12AF6, and a 12AE6A tube with a separate driver and power output transistor. The front-end and IF circuits are performed with tubes, while the audio circuits have transistors (**Chart 12-1**).

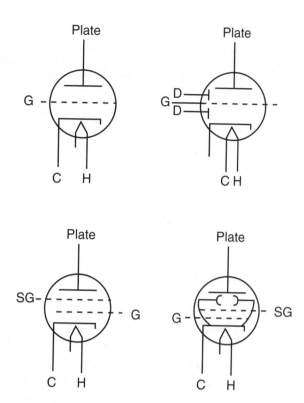

FIGURE 12-2. The different tube symbols found in the car radio.

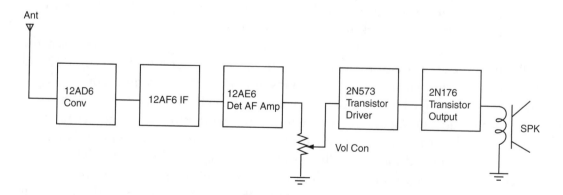

FIGURE 12-3. Block diagram of the AM car radio found in the hybrid car radio.

Use	Tubes							
RF	12AF6	12BL6	12CX6	12DZ6	12FX8			
Converter	12AD6	12FX8						
IF	12AF6	12BL6	12EK6					
Det-AVC-Amp	12AE6A	12DK7	12DL8	12DS7	12DU7	12DU7	12FN6	12J8
Output	2N155	2N176	2N401	2N573	2N1007	DS501	DS503	
Push-Pull Output	AR6	AR9	AR13	2N176	2N1033	2N1227	DS501	
	DS503	8L201B						

CHART 12-1. The integrated 12-volt tube line-up with a transistor AF and output tubes.

The 12-volt tube circuits found in the early all-tube car radio are the same circuits that are found in the hybrid chassis, except for the audio output circuits. The driver and output transistors are solid-state devices that amplify audio signal. Here, the hybrid car radio includes both tubes and transistors. The power output transistor might work in a single-ended transformer output circuit or in push-pull operation. Interstage transformer coupling is found between the output transistor and the PM speaker.

In the early transistor driver and output circuits, the driver transistor (2N573) is coupled directly to the AF amp tube with the collector terminal through an interstage transformer. Notice that both the driver and output transistors are PNP types; the collector terminal is always less positive than the negative base and emitter terminals. The driver transistor has a negative 10.5 volts at the base terminal, and 10.7 volts at the emitter terminal, providing a forward bias of 0.2 volts. Resistor R10 (56 ohm) provides a forward bias between the base and emitter terminals, while R11 and C10 provide a decoupling network from the 12.5-volt battery source.

The power output transistor has a transformer interstage coupling between the collector of the driver transistor and the base of the output transistor (2N176). R12 and R14 provide forward bias on the output transistor. The collector terminal of the output transistor has a tapped autotransformer winding to the PM speaker (Fig. 12-4).

In some early auto radio output circuits, a fusible resistor in the emitter terminal was made with a bias resistor type of adjustment. In a Karmann Ghia model O8G, bias resistor R2 is adjusted for 660 milliamperes, with a 6.3-volt DC source. The output meter is placed in series with the collector terminal and the output transformer for a correct setting of the bias control. Notice that this auto radio operates off of a 6-volt car battery source (Fig. 12-5).

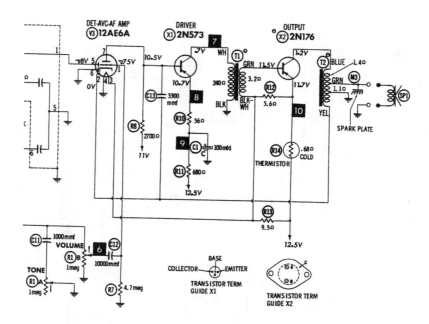

FIGURE 12-4. The 2N176 output transistor has a tapped transformer in the collector circuit that feeds the PM speaker.

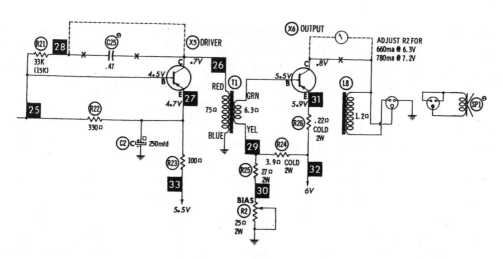

FIGURE 12-5. Adjust R2 for 600 milliamperes with a 6-volt source to properly set the bias resistor (R2) in a Karmann Ghia auto radio.

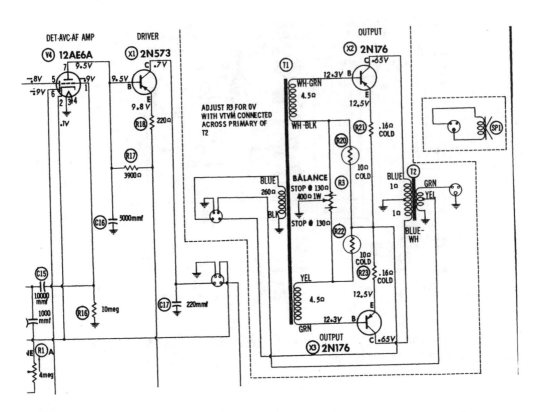

FIGURE 12-6. Two output transistors (2N176) work in a push-pull operation into a center-tapped output transformer (T2).

The driver transistor (2N573) collector terminal is tied to the interstage transformer that is coupled to two power output transistors (2N176) in a push-pull operating circuit. A separate secondary winding is connected to each base terminal of X2 and X3. The collector terminal of both output transistors is tied to each primary winding end of the output transformer with the center tap at common ground. R21 and R23 (0.16 ohm) resistors provide forward bias with a +12.6-volt source feeding the base and emitter circuits from a 12.6-volt battery source. The secondary winding of T2 is connected to a 6X9-inch 3-4-ohm PM speaker (**Fig. 12-6**).

In the Allstate 12-volt car radio model 6268, two push-pull output transistors (2N176) are found in the output circuits. Notice that the driver and both output transistors are a PNP type. The interstage transformer (T1) is coupled directly to the 12-volt AF amp tube (12AE6) and the secondary winding to the base of the driver transistor (2N176). Another interstage transformer (T2) couples the output of the driver transistor to both base terminals of the power output transistors. Bias resistors R24, R25, R26, and R27 connect to the 12.6-volt battery source. Both primary windings of the output transformer (T3) are tied to each collector terminal and center tapped at ground potential. The transformer secondary winding drives a 6X9- or 7-inch PM speaker (**Fig. 12-7**).

The most common trouble found in the transistor output circuits is a shorted or leaky output transistor with a burned or open bias resistor. Always remove one end of the bias resistor and check for correct resistance with the DMM. These bias resistors are very low in resistance, and can burn open or have a change in resistance. Sometimes if the car radio is left on too long, the power output transformer can become burned with shorted windings. Double-check the resistance of the primary winding of the output transformer after replacing a shorted or leaky power output transistor and bias resistor.

After operating for a few hours, the output transistor might become leaky, causing intermittent and distorted music. Spray the output transistor with a can of coolant spray and notice if the sound returns to normal; replace the output transistor if distortion is still present. Loss of audio and loss of stations with a hum in the speaker can be caused by an open AF amp or driver transistor. Replace the output transistor with a dead radio and a loud hum in the speaker. A leaky AF or driver transistor can cause a dead speaker.

If the car battery terminals are wired up backward or if the battery is charged up backward, the power output transistor will blow when the car radio is turned on. When the radio fades out, suspect an open bias resistor control in the emitter terminal of the power output transistor. Replace the power output transistor when the radio is turned on and has a cracking noise. Check the open audio output transistor and make sure that the heat sink is grounded when there is a popping noise in the speaker, and then nothing.

A red-hot power output transistor can cause a loud hum in the speaker. Replace the leaky output transistor for distortion in the speaker. Often the power output transistor becomes leaky between emitter and collector terminals. Suspect a shorted power output transistor when the transistor runs red-hot with a dead radio, and when the car radio pulls heavy current.

Replace both audio push-pull output transistors with a weak and distorted symptom. Sometimes the leaky or open driver transistor must be replaced when the output transistor is found leaky or shorted. Suspect the audio power output transistor when, after operating for a few minutes, the popping noise begins and the car radio goes dead. Replace the audio output transistor when there is static noise in the speaker. A noisy preamp or driver transistor can cause a frying noise in the speaker.

The Early AM and Multiband Car Receivers

Many different American manufacturers built the early AM/FM and multiband car radios, besides foreign manufacturers, like Blaupunkt and Volkswagen. In the early multiband radio, the car radio operated from a 6-volt battery and used the same tube

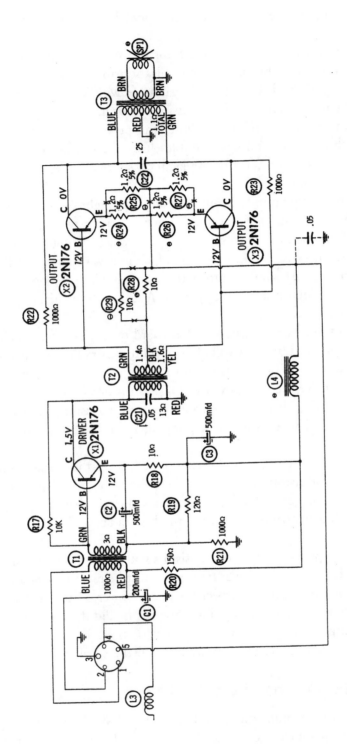

FIGURE 12-7. Push-pull audio output transistors (2N176) are found in an Allstate car radio audio output circuit

lineup as the AM band radio. A multiple-switch contact with a different coil in the RF and oscillator sections provided multiband reception with permeability tuning on the AM band. The 6-volt power supply contained a four-prong vibrator and a 6X4 rectifier tube (**Fig. 12-8**).

In the early Blaupunkt Hamburg model, the power supply operated from either a 6- or 12-volt car battery. The AM broadcast tuning range was from 540-1610 kHz, and the long wave band from 150-300 kHz. A sliding type of switch placed the different bands into operation in the RF and converter circuits. The RF amp tube used an EF89, and the mixer stage used an ECH81 tube. The IF frequency was 460 kHz. The IF amp

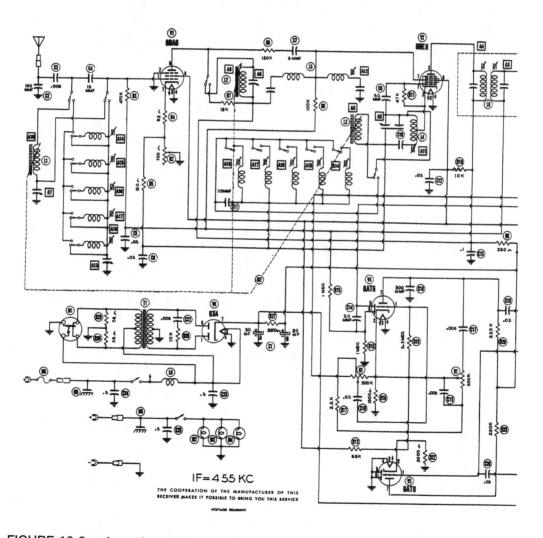

FIGURE 12-8. An early multi-band radio was found in a Volkswagen car radio operating on a 6-volt battery.

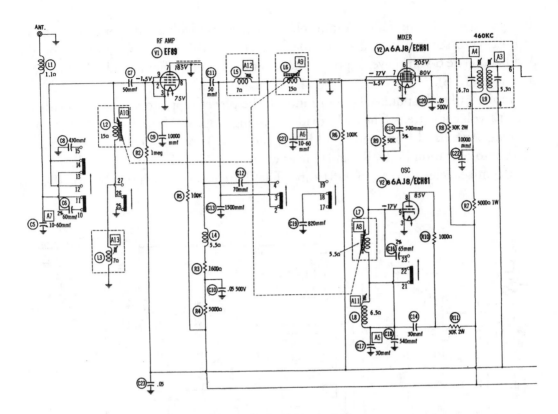

FIGURE 12-9. The Blaupunkt German car radio can be operated from a 6- or 12-volt battery with a sliding switch in the standard and long-wave band car radio.

tube used an EF89, Detector-AVC-AF amp (EBF89,) and a EL84 as the output tube. The heater circuits were wired in parallel from the 6-volt battery source (**Fig. 12-9**).

In the early AM/FM Motorola car radios, T1-400 or DS-41 transistors were switched into either the AM or FM band with a sliding switch assembly to the RF and antenna-input circuits. A T1-401 served as mixer and a T1-387 or DS-41 was the FM oscillator transistor, with the first, second, and third IF amplifier used a 4857 or a DS-66 transistor. The AM RF amp and converter stage used an 829B transistor. Another 829C transistor operated the AM IF amp stages at 262 kHz, while the FM IF frequency was 10.7 MHz. Permeability tuning was found in both front-end circuits (**Fig. 12-10**).

Two fixed diodes formed the ratio-detector circuits that provided analog or audio signal to the audio circuits. The same AM/FM slide switch would switch the different circuits into the volume control and the first AF amplifier (54D) transistor. Another AF amp transistor (54C) was directly coupled to the audio output transistor (SP1556-2). The single-ended output transistor collector terminal operated in a tapped autotransformer. The output transformer was tapped to match a 5X7 8-ohm PM speaker.

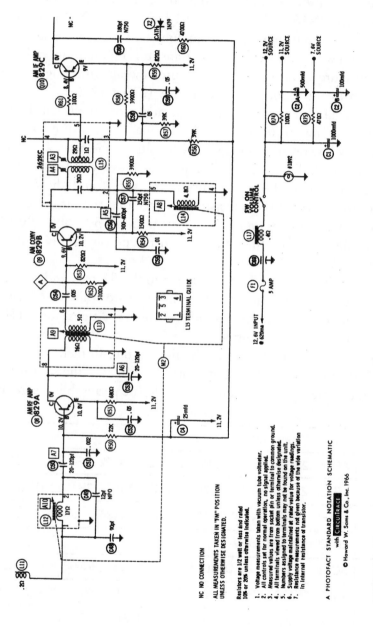

FIGURE 12-10. Permeability tuning is found in the early AM/FM Motorola car radio.

Early All-Transistor Auto Radios

The early all-transistor radios consisted of an RF Amp, converter, IF Amp, driver, and output transistors. An All-American UP-220 model consisted of a 2N1637 as RF Amp, 2N1639 as Converter, 2N1638 IF Amp, 2N591 Driver, and 2N301 as Power Output transistor. The power output transistor was mounted on the outside, at the rear of the car radio chassis. You will find many different transistors found in the various car radios of different manufacturers. Car radios manufactured by each firm might have the same transistor lineup in the different years and models (**Chart 12-2**).

One of the first all-transistor car radios operated from the 6-volt battery. These 6-volt operating transistors were all of the PNP type. Most of the base and emitter terminals had a very low positive voltage of 4.6 volts on the base, and had 4.8 volts on the emitter terminal. The collector voltage was lower yet, from 0 to 1.4 volts. A 6-volt source was fed to the audio output transistor, and the 5.2-volt source provided voltage to the rest of the transistors. Notice that the car radio pulls around 750 milliamperes with a 7.5-amp fuse protection. A 300-500-100-uF 16-volt electrolytics provided a proper filter network in the power supply and decoupling circuits (**Fig. 12-11**).

The antenna-loading coil couples the antenna to the RF coil, and C10 couples the antenna circuit to the base terminal of a RF Amp transistor (2N1637). The amplified RF signal is capacity-coupled to the base of the converter transistor (2N1639). The oscillator coil is found in the emitter circuit of the converter transistor with the secondary winding in the collector leg of the first IF transformer.

The 5.2-volt source feeds the base and emitter terminals of the converter transistor. All three input stages are tuned with a three-gang ferrite permeability core or rod. The

Use	Types of Transistors
RF Amp	AF127 2N637 2N1788 2N27 2SA72 2SA322 DS25 DS51 M76
Converter	AF127 2N27 2N462 2N1789 2N2089 2SA72 2SA322 DS25 DS52 M77
IF Amp	2N460 2N641 2N642 2N1638 2N1790 2N2672 2SA72 DS53
Det-AVC-Amp	2N139 1N295 DS-27
Driver	AC126 DS66 DS26 DS46 2SB186 2SB25A 2N408 2N466 M54 54D
Output	2N155 2N176 2N235 2B301 2SB54 DS501 DS520 SP8801 8P404A
Push-Pull	AD148 DS520 2N176 2SB254 SP404 SP880 SP1556 A-4002

CHART 12-2. The type of transistors found in the various stages of the all-transistor car radios.

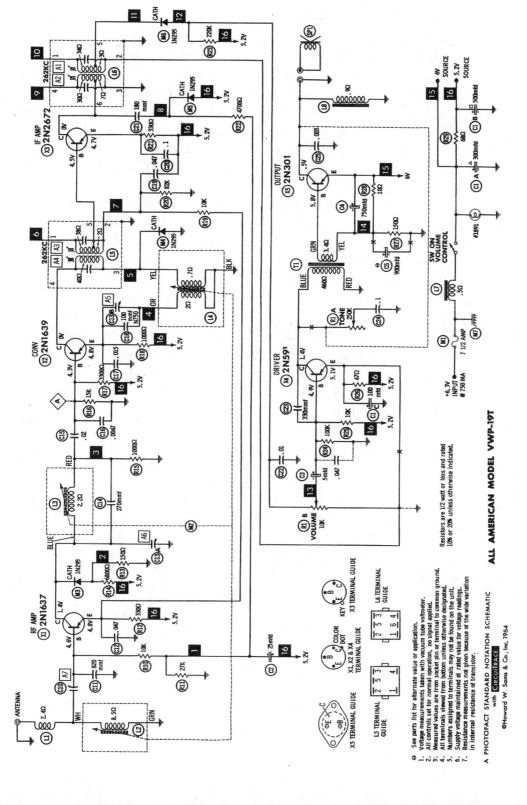

FIGURE 12-11. A complete schematic of an All-American transistor model VWP-197 car radio.

IF Amp transistor (2N2672) has the base terminal connected directly to the first IF transformer, and the intermediate frequency is amplified by the IF transistor.

A fixed diode (1N295) detector detects the audio signal from the second IF transformer, and appears across the 10K-ohm volume control. A 5-uF electrolytic couples the audio from the volume control to the base of driver transistor (2N591). Here, the audio signal is amplified and drives the primary winding of the interstage transformer (T1). The secondary winding connects directly to the base of the output transistor (2N301). The plate load coil L8 has a 15 mH impedance of 0.9 ohms to common ground. A five-inch 8-ohm speaker plugs into the secondary winding off of L8.

The 12-volt all-transistor car radio is operated from a 12.6-volt car battery with a RF transistor (DS-51), Converter (DS-52), IF Amp (DS-53), first and second AF amp with a DS-46, and the output transistor (DS-501). Here in the Opel model 980886, two separate AF amp transistors are connected in series to drive the output transistor. Notice that both the AF amps are NPN-type transistors.

The output transistor (DS-501) has a fusible resistor in the emitter circuit, and usually goes open when the output transistor becomes leaky or shorted between collector and emitter terminals. The output transistor is mounted on a heavy large heat sink bolted to the rear of the metal chassis. The metal body of the output transistor is the collector terminal with an insulated guide pin. Insulation with silicone grease applied between the insulator and metal transistor helps dissipate heat from the transistor (**Fig. 12-12**).

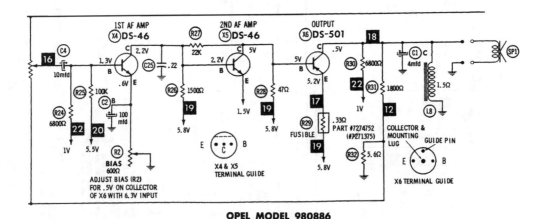

OPEL MODEL 980886

FIGURE 12-12. Replace the DS-501 output transistor with a guide pin and place the insulator with silicone grease on each side of the mica insulation.

THE POWER SUPPLY

6-Volt Vibrator Power Supply

The 6-volt vibrator power supply operates from the 6-volt car battery with a vibrator or interrupter working in the primary circuits of the power transformer. The stepped-up secondary winding produced a 360- to 400-volt AC source fed to the 6-volt rectifier tube (6X4). The early 6-volt transformer was center-tapped at 400 volts AC. Most of the 6-volt vibrators had a four-prong terminal and was inserted into a four-prong socket. You might find a 6-volt five-prong vibrator with the center terminal grounded. (**Fig. 12-13**). Notice that two 100- to 150-ohm resistors are found across the primary winding of the power transformer.

The high-voltage buildup in the secondary winding is applied to a couple of diode plates within the rectifier tube. To prevent a constant hash noise within the audio, a bypass capacitor (0.0068 uF) eliminates the noisy power supply. Notice that the 400-volt AC transformer is center-tapped to ground. You might find many different sizes of hash-type capacitors in the different 6-volt vibrator power supplies from 0.0035 uF, 0.006 uF, 0.0068 uF, and 0.008 uF at 1600 volts DC. Do not replace the hash noise capacitor with a smaller working voltage.

The DC-rectified voltage is found at the cathode terminal (7) and fed to a resistor-capacitor (RC) filter network. Two 10-uF electrolytics provide filtering action with a 3.3K-ohm resistor. The 190-volt DC filtered source supplies voltage to the output tube and a 155-volt DC source is fed to the radio circuits. You may find the vibrator, DC power supply, and audio output tubes with circuit components in a separate shielded unit (**Fig. 12-14**).

A 7.5-amp fuse protects the vibrator power supply and radio circuits within the 12-volt vibrator power circuits. L8 is a hash choke filter to help eliminate noise, while L6 and L7 are filament choke coils. The 12-volt vibrator is fed from 12 volts at pin 6 and the primary center tap of the power transformer (T1). Notice that two 150-ohm resistors are connected across the primary winding. The 400-volt AC center-tapped secondary winding is fed to the diode plates of a 12X4 rectifier tube.

A 0.01-uF 1600-volt capacitor and a 10K-ohm resistor prevent filter hash noise from the audio circuits. The 180-volt DC source is fed from the cathode terminal of the 12X4 tube to the audio output tube (12AQ5). Again, two 10-uF 250-volt electrolytic capacitors provide filter action with a 3.3K-ohm 1-watt resistor. A +135 volt source is fed to the screen grid of the output tube and to the other radio circuits (**Fig. 12-15**).

You might find a little different vibrator power supply circuit in the 6- and 12-volt vibrator power supplies. The 6-volt vibrator supply might be contained in a separate

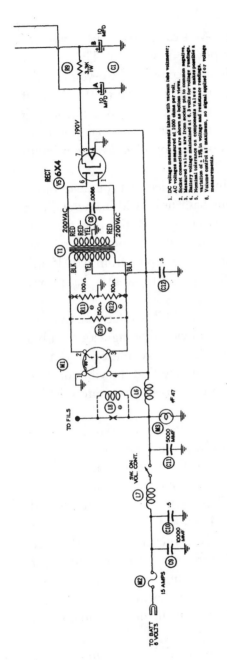

FIGURE 12-13. The 6-volt vibrator and power-supply circuits with a 6X4 rectifier tube and RC filter network.

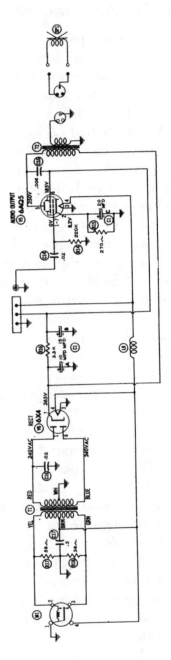

FIGURE 12-14. Here the vibrator, rectifier tube (6X4), and output tube (6AQ5) are found in a separate shielded unit.

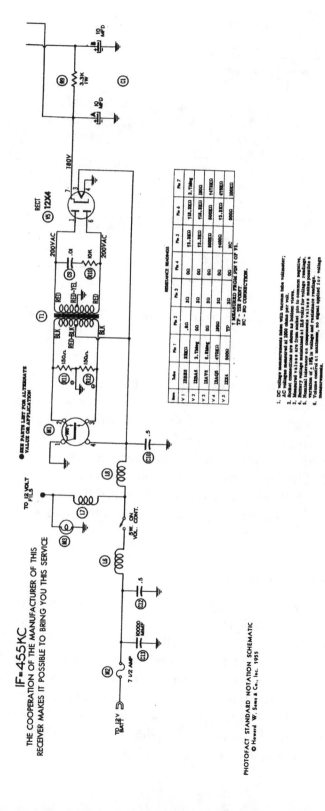

FIGURE 12-15. The 12-volt vibrator power supply with a 12X4 rectifier tube.

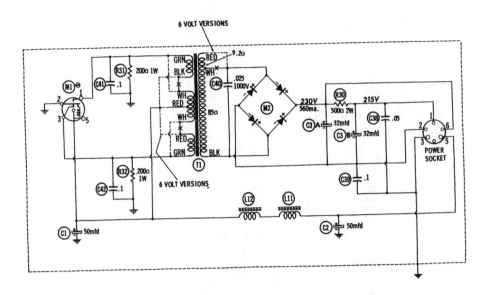

FIGURE 12-16. A bridge diode rectifier circuit is found in a Blaupunkt Hamburg car radio with a vibrator power supply.

metal unit. The primary winding has three separate windings instead of a center-tapped winding found in American car radios. The secondary transformer voltage is 460 volts with no center tapped to ground. Instead of a 6-volt rectifier tube, a full wave bridge diode circuit is formed to produce 230 volts DC.

RC filter action with two 32-uF electrolytic capacitors and a 500-ohm 2-watt resistor provides the dc filtering action to a six-prong power socket in the Blaupunkt Hamburg car radio. The 230-volt source is fed to the 6BQ5/EL84 power output tube, and a 215-volt source is fed to the screen grid of the output tube and other tube circuits (**Fig. 12-16**).

Hybrid 12-Volt Power Supply

The integrated 12-volt power supply might consist of only a few components such as the 7.5-amp fuse, spark plate M4, pilot light, on/off switch, filter choke (6), and "A" lead. The RC network consists of three filter capacitors (700-200-100-uF 16-volt) to filter out the hum and motorboating noises (**Fig. 12-17**). The spark plate consists of a piece of metal plate with insulation between plate and metal cabinet to form a spark type of capacitor.

The power on/off switch is found before the pilot light and filter choke (L6). The power on/off switch might be a SPDP type of switch that supplies 12 volts to the radio circuits, while the other switch turns on the pilot light. If a SPDT switch is not available, connect the pilot light wire to the same switch terminal as the radio. Excessive

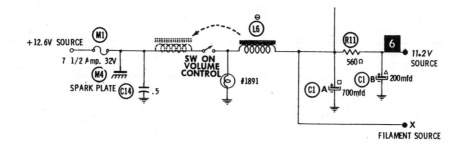

FIGURE 12-7. Push-pull audio output transistors (2N176) are found in an Allstate car radio audio output circuits.

hum and input line noise is filtered out with a 0.5-uF bypass capacitor and filter choke. A selenium type of rectifier might be found in some car radios with the cathode to the positive 12-volt side and the anode terminal to ground.

Check the filter choke when hum is found in the speaker after all filter capacitors have been replaced. A scorched or burned choke coil or transformer can be replaced with filter chokes from 5- to 15-Henries at several antique electronics mail-order firms. If the choke coil is not available, simply wind on the same amount of turns of wire found on the original choke winding. Count the windings as you remove the old burned wiring. You can use a wire gauge to determine the same size of enameled wire. These choke coils are often around 500 uH or less and can be measured with an inductance meter found on some DMMs.

For instance, in a Karmann Ghia (Volkswagen) model OBG, the line choke is around 500 uh with a 0.2-ohm resistance that will pass 2 amps of current. If the inductance meter is handy, wind on the same size of wire until the DMM ohmmeter reads 0.2 ohms resistance. In a Mopar auto radio 110 model, the filter line choke is 0.0048 Hy with a DC resistance of 0.2 ohms, and passes a 1-amp current rating (**Chart 12-3**).

In a 12-volt Edsel model O4BE hybrid car radio, the 12.6-volt DC supply source feeds to the emitter terminal of a power output transistor (2N1227). A 12.6-volt source feeds to the emitter terminal of the driver transistor (2N1287), while the 10.5-volt source feeds voltage to the other tube plate terminals. Notice that the power switch is an SPDT switch on the rear of the volume control to provide 12.6 volts to the radio circuits and to the separate pilot light (57).

L7 is a line choke coil and M2 is a spark plate of around 200 pF to eliminate noise on the 12-volt line. The RC filter network consists of a 220-ohm resistor (R22) and a bank of electrolytics (450-100-50-uF 16 volts). Check all three electrolytics with the ESR meter if hum appears in the sound (**Fig. 12-18**).

Car and Model	Current Measured	DC Resistance	Inductance	Fuse
American Motors Model 8990543	1A	0.3 Ohms	0.004 Hy	5A-32v
Moper Model 101-104 & 110	1A	0.2 Ohms	0.0048 Hy	7.5A-32v
Automatic Model NY-3100	1.2A	0.9 Ohms	6 Mhz	7.5A
Motorola Model RX	1.1A	0.6 Ohms	0.001 Hy	9A-32v
Karmann Gia Model OBG	1.1A	0.2 Ohms	500 uF	2A-32v

CHART 12-3. A filter choke current, DC resistance, inductance, and protection fuse in the different car radios.

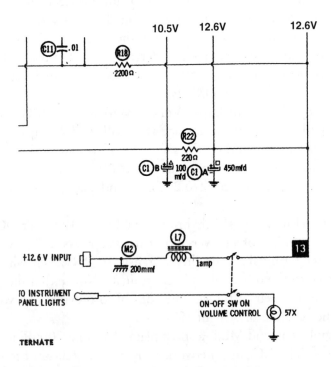

FIGURE 12-18. In the Edsel 04BE 12-volt car radio, the filter network consisted of an RC network of 100-450-50 uF 16-volt electrolytics.

All-Transistor Power Supply

Several output voltage sources were found in the 6-volt all-transistor Opel model 980886 power supply. The 1-volt source is tied to the base terminal of the first AF Amp transistor (DS-46), while the 1.5-volt source was connected to the emitter terminal of the second AF amp transistor (DS-46). A 5.5-volt source was fed to the base resistor of the first AF amp, emitter terminal of RF Amp (DS-51), emitter and base terminals of the converter transistor (DS-52), base, and emitter circuit of the first IF Amp (DS-53), the anode terminal of the AVC diode (DS-27), and the AF detector diode M5 (**Fig. 12-19**).

In the early 12-volt all-transistor car radio, the radio is protected by a 7.5-amp fuse with the on/off switch before the pilot light (1891) and choke coil (L7). The

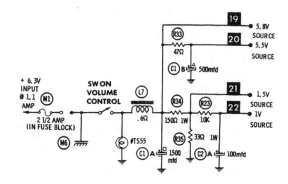

FIGURE 12-19. The early all transistor 6-volt power supply consisted of several different voltages with 1500-500-1000 uF 16-volt filter electrolytics and voltage-dropping resistors.

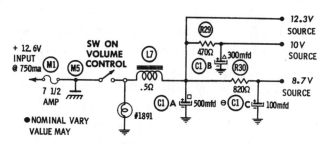

FIGURE 12-20. An early all transistor radio has a 500-300-100 uF at 16 volts in an RC filter network.

filter choke has a DC resistance of 0.5 ohms and an inductance of 10 mH. An 8.7-volt source feeds the transistor driver stage, while the 10-volt source is fed to the RF Amp, converter, IF, and detector circuits. Output transistor (8P404) is connected to the 12.3-volt source before the filter RC network. Filter capacitor C1 (500-300-100 uF) provides filtering action with an 820-ohm voltage-dropping resistor. The 10-volt source has a 300-uF decoupling electrolytic after a 470 ohm resistor (**Fig. 12-20**).

POWER-SUPPLY PROBLEMS

Vibrator Power-Supply Problems

A bad vibrator or shorted hash capacitor within the high-voltage secondary of the power transformer might cause a blown fuse. Replace the high-voltage capacitor first. A weak rectifier tube can produce a really low or dead DC voltage. A shorted rectifier tube can blow the main fuse. Replace the vibrator when you cannot hear it vibrate, or if it sputters as it tries to operate. Touch the outside metal case of the vibrator to notice if it is vibrating.

Dried-up filter capacitors can cause excessive hum in the speaker. A bad filter capacitor can cause a motorboating sound. Replace the dual electrolytic capacitors with two separate capacitors. Check all filter capacitors with the ESR meter.

Although the four-prong 6- and 12-volt vibrators are difficult to locate, try to repair them if possible. Check the continuity of the field coil winding with the low-ohm scale of the DMM. If the coil is open, remove the outside metal cover and take a quick peek at the coil pin terminals to see if the wire lead might be broken off of the vibrating coil. Use a fingernail file and sandpaper to clean up the vibrating point contacts. Sometimes the reed blade or vibrating points can be bent with a pair of long-nose pliers to make a better contact. Take it easy, as this is a very delicate job; but it might place that antique car radio back into service once again.

After checking the vibrator, test or substitute another rectifier tube. When a glass OZ4 rectifier tube is found in the car radio, do not think that it is bad when blue lights appear inside the glass area; this is a gaseous type of rectifier tube. Measure the voltage on the filter capacitors and compare with the schematic. Old dried-up filter capacitors can cause low voltage with hum in the sound. Suspect a bad or open on/off switch for loss of sound or for tubes not lighting up.

To rebuild the low-voltage power supply, replace all electrolytic filters and hash capacitors. Check the hash resistors for correct resistance by removing one end from the circuit and checking the low ohmmeter range of the DMM. Test the rectifier tube for an emission check on the tube tester. Recheck all fixed resistors in the filter circuits for burns or a change in value. If the vibrator sometimes vibrates and other times is dead, replace it or try to repair it. Check all hash coils and chokes for burned windings. Suspect a shorted choke transformer when hum can still be heard after replacing the filter capacitors (**Fig. 12-21**).

Of course, the vibrator transformers are no longer available, but they may be substituted from another old car radio chassis. Check the primary and center-tapped winding with the ohmmeter. Each winding from the center tap to the outside winding should be close in resistance. Likewise, measure the secondary in the same method. Sometimes the transformer leads break off where the insulated connecting wire is soldered to the enameled coated winding. A poor soldered joint at this connection might appear open. Simply remove the outside wrapping to get at these connections, and resolder the bad connection. Scrape the enameled wire with a pocketknife, and place a dab of solder paste on each wire to make a better connection.

If the transformer winding is open or the winding is badly charred, try to exchange it with another transformer from an old car radio with a 6- or 12-volt input winding.

If the old transformer high-voltage winding is a lot higher than the original, experiment by inserting another size of resistor in series with the RC network, to help lower the DC output voltage. Make sure the old transformer will fit into the required space before connecting all transformer wires.

Burned filament, heater, or hash line chokes can be repaired by winding the same size of wire over a wooden dowel of the same diameter as the original choke coil

FIGURE 12-21. The 6-volt car radio with several different vibrator components found in the car radio.

(**Chart 12-4**). Often these choke coils are wound with larger coil wire than magnet coil wire found at the electronics mail-order supply stores. These filament chokes are air wound with no metal laminations. Choose an 18 or 16 solid hookup wire to make the new choke coil. Wind on the same amount of turns as the old choke coil has. Apply clear glue, silicone, or epoxy cement over the two sides of the coil to hold the wire in a form of a line choke coil.

12-Volt Power Supply Problems

A quick method to determine if the filter capacitors and the 12-volt power supply are normal, take a voltage measurement across the large filter capacitors and check it against the schematic. If the voltage is really low, suspect an increase in the RC network and poor filtering action. Old dried-up filter capacitors can cause excessive hum in the

Make Model	Inductance	Resistance
Automatic	1.3 Microhenries	0 Ohms
Allstate 5033	9 Microhenries	0.1 Ohms
Ford 66MF	123 Microhenries	0.2 Ohms
Hudson 236486	11 Microhenries	0.2 Ohms
Loncoln	11 Microhenries	0.4 Ohms

CHART 12-4. The inductance and resistance of the filament line choke coils found in the early auto radio.

power supply. A noisy motorboating sound in the speakers can be caused when the top is blown off of an electrolytic, with black or white substance oozing out around the terminals of the filter capacitor. A high-pitched whistle, which may disappear when the volume control is turned upward, can be caused by a 500-uF filter capacitor. A dead audio symptom can result from a bad 12.6-volt connection to the power supply.

A bad electrolytic or choke transformer can cause excessive hum in the speaker. Replace a 500-uF or 1000-uF electrolytic with hum in the speaker when the volume control is turned all the way down. A bad positive lead off of a 500-uF 16-volt electrolytic can cause a loud popping noise with the volume turned down. A defective decoupling electrolytic (50-100 uF) can cause a loss of volume or weak station reception.

Replace the 500-1000-uF capacitor with a motorboating sound in the speaker. A bad 1000 uF electrolytic can also cause motorboating and chirping sounds. Check the 500-1000-uF electrolytic for a squeaky noise while the volume is turned up. A main filter capacitor can cause a really loud whistling sound in the speaker. Check the 50-uF electrolytic for a screeching noise in the speaker.

HYBRID AND TRANSISTOR AUDIO OUTPUT TRANSFORMERS

The transistorized audio car radio circuits might have a tapped output autotransformer or have only one winding to match the speaker impedance and the transistor output. A single-winding transformer primary coil is a fraction of the entire winding, and the secondary steps down to match the speaker impedance. The tapped winding might have an impedance of 3-4 ohms or 8 ohms to match the speaker impedance resistance (**Fig. 12-22**).

The autotransformer can be damaged when the output transistor draws heavy current or becomes shorted through the primary winding. These autotransformers are only found in single-ended transistor audio output circuits within the hybrid and all transistor car radio circuits.

A Chevrolet model 98788 car radio has a single output transistor (DS-503) with a tapped auto-output transformer. The total winding is 8 ohms with a 3-4-ohm tapped output impedance.

FIGURE 12-22. The hybrid audio output transformers might have a center tap winding fed to the PM speaker.

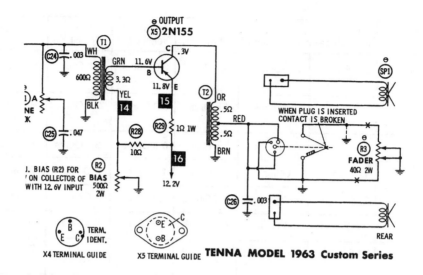

FIGURE 12-23. The Tenna model 1963 auto radio had a 30-ohm impedance transformer tapped at 8 ohms impedance with a 0.5-ohm resistance.

The bottom half of the autotransformer is tapped at 1.3 ohms to match a 3.2-ohm 6X9 PM speaker. In a Tenna model 1963 custom series early car radio, the audio output transformer has a 30-ohm primary impedance and the secondary tap is at 8-ohm impedance. The total resistance is 1 ohm, and the tap is at 0.5-ohms resistance for the speaker connections (**Fig. 12-23**).

These auto-output transformers are no longer available. Simply wind your own or take one from another old car radio with the same impedance. When your transformer is burned or charred, remove the damaged transformer from the radio by drilling out the metal rivets.

Cut off the outside wrapping from the transformer. Remove the metal laminations. Count the number of turns of charred wire that you unwind from the transformer core. Write down the number of turns where the tap is found in the winding. Make sure what lead goes where, when removing the first winding. Likewise, keep counting the number of turns as the whole coil is unwound, and write it down. Most of these coils are wound with 26- or 28-gauge enameled wire. Check the size of the wire on a wire gauge if one is handy.

Now wind on the new winding with the same gauge of enameled magnet wire. Place a layer of masking tape over the core area. Wind on the primary winding and twist a small loop of wire for the tapped coil. Bring the tapped coil wire out to one side of the transformer. Clean off the enamel from the wire with a pocketknife and dip the end into solder paste to make a good soldered connection. Solder on a flexible piece of hookup wire and bring it out to one side of the coil winding. Keep each wire placed alongside the next winding in a closely wound (CW) fashion.

Make Model	Primary Resistance	Secondary Impedance	Total Ohms	Tapped at Voice Coil	Voice Coil SPK	Output Tr
American Motors Model 93MR	21 Ohms	3-4 Ohms	3.1 Ohms	0.6 Ohms	3.2 Ohms	Hybrid 2N176
Automatic Model C349	14 Ohms	3-4 Ohms	1.6 Ohms	0.6 Ohms	3.2 Ohms	Hybrid 4247
Ford 84BF	8 Ohms	3-4 Ohms	1.4 Ohms	0.8 Ohms	3.2 Ohms	Hybrid 2N155
Ford 4TBT	12 Ohms	8 Ohms	2 Ohms	1.3 Ohms	8 Ohms	Trans DS520
Riverside	24 Ohms	8 Ohms	1.4 Ohms	.8 Ohms	8 Ohms	Trans 8P404R
Soundex	25 Ohms	3-4 Ohms	0.3 Ohms	0.3 Ohms	3-4 Ohms	Tran 2N301
Motorola TM326M	24 Ohms	8 Ohms	2.6 Ohms	1.3 Ohms	8 Ohms	Tran SP1556-2

CHART 12-5. The early car radio autotransformer tapped for the speaker connection with the various output transistors.

If the transformer inductance is given on the schematic, check for correct inductance with an inductance meter, or check the low-ohm scale of the DMM. After the coil is finished, place masking tape over the outside winding (**Chart 12-5**). If the turns are not known, wind on enough turns of 26-gauge enameled wire to equal the resistance of the tap and the total transformer resistance.

Some interstage and audio output transformers are available from Antique Radio Supply mail-order house for the tube and transistor chassis. Transistor interstage and transistor output transformers are available from some electronics dealers and firms. The only repair that can be made on these types of transformers with an open winding is to check where the flexible lead connects to the enameled winding. A poorly soldered joint can occur at these connections. Carefully remove the outside wrapper of the transformer where it is soldered together. Inspect the bad joint or the poorly soldered connection. Apply solder paste on the connection and then resolder. Now check the winding with the low-ohm range of the DMM.

CIRCUITS AND STAGES OF THE EARLY CAR RADIO

Early Tube Radio Circuits

The standard tubes found in the early auto 6-volt car radios were a 6BA6, 6BE6, 6AV6, 6V6GT, and 6X5GT rectifier. Some 6-volt tube lineups might consist of a 6SK7,

6SA7, 6SQ7, 6AQ5, and 6X5 rectifier tube. The miniature tubes found in the early car radios were a 6BA6, 6BE6, 6AV6, 6AT6, 6AQ5, and 6X4 tube. The auto radio with push-pull output tubes might consist of two 6AQ5 or 6V6GT tubes (Chart 12-6).

The early Buick car radio had a ganged-tuning capacitor of 22-527 uF tuning the RF, oscillator, and mixer circuits. The RF amplifier consisted of an 6K7, 6SA7 oscillator, 6SK7 as IF Amp, 6SR7 as DET-AVC, and two 6V6GT tubes in push-pull output operation. The

Use	Tube Types		
RF Amp	6AD6	6BA6	6SK7
Converter	6BE6	6SA7	
IF Amp	6BA6	6BD6	6SQ7
Det -AVC Amp	6AT6	6AV6	6SQ7
Output	6AQ5	6AS5	6V6GT
Push Pull	6AQ5	6VGT	
Output			
Rectifier	6x4	6x5GT	

CHART 12-6. The early 6-volt vibrator car radio tube line-up in the different car radios.

IF frequency was 262 kHz. The 6-volt vibrator sat in a five-prong socket. You might find a seven-inch speaker with a 3.7-ohm voice coil (**Fig. 12-24**).

An early Automatic car radio tube lineup might consist of a 6BA6, 6BE6, 6BA6, 6AT6, 6AQ5, and a 6X4 rectifier tube. The RF Amp (6BA6) and converter (6BE6) circuits were tuned with a three-ganged-tuning capacitor (25-397 uF, 27-386 uF, 24-201 uF) with an IF frequency of 455 kHz. The 6-volt vibrator, transformer, rectifier tube (6X4,) and power output (6AQ5) were mounted on a separate chassis (**Fig. 12-25**). The power output tube drives a five-inch PM speaker.

The RF Amplifier Tube Circuits

The RF amplifier tube picks up the radio signal from the antenna receptacle and the RF input coil. This RF coil is tuned by a variable capacitor. Capacitance tuning is accomplished by a variable tuning capacitor of a given circuit. The input-loading coil is found between the antenna and the RF coil. Sometimes this coil has vibrated loose and it might lie in the bottom of the receiver, leaving only a rushing sound in the speaker. In some antenna circuits, a small trimmer capacitor can be varied to tune the car antenna to the radio input circuits.

The tuned RF signal is found on the grid terminal (1) of the RF Amp (6BA6) and is amplified to another set of coils out of the plate terminal 3. Here again, the RF coil circuits are tuned with the ganged variable capacitor and tied directly to pin 7 of the converter tube (**Fig. 12-26**).

Within the Standard RF Amp tube (6K7), the RF input signal is tuned by a variable ganged-tuning capacitor and fed to the grid cap of the RF Amp 6K7. The RF output

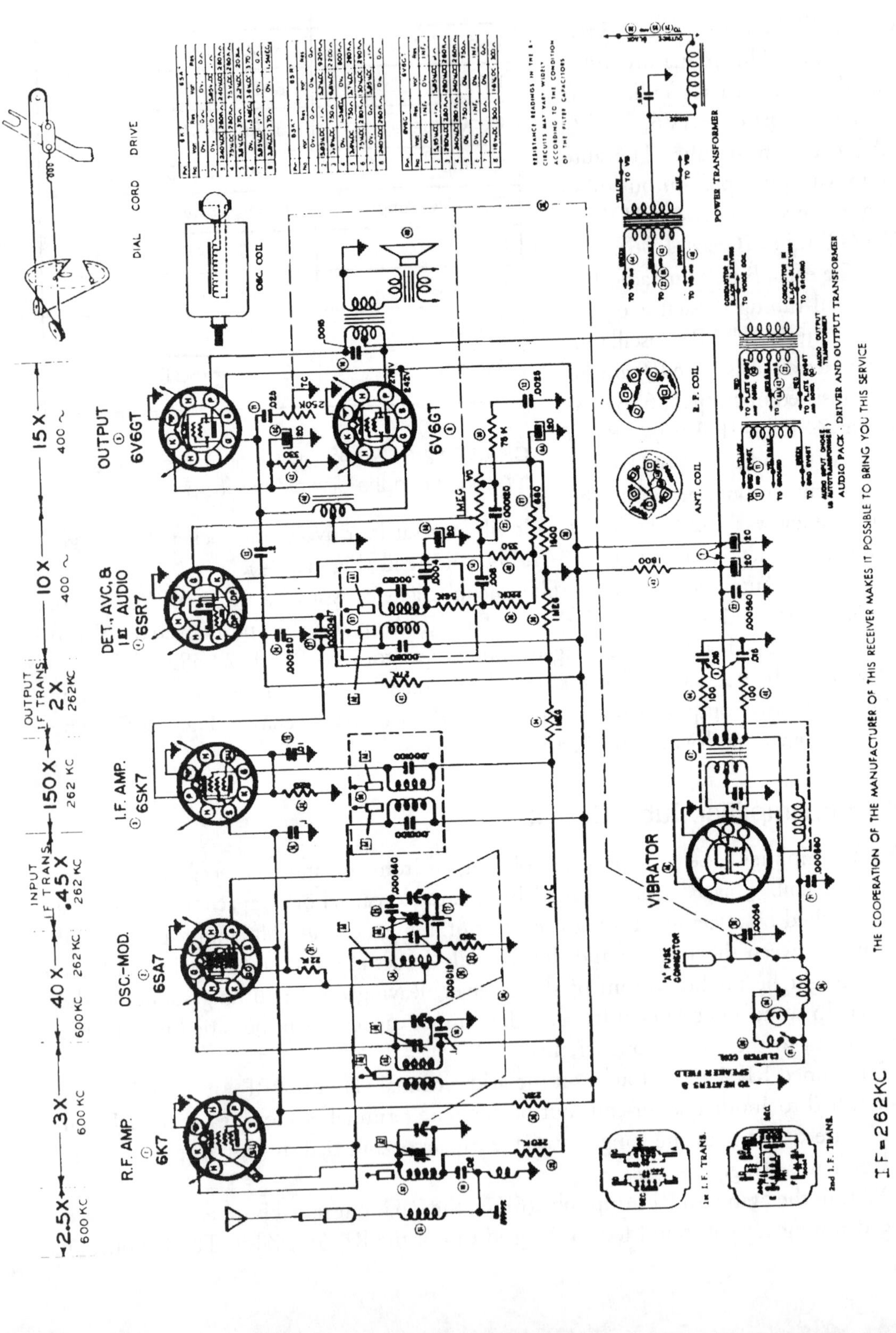

FIGURE 12-24. The standard tube line-up and circuits found in early Buick car radio.

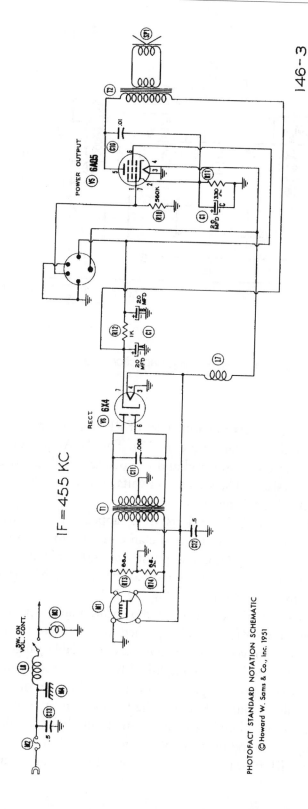

FIGURE 12-25. Here the vibrator, rectifier, and power output tube (6AQ5) are found in a separate early car radio chassis.

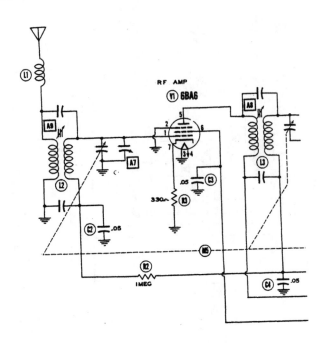

FIGURE 12-26. Notice that the variable tuning capacitors are found in the very early antique 6-volt car radios.

signal is amplified, and it connects to another set of RF coils out of the plate terminal of the 6K7 to the grid of the oscillator tube (6SA7). The RF Amp plate voltage is very high at 240 volts DC. The suppressor grid and cathode terminals are tied to common ground with a 390-ohm resistor. The screen grid voltage is at 75 volts with a 22K-ohm resistor tied to the 240-volt DC source.

The antenna-loading coil (L1) connects to a tapped RF coil, with a trimmer capacitor tied to common ground. A7 tunes the antenna coil to the coil input circuits of the RF stage. Rotate the tuning knob and pointer to 1400 kHz on the dial assembly and adjust A7 for maximum volume in the radio. Oftentimes a hole is found close to the antenna receptacle for this adjustment (**Fig. 12-27**).

Later on, the RF stage in the 6-volt car radio incorporates a permeability circuit instead of a variable capacitor. A ferrite rod is pushed in and out of the RF coil to tune the RF stage. The RF and converter AM circuits use a variable ganged-tuning ferrite tuning system. Permeability tuning is accomplished by varying the frequency of an LC circuit by changing the position of a magnetic ferrite core within the coil or inductor. The variable ganged-permeability tuning is found in the RF and converter or oscillator stages.

The RF radio signal is picked up by the car antenna, coupled through a loading coil, tuned by a trimmer capacitor, and fed through a tuned RF coil that is capacity-coupled to the grid (pin 1) of the RF tube (6BA6). The amplified RF signal is found at the plate

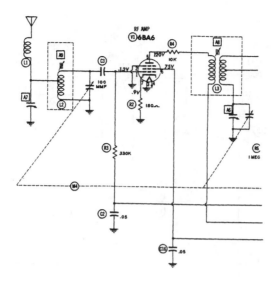

FIGURE 12-27. Adjust the antenna to the car radio input circuits with a trimmer capacitor (A7) in a 6-volt car radio.

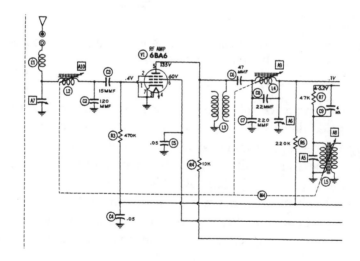

FIGURE 12-28. Here the radio station is picked up by the outside antenna, tuned by permeability tuning to the RF and converter tubes.

terminal 5 and coupled to another RF tuned circuit. Notice that the plate voltage on the RF tube (6BA6) is only 135 volts compared to the standard RF tube series. A 60-volt source is found on the screen grid (pin 6) with the suppressor grid and cathode terminal to common ground (**Fig. 12-28**).

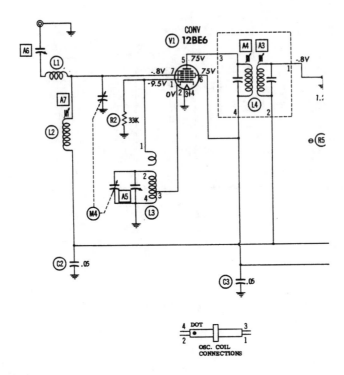

FIGURE 12-29. Notice that no RF amp circuit is found here as the converter accepts the RF signal and mixes it with the local oscillator signal.

You might not find a RF tube in the early 12-volt car radio, because a converter stage (12BE6) is tied directly through a loading coil to the antenna input receptacle. L2 is tuned by a two-ganged tuning capacitor (M4). The RF signal is found at pin 7 of the converter tube and amplified out of the plate terminal 5. Both the plate and screen grid voltage operate at 75 volts DC (**Fig. 12-29**).

In the early 12-volt tube car radio, the RF tube might be a 12FX8 that picks up the RF antenna signal and amplifies it to the grid terminal 6. Again, the permeability antenna coil has a trimmer capacitor (40-130 pF) that tunes the antenna to the radio input circuits.

The amplified RF signal is found on plate terminal 8 of the RF tube (12FX8) and capacity-coupled to another permeability RF coil to pin 9 of the converter tube (12FX8). Notice that one half of the first tube serves in both RF and converter circuits. Here the RF amp plate voltage operates at 10.2 volts with 10.2 volts measured on the screen grid terminal 1. The suppressor grid and the cathode terminal 7 are fed to common ground (**Fig 12-30**).

The tube lineup in a 12-volt vibrator car radio might consist of a 12BA6, 12BE6, 12AV6, 12AQ5, 12X4, and OZ4 as rectifiers. A few of these 12-volt tube radios do not

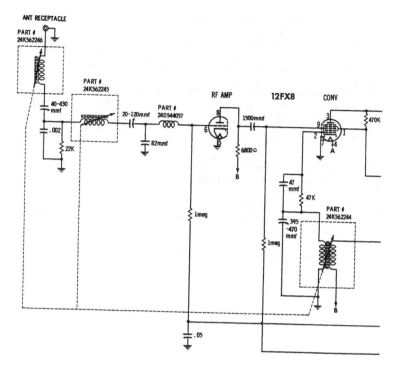

FIGURE 12-30. The triode section of a multi 12-volt tube (12FX8) serves as the RF and converter circuits.

have a separate RF amp tube, as the Rf signal is fed directly into the convert circuits. The early car receivers were capacity tuned, while later on, permeability tuning was found in a three-ganged circuit. The IF frequency consisted of 262 kHz and 455 kHz. A variation of different tubes was found in the 12-volt vibrator radio circuits (**Chart 12-7**).

Converter Tube Circuits

The converter tube (6BE6) circuit within the early 6-volt car radio acts as a mixer and oscillator stage. The RF tube amplified signal is coupled to the grid terminal 7 and mixed with the oscillator circuit found in the cathode and grid circuits of the 6BE6 oscillator tube. The cathode terminal 2 goes to common ground through the secondary winding of the oscillator coil. A 155 DC voltage is found on the plate terminal 5, and 60 volts on the screen grid pin terminal 6 from a voltage-dropping resistor. Both the RF input coil and the oscillator coil are tuned with a variable tuning capacitor (**Fig. 12-31**).

The mixed radio output signal from the converter stage results in a difference of the incoming tuned signal and the oscillator signal, producing the intermediate frequency

Stage	Types of Tubes
RF Amp	12BA6 2AT7 12BL6 12CX6
Converter	12BE6 12AD6 12AT7
IF Amp	12BA6 12AV6 12CN5 12FR8
Det-AVC Amp	12AV6 12AL5 12CR6 12FR8 12J8
Output	12AQ5 12AB5 12V6GT
Push = Pull Output	12AB5 12v6GT in Push=Pull Operation
Rectifier	0Z4 12x4

CHART 12-7.   The 12-volt tube lineup found in a 12-volt vibrator power circuit.

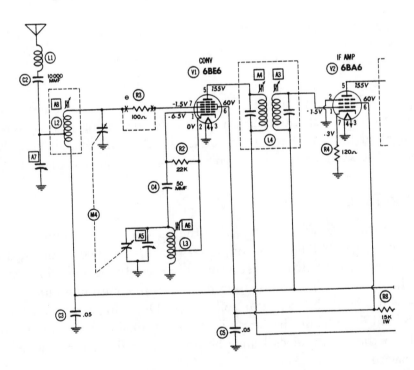

FIGURE 12-31.   The RF and oscillator coil are tuned by a variable-gang capacitors in the 6-volt converter tube (6BE6) circuits.

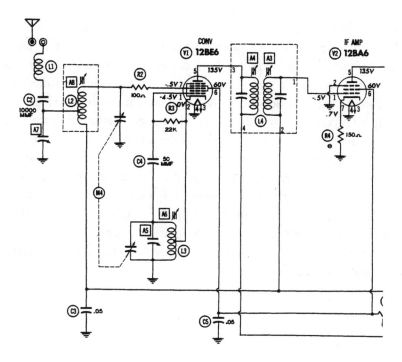

FIGURE 12-32. The early 12-volt car radios had a 262 kHz IF circuit.

(IF). The early IF frequency was 262 kHz, while the latter 6-volt car radios have a 455-kHz frequency. The IF signal out of pin 5 of the 6BE6 tube is transformer coupled to the first IF Amp tube (6BA6).

A converter tube (12BE6) within the early 12-volt car radio is coupled to the RF amp tuned coil and to the grid of the converter tube circuits. In some car radios, the converter stage is connected to an RF coil that has permeability tuning and is connected directly to the loading coil and outside antenna receptacle. Likewise, the oscillator coil is found in the cathode circuit (pin 6) of the converter tube. A three-ganged permeability tuning tunes in the favorite station within the RF and converter oscillator circuits.

The plate voltage for the converter tube is applied through the primary winding of the IF transformer to pin 5 of 135 volts. The screen grid voltage is at 80 volts from a resistor-capacitor decoupling voltage network. The converter IF frequency is transformer-coupled by L4 to the IF stages. The IF frequency might be at 262 kHz and 455 kHz in the later car radios (**Fig. 12-32**).

Octal Tube RF and Converter Circuits

In the early car radios with octal radio tubes, the RF stage consisted of a 7A7 tube with the grid terminal 6 tied to the grid filter choke coil. The outside antenna receptacle was soldered to the antenna choke coil and antenna trimmer capacitor. Both the RF coils in the grid and plate circuits have permeability tuning. The cathode terminal 7 and suppressor grid terminal 4 are tied directly to common ground, while the plate terminal 2 is capacity-coupled to the RF circuit in the input of the converter tube (7Q7). A three ganged-permeability tuner tunes in the RF and oscillator circuits (**Fig. 12-33**).

The RF signal is coupled from the RF coil to the grid terminal of the converter tube (7Q7). The oscillator circuits of the converter tube are found in the grid-4 and cathode-7 terminals. The oscillator trimmer capacitor (A5) is adjusted for maximum output with the slug or ferrite rod out of coils, and the dial setting around 1600 kHz. The oscillator signal is mixed with the incoming RF signal and the difference is the IF frequency of 260 kHz. The plate voltage of the RF 7A7 tuber is around 92 volts and the screen grid at 58 volts DC. Likewise, the plate voltage of the 7Q7 converter stage is 200 volts, and the screen grid voltage is at 58 volts DC. The converter IF signal is transformer-coupled to the first IF tube (7A7).

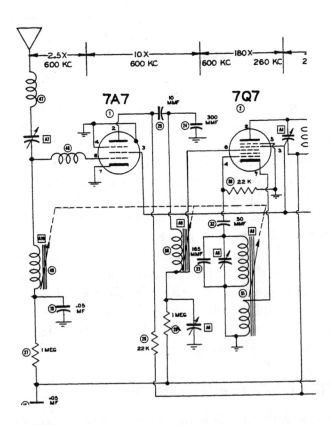

FIGURE 12-33. Permeability tuning is found in the RF loctal and converter tube circuits.

The Intermediate Frequency (IF) Stages

The IF stages within the early 6-volt car radio consisted of a 6BA6 IF tube with the first IF transformer-coupled directly to the grid terminal number 1. A 120-ohm resistor is found in the cathode circuit at pin 7. The amplified IF signal (455 kHz) was transformer-coupled to the detector tube (6AV6) from plate terminal 5 of the converter tube. The plate voltage on pin 5 was 105 DC volts, while the screen grid voltage was around 75 volts from a 15K-ohm decoupling resistor voltage network (**Fig. 12-34**).

A 12-volt vibrator car radio IF circuit consists of a 12BA6 IF amplifier tube with the incoming signal at pin 1. The 150-ohm resistor provides bias to the IF amp tube out of pin 7. The second IF transformer is coupled to the plate terminal 5 of the 12BA6 tube and transformer-coupled to the detector pin terminals 5 and 6 of the Detector-AVC AF Amp tube (12AV6). The early 12-volt car radios had a 455 kHz IF frequency.

In the early standard 6-volt tube car radio, the first IF transformer was coupled to the plate terminal of the oscillator-mixer tube (6SK7) and transformer-coupled to the grid of the IF tube (6SK7). The second IF stage is coupled from the plate of the 6SK7 tube to the detector terminals of the Detector-AVC-AF audio amp tube (6SR7). The plate voltage on the IF tube is around 240 volts with a screen voltage of 75 volts from a resistor-capacitor decoupling filter, voltage-dropping network. The early intermediate frequency of the standard tube car radio was at 262 kHz.

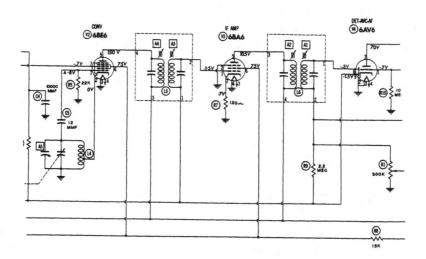

FIGURE 12-34. The early 6-volt IF tube (6BA6) has an input IF stage and output IF stage with L5 and L6 in a 6-volt car radio.

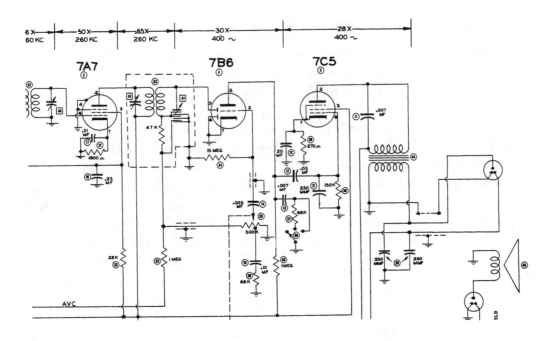

FIGURE 12-35. A Loctal 7A7 tube serves as the IF amplifier and a 7B6 works as a diode-detection triode amplifier with signal coupled to the 7C5 output tube.

Octal Tube IF And AVC Circuits

The first IF transformer couples the IF signal from the plate terminal of the converter tube (7Q7) to the grid terminal of the first IF tube (7A7). The first IF tube cathode terminal provides bias through a 1.8K-ohm resistor with a 0.01-uF bypass capacitor. A 50- or 100-volt bypass capacitor can be used here, as the voltage is only 3.7 volts DC on the cathode terminal. The second IF transformer is connected to plate terminal 2 of the IF Amp (7A7), and transformer-coupled to the diode plates of the Detector-AVC-Amp tube (7B6). The secondary winding is connected directly to pin 5 of the diode detector tube.

Besides detecting the radio IF signal, the 7B6 tube provides AVC action, and amplifies the audio applied to grid terminal 3 with a variable 300K-ohm volume control. A 0.005-uF coupling capacitor is found between the center volume control terminal and pin 3 of the 7B6 tube. The audio is amplified by triode section of the (7B6) tube and is capacity coupled with a 0.05-uF capacitor from the plate terminal 2 of the 7B6 to the grid terminal of the output tube (**Fig. 12-35**). Plate terminal 2 of the triode section has a 70 DC applied voltage from a 1-megaohm load resistor.

The 6-Volt and 12-Volt Detector and AVC Circuits

In the early tube detector-AVC and triode amplifier circuits, you might find a 6AT6, 6AV6, or 6SQ7 tube. The second IF transformer is connected to the input detector terminal 6 of V4 (6AT6). The IF signal is rectified and turned into an analog or audio signal. A negative 0.3 volts is found on both detector diode and the AVC diode elements inside the 6-volt tube (6AT6). The cathode element 2 is grounded and the audio appears at plate terminal 7. The audio signal is controlled by a 250K-ohm tapped volume control to a 0.1-uF coupling capacitor to grid terminal 1 of the output tube. A high resistance grid leak resistor of 4.7 megaohms is tied to the grid terminal and common ground (**Fig. 12-36**).

Within the early 12-volt vibrator tube circuits, a 12AV6 tube serves as a IF detector-AVC and triode amplifier. The second IF transformer couples the IF signal directly to the detector terminals 5 and 6 of the 12AV6 tube. A 500K-ohm volume control is found at the bottom leg of the secondary winding of the IF transformer. The center tap of the volume control controls the audio signal coupled through a 0.005-uF capacitor to the triode grid terminal 1. R7 and C6 consist of a resistor and capacitor module

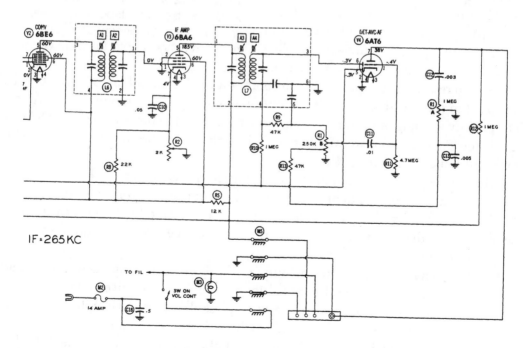

FIGURE 12-36. The 6-volt Detector, AVC, and amplifier tube (6AT6) detects the IF signal and amplifies it in the early car radio.

component connecting the audio to the grid of the output tube. The voltage on pin 7 of the 12AV6 tube is 65 volts through R7B (470-K ohm) resistor inside the plastic enclosed module component (**Fig. 12-37**).

Early Auto Radio Output Circuits

The early 6-volt car audio output tubes might consist of a 6AQ5, 6AS5, 6V6GT, or two 6AQ5s and two 6V6GT tubes in a push-pull audio output circuit. A 12-volt tube line-up might be a 12AQ5, 12AB5, 12V6GT, or two 12AB5s and two 12V6GT in a push-pull audio output circuit. The Loctal audio output tube might have two 6V6GT and two 7C5 tubes in a push-pull audio output circuits (**Fig. 12-38**).

The single audio output tube (12AQ5) was capacity-coupled to the plate circuit of a 12AV6 tube with a 0.005-uF coupling capacitor. The grid resistor was a regular

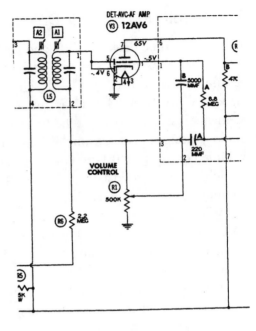

FIGURE 12-37. The 12-volt Detector, AVC, and first Amp tube (12AV6) has a module made up of capacitors and resistors connected to the plate circuit.

470-K ohm resistor tied to pins 1-7 of the output tube. A plate voltage of 170 DC volts is applied to pin 5 with a 135-volt source at the screen grid of output tube (12AQ5). The primary winding of the output transformer is coupled to pin 5 with a 180 DC volt supply source. The secondary winding of T2 is applied to a 4X6 inch PM speaker with voice coil impedance from 3-4 ohms.

Most of the problems found in the audio output tube chassis are in the output tube. This tube can cause intermittent sound, no audio, hum, and distortion in the speaker. A microphonic or cracking noise in the speaker can result from loose particles or elements inside the output tube. Test the tube on the tube tester or substitute another tube to determine if the output tube is okay.

A leaky coupling capacitor to the output grid terminal 1-7 can cause a weak and distorted sound. Check the grid terminal for any signs of a positive voltage; the grid terminal voltage is always negative. It's best to replace all bypass and coupling capacitors when trying to restore the antique car radio. Check the resistance of the cathode resistor from pin 2 to ground for distorted sound. A leaky electrolytic across the 180-ohm cathode resistor can cause a weak and distorted sound in the speaker. A bad combination of the cathode resistor and bypass capacitor in the cathode circuit can sometimes cause a motorboating sound.

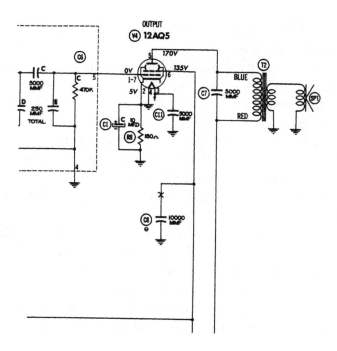

FIGURE 12-38. A beam-powered 12AQ5 tube amplifies the audio signal and transformer couples it to the PM speaker.

A dead car radio with a low hum in the speaker might be caused by a dead audio output tube and burned cathode resistor. Low or weak volume can occur with a leaky grid coupling capacitor of the output tube. Replace the audio output tube if the speaker begins to crack and pop when the radio is first turned on. Replace the output tube if it makes a cracking sound noise in the speaker when tapped.

An open primary winding of the output transformer can cause a dead chassis without hum. Suspect a short, and a red spot on the plate terminal of the output tube, with a loud hum in the speaker. A blue gas-like light at the bottom area of the output tube may prove the tube is gassy and should be replaced if the sound is somewhat distorted. Intermittent audio can result from a bad power output transformer to the PM speaker.

Front-End Transistor Circuits

Practically all of the early transistor car radios contained permeability tuning within the front-end circuits. Most RF transistor problems were caused by the RF transistor becoming leaky or open. A defective RF transistor can cause really weak local station reception. Suspect a bad RF transistor when the auto radio becomes intermittent after three or four hours of operation. Try using a couple of applications of coolant spray on the RF transistor to make it act up.

Intermittent sound can result from a defective coupling capacitor off of the base terminal. An open RF transistor can cause a low hum and a loss audio with no stations received. When the car hits a bump and the car radio quits playing, replace the RF transistor, and solder up all the connections around the RF transistor.

A RF coil wire broken off of the permeability coil assembly can cause really weak radio station reception. If only a local station can be tuned in and the RF transistor tests fine, replace it anyway. Suspect a broken lead off of the antenna-loading coil if there is only a loud rushing noise and no stations are tuning in. Check the antenna-receptacle socket for a broken lead wire.

Replace the RF and converter transistors if there is a loss of RF but the audio circuits appear normal. Check for a defective bypass capacitor (0.47 uF) off of the emitter terminal of the RF transistor for low volume and the radio cutting out. A bad oscillator trimmer capacitor in the converter stage can cause intermittent reception. Replace the RF transistor for intermittent and microphonic noise in the speaker. Readjust the antenna-trimmer capacitor for stations poorly tuned in at the high end of the dial.

Suspect a defective converter transistor with a loud rushing noise in the speaker with no stations tuned in. A bad tuning network on the converter transistor can cause dead stations on the high end of the dial. Check the converter transistor for a dead radio with only hum. For loss of AM reception, check for broken wires on the AM tuning assembly. A leaky converter transistor can cause a dead AM section.

Replace the bad converter transistor when the car radio cuts in and out with a low volume. A bad oscillator transistor can cause a loss of rush. Loss of rush or station reception can result from an open converter transistor. A leaky converter transistor can cause a dead radio and a rushing sound when the volume is turned up.

If the car radio is very weak with some stations, and changes the location of other stations on the dial assembly, suspect a broken ferrite rod or core inside the converter permeability-tuning coil. Check for a bad converter transistor with only a hum and a dead car radio. A dead radio might be caused by a wire torn off of the 0.01-uF capacitor to ground in the converter section. Check the converter transistor when the volume cuts in and out on low volume.

A shorted diode or an increase in resistance within the power-supply voltage can cause intermittent AM radio reception. Replace the converter transistor for a loss of RF or stations when the audio section is normal. Spray coolant on the converter transistor when the radio is first turned on but brings in no stations.

Low or no voltage to the IF circuits can cause a weak or dead car radio. Suspect a bad second IF transistor when the radio is intermittent and then dead, and when the radio begins to play when the base of the IF transistor is touched. Loss of RF or audio can be caused by an open first IF transformer winding. An open winding in the second IF transformer can cause the intermittent front-end with a normal audio section. Check for an intermittent IF transistor for intermittent radio reception.

A leaky IF transistor can cause a weak local station reception with hum. The loss of rushing noise with dead reception can result from an open IF transistor. A badly soldered connection on the collector terminal of the

FIGURE 12-39. A bad IF transistor with an open primary winding can cause a dead radio with only a rushing sound.

IF Amp transistor can cause a very weak local radio station reception. An open IF Amp transistor can cause the dead car radio with only a rush when the volume control is rotated (**Fig. 12-39**). Replace the second IF transistor when the car radio cuts up and down.

A leaky IF transistor can cause a dead car radio without a rushing sound. A badly soldered IF connection on the IF transformer can cause intermittent radio reception. Suspect a bad primary winding of the IF transformer coil touching the inside metal case when the car radio becomes intermittent and operates for several weeks before it acts up again. Replace the IF transistor if the car radio plays but then goes dead when the car sits out in cold weather.

Replace the first IF Amp transistor with an open primary winding that results in only one station able to be heard. Simply move the IF transformer if a bad connection inside the metal-can area causes intermittent reception. Check the power-supply voltage source, decoupling capacitor, and resistor for loss of AM reception.

Suspect a defective diode in the second IF transformer circuit for poor radio reception. A broken capacitor inside the IF transformer can result in only one station being heard on the low end of the dial. A poorly soldered connection on the cathode side of the signal diode can cause intermittent AM. Check for a dried-up or open filter and bypass capacitor in the IF section when the radio begins to oscillate and create a tuning sound. Replace the AVC limiter diode for some distortion on local radio stations.

VARIOUS PROBLEMS, REPAIRS, AND REPLACEMENTS

Tube Problems

Most tube problems consist of an open filament or heater, with a dead radio, or weak reception with a bad corroded cathode element and a weak flow of electrons. When the cathode element has many ions built up on the heater-cathode element, the vacuum tube will test weak in emission and produce weak or no radio reception. Sometimes the cathode element will flake off and fall into the grid element, producing a shorted tube. The vacuum tube can produce intermittent or cracking noise in the speaker with loose elements inside the tube or with poor pin connections. Gas built up inside the tube by the various elements can produce a soft tube. Shorted elements within the tube can cause intermittent sound and provide damage to the voltage-dropping resistors and transformer windings.

All tubes should be checked in a tube emission tester for shorts and good emission or be subbed with another tube of the same number. Since tubes are now hard to obtain and are a lot more costly, try to locate a tube tester at the TV shop or possibly one you can borrow. A lot of tubes found in the antique car radio can be purchased at the antique electronics supply or found listed in radio and electronics magazines for sale. Check the back of this book for a list of electronics mail-order dealers.

Checking The Fuses

When the car radio quits all of a sudden, or the radio begins to play and then stops, suspect an open fuse. Often the fuse blows when it is too small in amperage or when a component inside the radio is pulling too much current. Most fuse problems are related to a shorted or leaky audio output transistor. The fuse would blow in the early vibrator car radio with an overload part in the rectifier section, or with a bad or stuck vibrator. Leaky or shorted electrolytics can cause the main fuse to open (**Fig. 12-40**).

The car radio fuse should be replaced with the exact amperage as the one taken from the fuse holder. Remove the old fuse and take a peek at the amp sign on the fuse or fuse holder. Also look at the amps marked on the

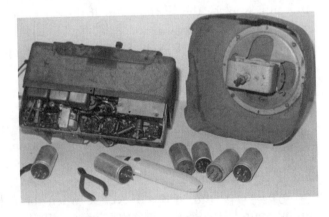

FIGURE 12-40. An early Motorola 6-volt car radio with vibrator and tubes in one case that fastens to the firewall of the automobile.

Radio Voltage	Type of Fuse
6v Vibrator Radio Supply	14A- 15A- 20A
12v Vibrator Supply	4A Oriegn- 5A- 7.5Amp
Hybrid Radio 12v	2.1A-4a-32v -5A 32v -7.5A-b3v
All Transistor Radio 6v	2A- 2.5A -5A -7.5A
All Transistor 12v Radio	1.2A-2A- 2.5A -4A -5A -6A -7.5A

CHART 12-8. A fuse chart of the different sizes of fuses found in the early car radios.

metal end area of the fuse. Do not wrap tin foil around the fuse instead of replacing with the correct amperage; you can destroy critical components inside the car radio and cause it to smoke. Most fuse holders will only hold a certain type of amp fuse as they appear in different lengths. Check **Chart 12-8** if the fuse is not marked or if it is missing.

Filter Capacitor Problems

Filter capacitors provide capacitance reactance to a power-supply filter network, while blocking direct current. Electrolytic capacitors are made up of sheets of tin foil and a dielectric-constant type of paste. After many years of operation, the filter capacitor can dry up, have a lower capacity, or become leaky and hence short out the DC source. Although the large two or three electrolytic filters in one metal might not be available today, they can be replaced with single radial electrolytic capacitors. For instance, a filter network found in a 6-volt car radio power supply of a dual 10 uF at 450 volts can be replaced with two 10-uF radial 450-volt capacitors that are still available. Simply tape them together and reconnect the terminals to the correct filter network.

The 12-volt vibrator power supply with a 20-20-10 uF electrolytic at 450 volts can be replaced with individual radial electrolytics of 22-22-10 uF at 450 volts. An early hybrid car radio with a 700-200 uF at 16 volts can be replaced with individual radial electrolytics of 1000-1000 uF at 16 volts.

It's always wise to replace old capacitors with those having a larger capacity and voltage than the original ones. Make sure the working voltage is higher than the old capacitor. If in doubt, measure the voltage across the filter capacitor and replace it with a higher working voltage. Filter capacitors cause more hum problems within the speaker than any other capacitor within the power supply. Tape the two or three electrolytics together with masking or plastic tape to replace a dual or triple metal-can electrolytic (**Chart 12-9**).

Make and Model	Capacity	Voltage	Power Supply
Allstate 6240	10-10-10 UF	250-250-25 VDC	6v Vibrator
Buick Model 981902	20-20-20 uF	400-400-250 VDC	12v Vibrator
Edsel 04BF	450-100-50 uF	16-16-16- VDC	12v Vibrator
Karmann Gia	500-250-250 uF	8-8-8 VDC	Foriegn All Transistor

CHART 12-9. Chart of electronic filter capacitors found in the various early antique car radios.

Audio AF-Driver and Output Transistor Problems

Suspect an open or leaky driver or AF transistor in the case of a weak audio and distortion in the speaker. A dead radio can result from a leaky driver transistor. Check the voltage on a directly driven AF transistor for a leaky or open AF transistor. An open AF transistor can cause loss of audio in the speaker. Spray coolant on the AF or driver transistor with intermittent reception.

An intermittent junction inside the AF transistor can cause intermittent reception. The loss of audio with a hum in the speaker can result from an open AF transistor. A shorted AF amp transistor can cause distorted audio. A leaky electrolytic capacitor can also cause distortion in the sound.

An open bias control on the first AF amp transistor can cause the radio to just fade out. The shorted coupling capacitor to the base terminal of the audio AF transistor can cause a loss of radio sound.

A microphonic sound in the speaker can result from a noisy AF transistor. Suspect the AF transistor when there is a low hissing sound in the audio; spray coolant on the transistor and notice if the hissing sound is no longer heard. When noise cannot be adjusted out with the volume control, check for a noisy AF amp transistor. The AF amp transistor might be damaged with a directly driven shorted output transistor. Replace the AF and output transistor for a weak and distorted sound in the speaker.

A motorboating sound in the audio circuits can be caused by dried-up filter capacitors (500-300-100 uF 16 volts). Stations might become weaker as the radio operates with a defective coupling capacitor to the base of the AF transistor. The sound in the speaker may appear intermittent when a 100-uF electrolytic is moved in the base circuit of the AF amp transistor.

The defective audio output transistor can cause intermittent and distorted sound. The shorted output transistor can cause low and distorted volume in the audio. Replace the DS-67 output transistor for weak radio reception. The shorted output transistor can pull heavy current and keep blowing the fuse. A leaky output transistor can also cause weak radio reception. Really low volume on all stations can result from a leaky power

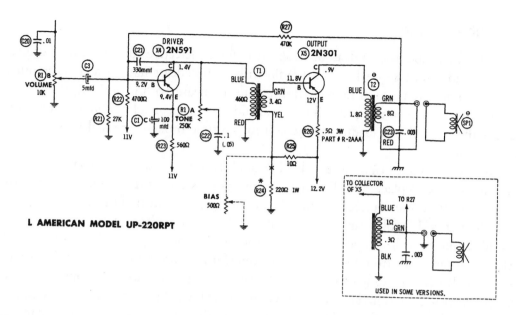

FIGURE 12-41. A leaky output transistor (2N301) can cause really low volume on all stations.

output transistor with a 0.5-ohm resistance between emitter and collector terminals. Replace both output transistors with weak radio reception (**Fig. 12-41**).

Suspect a leaky output transistor with a loud motorboating sound when the volume control is turned down. Replace both push-pull output transistors with a loud motorboating sound. A loud howling noise in the speaker can result from a red-hot output transistor.

Having one push-pull transistor open and the other one leaky can cause distorted output. After the radio operates for several hours and then becomes distorted, replace the power output transistor. A defective output transistor can cause low volume and distorted music. A shorted output transistor and burned 0.33-ohm bias resistor can also cause distorted and weak audio. Replace both AF and power output transistors for distortion and a blown fuse.

Replace both power output transistors when the fuse keeps blowing. The ungrounded heat sink can cause hum and motorboating in the output circuits. A bad plug connection between the separate amp and the radio section can cause intermittent hum.

An open bias resistor in the emitter terminal of the output transistor can cause really weak volume with no click in the speaker. A shorted output transistor and burned output transformer can cause a low hum with no other noise. A bad fusible resistor (0.47 ohms) in the emitter circuit of the output transistor can cause the car radio to have no sound for about one hour of operation. Replace the bias diode and output transistor for a dead car radio.

RF Tube Problems

First check all tubes with a dead or weak car radio. A defective RF tube can be shorted or leaky and can cause weak or dead radio with no broadcast stations able to be tuned in. Sometimes even a local radio station cannot be heard with a weak RF tube. Place a screwdriver blade on the grid coil terminal and try to tune in a local broadcast station. Check the RF coil and loading choke coil for possible broken terminal leads or open coil windings.

When no sound or else just a clicking noise is heard, suspect a defective RF tube or components. Replace the RF tube when no stations can be tuned in. Check the RF tube with intermittent reception and if only one station can be heard.

A defective RF tube can cause a really weak local broadcast station. Intermittent sound can be caused by a poorly soldered lead on the coupling capacitor. A very weak AM signal can result from by a bad first IF transformer connection. A coil wire broken off of the RF coil winding can cause really weak receptions of local stations. Poor AM reception with no stations able to be tuned in on the lower half of the dial can be caused from poor alignment of the trimmer capacitor on the antenna coil. A broken connection on the antenna choke coil can cause loss of AM reception. Loose elements or particles within the RF tube can cause microphonic reception. Clean out the dust particles with a thin piece of cardboard, or blow the dust free.

A bad antenna jack can produce intermittent radio reception. Suspect a broken ferrite rod or core inside the RF coil when a local station is weak or if the volume unexpectedly changes. A bad trimmer capacitor in the RF stage can cause intermittent and dead radio reception. Suspect a dirty variable capacitor when tuning through the stations and only a scratching noise is heard.

Mixer Tube Problems

When the car radio will not tune in a station in a 6- or 12-volt car radio, suspect a defective converter tube or corresponding part. Check for a defective converter tube or components when the car radio chassis is dead on the high end of the dial. After checking the tube and voltage measurements, double-check the oscillator trimmer capacitor. A bad converter tube socket can cause intermittent sound. A loud rushing noise with some hum and no tuned-in stations can result from a bad converter tube or surrounding parts.

The fading of an AM station can result from a trimmer capacitor with moisture between the metal plates. Check the oscillator coil for a broken lead with only a rushing noise in the speaker. A poorly soldered connection on the oscillator coil can cause loss of AM reception. Notice if a lead is broken off of the AM tuner without tuning in any stations. Suspect a broken oscillator ferrite rod or core inside the permeability coil if the car hits a bump and causes the station to fade out or appear at another point on the radio dial. Hum and a loss of stations are caused by a broken wire on the oscillator permeability coil.

Tube Detector and Audio Amp Problems

A weak 6AT6 or 12AV6 detector amp tube might cause a weak audio station reception. Measure the voltage on the plate terminal 7 when no audio is heard. Place a screwdriver blade on the center terminal of the volume control and you should hear a loud hum in the speaker. If not, check the audio output circuits.

The open coupling capacitor can cause very weak, intermittent, or no audio in the speaker. When the music or voice of the radio station is distorted, suspect a leaky capacitor between the first amp tube and the output tube. Replace the leaky coupling capacitor with a positive voltage on the grid of the output tube (6AQ5 or 12AQ5). A bad detector tube can cause some distortion when a local station is tuned in.

In some early car radios, both the coupling capacitors between the volume control and grid terminal of the detector tube (12AV6), and the coupling capacitor (0.005 uF) between the plate and grid of the output tube, are found in one component or molded module. Capacitor component C6 has four different capacitors in one module with leads connecting to the preamp and output circuits.

When one of these resistors or capacitors goes open or increases in resistance inside the molded component, replace with separate outside parts. Choose the same type of resistors and capacitors that are inside the module. Solder each component together as shown in the schematic and connect back to the radio circuits. Because these components are no longer available, you can make your own with regular components (**Fig. 12-42**).

Push-Pull Transistor Transformer Repair

In a Motorola CTA62 car radio, the push-pull output transistor was shorted and showed signs of overheating the output transformer. Both output transistors were replaced with excessive distortion in the speaker. The total resistance of the output transformer winding was only 0.1 ohms. The output transformer was removed and dismantled. All metal laminations were removed from the transformer core. Now the turns of the winding could be counted while they were unwound from the transformer core.

The coil form was wound full with number-26 enameled wire at the same time. Both windings were wound together. It is much easier to wind the coil winding when the core is removed from the laminations. The outside winding of the core was wound so that it did not short against the metal of the laminations when replaced. The total resistance on a regular DMM was 2.8 ohms. A flexible wire was soldered to the tapped coil for the ground connection. Both enameled wire ends were scraped clean and dipped in solder paste to make a good connection (**Fig. 12-43**).

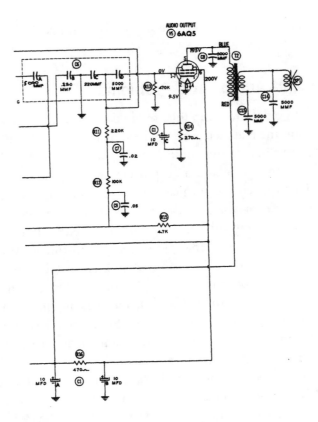

FIGURE 12-42. If the coupling and RC network module is no longer available, solder up the unit with several different resistors and capacitors.

After the transformer was completely assembled, the coil ends were soldered to each output transistor collector terminal. The auto radio was fired up and the sound was clean and clear. Readjustment was made on the balance bias control to correct any distortion. With this repair, another antique car radio was placed back into action.

Antique Speaker Repair

Most of the old PM speakers have a broken or torn speaker cone. Of course, these speakers can be repaired and reconed at several different places around the country. If the voice coil wires are broken at the cone area or where they attach to the speaker terminals, repair them with flexible Litz wire. Be very careful when cleaning up the flexible wire at the speaker cone area so as not to break them off or pull out the voice coil wires. Clean up the broken ends and solder the new wire to the old. Dipping wire ends in or applying solder paste can help to make a good connection. Use speaker or

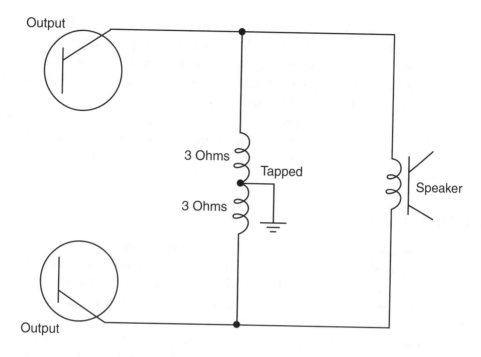

FIGURE 12-43. The burned center tapped output transformer was rewound with number-26 enameled wire in a Motorola CTA62 model.

FIGURE 12-44. An old antique speaker that mounts on the firewall of the automobile with a torn cone.

model airplane glue to fasten down the outside rim of the speaker cone. Likewise, reglue the spider assembly of the speaker for a blatting sound (**Fig. 12-44**).

The defective auto radio speaker might have an open voice coil, warped cone, and holes poked into the paper cone. Check the continuity of the voice coil with the low-ohm range of the DMM. Extreme weather conditions and water seeping into the car speaker can cause the cone to warp. The speaker cone can rub against the magnet pole piece and cause a muffled or tinny sound. Push down on the speaker cone area and notice if the cone assembly is dragging against the metal pole magnet. Test the intermittent speaker sound problem in the same manner. If intermittent, the sound will cut up and down.

Dial Cord Repair

Although some dial cord is still found at some antique electronics firms, you can use a rayon or cotton fishing line to wind that broken dial cord assembly. Try to select the same diameter of dial cord as the old one. The no-stretch black of nylon braid over a fiberglass core is still available in either the standard thin and mid-heavy at several electronics stores or mail-order firms.

Notice how the cord is wrapped around the tuning shaft and the various pulleys. Draw a rough sketch of how the cord is wound if no diagram is available. Some of these dial-stringing charts are rather simple while others are very difficult. Usually four turns around the tuning shaft is sufficient for the tuning of most dial cords (**Fig. 12-45**).

Tie the dial cord into an end loop and fasten it to the metal lip within the metal tuning drum. String the dial cord around the various pulleys and tuning shaft, and end up at the other end of the dial drum assembly. Tie a knot to one end of the small metal spring and stretch the spring to apply pressure on the dial assembly. Clip the end of the spring into the metal clip on the drum assembly. Rotate the tuning dial shaft and notice if there is no dial slipping on the tuning shaft. Sometimes if the dial cord is slipping around the tuning shaft, clip a couple of turns off of the metal spring and rehook the spring in the small hole or clip of the drum area.

Tune in a local broadcast station to set the dial pointer. For instance, if the local station is at 1400 kHz on the dial, tune in that station and set the dial pointer at the same frequency. The dial cord might go under a strip of metal at the back of the dial pointer. Place a drop of glue on the cord at the dial pointer, at the coil spring end, and at the dial cord fastened to the drum assembly (**Fig. 12-46**).

Old Volume Controls

Old volume controls have a tendency to make a scratching noise when the control is rotated. Some volume controls can become worn, and the audio might cut in and out

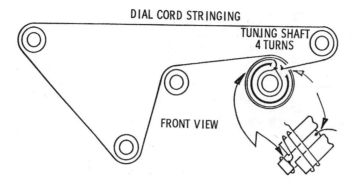

FRONT VIEW

FIGURE 12-45. Most early auto radios have a simple dial cord stringing showing the various pulleys and tuning shaft.

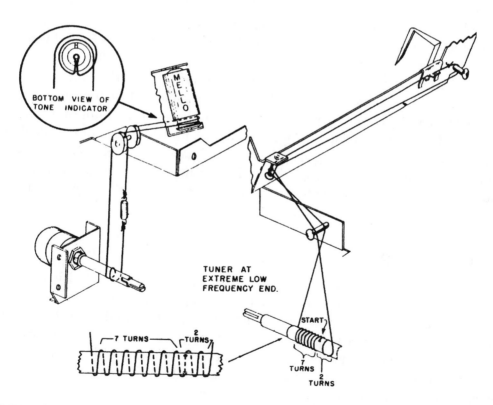

FIGURE 12-46. Some of the early car radios have a more complicated dial cord stringing arrangement.

Make of Radio	Volume Control Resistance
12v Tube	500k Tapped 100k 500k Tapped 150k 1Meg Tapped 180k
Loctal Tube	350k 500k
12v Hybrid	1 Meg Tapped 300k 2 Meg Tapped 500k 5 Meg Tapped 2 Meg
Early All Transistor	10k -20k -10k Tapped 2k- 13k Tapped 3k - 150k Tapped 75k
12v All Transistor	10k -12k -25k -50k - 100k
AM/FM All Transistor	10k tapped 5k -7.5k tapped 5.5k - 10k tapped6k -25k tapped 6k

CHART 12-10. A chart of the various volume controls found in the early antique car radios.

on the control at that point. Replace the old volume control when the music cuts up and down; these controls cannot be repaired with cleaning liquid sprayed inside the control area. A scratching noise when the control is rotated can sometimes be repaired by placing cleaning fluid down inside the control. Suspect a loose ground wire when the audio cannot be turned down.

The volume controls found in the antique car radio might have a resistance from 500K-ohm to 1 megohm within the 12-volt vibrator radio. A 1- to 5-megohms control is found within the 12-volt hybrid car radio. The early 6-volt transistor car radio might start at 10K to 13K ohm with a tapped connection to the tone control, while the 12-volt all-transistor car radio starts at 7.5K ohm and is tapped at 5.5K ohm or a 10K ohm tapped at 5K ohm (**Chart 12-10**). Most volume controls and switches are available at most electronics and mail-order firms. The control shaft length should be cut off to the same length as the old volume control.

Typical AM Alignment

The AM/FM signal generator and output meter was used to align the RF, Converter, and IF circuits. The output of the generator was connected to the base terminal of the converter stage for IF alignment through a 0.1-uF capacitor and the shield to common ground. Set the AM IF frequency at 262 or 455 kHz on whatever IF frequency is being used in the car radio. Rotate the AM dial to the high frequency end-stop. Connect the output meter or FET-VOM meter across the voice coil of the speaker. Adjust each core of the IF transformer for a maximum reading on the meter.

Likewise, set the signal generator to 1615 kHz and the radio dial setting to the same frequency. Connect the signal generator through a dummy load to the antenna receptacle. Adjust the trimmer capacitor (A5), found in the oscillator circuit of the converter stage, and RF trimmer capacitor (A6), found in the collector circuit of the RF amp transistor,

ALIGNMENT INSTRUCTIONS

Check for specified source voltage.
Use only enough generator output to provide a usable indication.
Suggested Alignment Tools: A1 thru A5, A9, A10   GENERAL CEMENT: 5009, 8728A, 8987   WALSCO: 2531X, 2536
                          A6, A7, A8   GENERAL CEMENT: 8290, 8868, 9087   WALSCO: 2528, 2587

	SIGNAL GENERATOR COUPLING	SIGNAL GENERATOR FREQUENCY	RADIO DIAL SETTING	INDICATOR	ADJUST	REMARKS
1.	High side thru .1mfd to point Ⓐ, low side to ground.	262KC (400~ Mod.)	High Freq. End Stop.	Output Meter across voice coil.	A1, A2, A3, A4	Adjust for maximum.
	Check setting of oscillator coil core (A5). Rear of core should be 1 3/8" from mounting end of coil form.					
2.	High side thru 82mmf to antenna receptacle, low side to ground.	1615KC	High Freq. End Stop.	Output Meter across voice coil.	A6, A7, A8	Adjust for maximum.
3.	"	600KC	600KC Signal	"	A5 A9, A10	Adjust for maximum. Repeat step 2.
	With radio installed in car and antenna fully extended, tune in a weak station near 1000KC and adjust A8 for maximum output.					

CHART 12-11.   A typical early AM car radio alignment chart and instructions of adjustment of the IF transformer, RF, and low-band adjustments.

to maximum on the output meter. Tune in a weak station around 1400 kHz and adjust the antenna trimmer capacitor for maximum output. Follow the manufacturer's alignment chart procedures for correct RF and IF alignment (**Chart 12-11**).

Filter Choke Replacement

Filter chokes that are wound on a core like the regular power transformer can be repaired if they are no longer available. Some filter chokes have an iron core with metal laminations. You may find some filter chokes that are still available from mail-order firms in the 5-15 Henries with a 50- to 90-milliamperes current rate. The filter chokes found in the hybrid car radios might range from 0.00 Hy to 500 microhenries with a DC resistance of 0.2 to 0.9 ohms. These filter chokes are found from 2 amps at 32 volts to 9 amps at 32 volts.

Try to replace the burned or charred filter choke with one taken from another old car radio. You can rewind the choke coil if one is not available. Remove the burned core from the metal laminations by removing the outside metal wrapping. Now pry loose each metal piece from the coil form. It is much easier to wind the new coil with the laminations removed.

Count the number of turns as you unwind the burned windings. Choose the same diameter of wire that was used on the old coil. Use a wire gauge to select the correct enameled wire for the new choke coil. Enameled magnet wire from a 21 to 26 gauge can be used to rewind the filter coke coils.

First, place a layer of masking tape around the core area if it's not already insulated. Wind the coils in layers and place a layer of cellophane tape between each layer. Wind the coil by placing the wire in a closely wound (CW) fashion. After the coil has been completely wound, place a layer of masking tape over the outside winding. Replace each metal piece to return the metal laminations back to form the choke coil or transformer. Check the inductance of the original filter on the schematic and compare the two measurements, if one is handy.

A defective filter choke can cause hum in the audio even after all filter capacitors have been replaced. Very often the filter choke becomes burned and the main fuse does not blow or open. This can happen when tin foil is wrapped around the open fuse and is not replaced with the required amp fuse. The choke coil wires become hot and they short out inside the choke coil. You can spot a bad filter choke with a burned outside wrapping on the choke coil.

Replacing the Dial Lights

Replace the dial light when no light bulb is lit, or even if the bulb does light up but has a black glass. Replace it now, while the car radio is apart on the bench, as one with a black glass will not last very long, even if it now lights up. The 6-volt vibrator power supply might have a 6-volt screw-in bulb. A 12-volt vibrator car radio may have a 12-volt (1891) bayonet bulb or lamp. The hybrid or integrated 12-volt car radio has a 47, 57, or 1891 dial light bulb, while the all-transistor car radio may have an 1891 or DS-27 found in the Delco car radio series. A foreign car, such as the Volkswagen, can have an 1892 pilot light. The 6-volt Volkswagen might have a number-47 pilot light.

Inspect the socket and wiring when the pilot light will not light up. Make sure that no hot wire is grounded at the end of the socket or that it isn't pinched when the covers are replaced. A pinched wire can short out the whole power supply system and cause the main fuse to blow. Most pilot light bulbs are still available and can be picked up at most electronics dealers and mail-order firms.

SOME CASE-STUDIES

Only a Rushing Noise and Local-Station Reception

In this example, the RF tube was initially suspected as the culprit in a Mopar 602 car radio; the 7A7 tube was removed and it tested normal. Only a local radio station at 1400 kHz could be heard, along with a loud rushing noise. When the RF loading coil was touched with a screwdriver blade, the local station became louder. One end of the loading coil was broken loose from the grid terminal 6 of the 7A7 RF tube.

No Sound—Only a Rushing and Humming Noise

All of the tubes were checked in an Allstate 6330 model car radio and tested well. When a screwdriver blade was touched to the 500K-ohm volume control, a loud hum was heard. This meant that the audio was normal. A quick voltage check on the converter tube indicated very little voltage on the plate terminal pin 5.

By checking the first IF primary winding terminals 3 and 4 of L4 indicated that voltage was going into the primary winding and not out.

No resistance or continuity of the primary winding was measured with the low-ohm scale of the DMM. Removing the IF transformer from the chassis, and taking the coil outside of the metal shield, found one coil lead was broken in two, off of terminal 3 of L4. Extending the coil wire with a piece of hookup wire restored the converter circuits (**Fig. 12-47**).

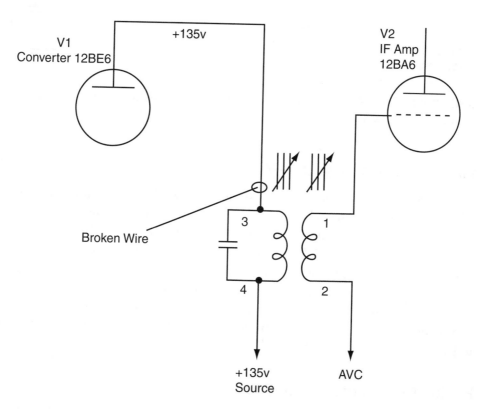

FIGURE 12-47.   A broken wire off of the IF coil winding inside the shielded can cause loss of sound in an Allstate 6330 model antique car radio.

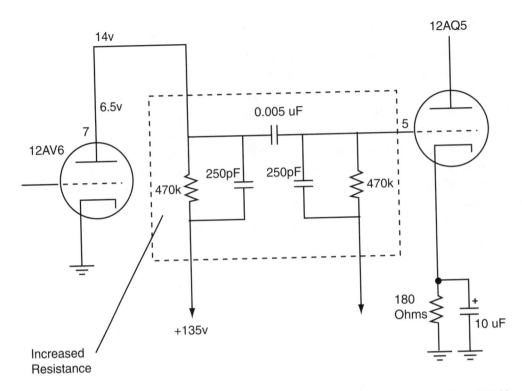

FIGURE 12-48. The weak and distorted Allstate receiver was repaired when the 470-K ohm plate resistor had increased to 3.9 megaohms.

Weak and Distorted Audio—Allstate

The sound of a regular local broadcast station was weak and distorted in an Allstate 6330 model car radio. After checking all tubes, voltage measurements were made on the AF amp tube and the audio output tube (12AQ5). The plate voltage on the 12AV6 tube (pin 7) was low at 14.7 DC volts. On checking the schematic, the plate load resistor was found molded inside of a component with other resistors and capacitors (**Fig. 12-48**). The voltage measured at pin 7 was around 137 volts. No doubt the plate load resistor had increased in resistance. Since the molded module was no longer available, the correct resistors and capacitors were soldered together to form another module, which solved the weak and distorted car radio.

A Really Weak Motorola GV800

In this example, the volume on the Motorola car radio became weak and weaker as it began to play. You could barely hear the music on any station, even though they all could be tuned in. There was no doubt that the problem was in the audio output

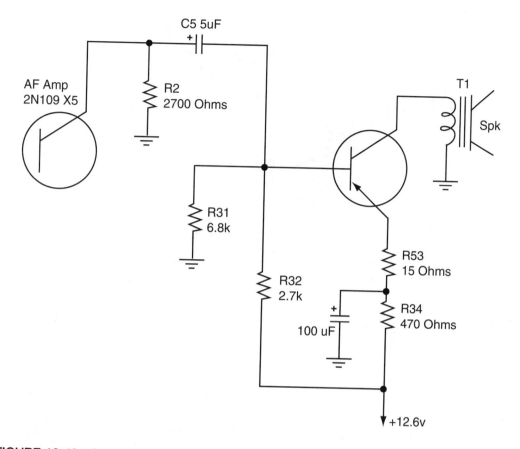

FIGURE 12-49. C5 caused the sound to become very weak in a Motorola 6V800 auto radio.

circuits. With an audio signal from the signal generator, the tone was good on the base of the driver transistor (2N270). When X5 or the AF amp (2N109) collector terminal was touched, the audio signal was very weak (**Fig. 12-49**).

Upon checking the schematic, a 5-uF electrolytic coupling capacitor was found between the AF amp and the driver transistor. The suspected C5 was shunted with another 5-uF electrolytic and with the volume popped up. Replacing C5 with another electrolytic (5-uF 10-volt) solved the weak car radio reception.

Hum and a Dead Lincoln Auto Radio

A Lincoln 85BH car radio was pulling heavy current when connected to the bench power supply. A quick transistor test on the push-pull output transistors (2N399) indicated that one was open and the other was shorted. When both were removed from the radio, the one that tested open again tested normal. Replacing both output

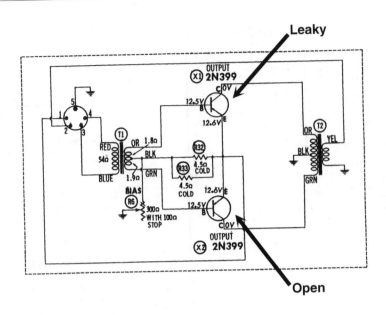

FIGURE 12-50. Both 2N399 output transistors were replaced in a Lincoln car radio for a dead radio with hum.

transistors and the 4.5-ohm bias resistors (R32 and R33) solved the problem of the dead car radio.

Bias resistor R6 (300 ohms) was adjusted for a clean and clear music sound, with 14.4 volts applied. R6 was adjusted for a meter reading around 100 milliamperes. Connect the DC milliamperes meter in series with the center tap (black lead) of the power output transformer (T2) to measure the correct current drawn by both transistors. (**Fig. 12-50**).

Only a Whispering Delco

In this example, a Delco 7303311 car radio appeared to be dead, but you could hear a click or whisper when the volume control was turned up. Output transistor DS-501 was checked and it tested normal; even the fusible resistor (0.47 ohm) bias resistor in the emitter terminal was good. A voltage check on the base and emitter terminals of the output transistor was higher than normal (12.3 volts). When a voltage measurement was made on the first AF amp (DS-66), the voltage was also quite high on the collector terminal. R2 (600-ohm) emitter resistor was found open. Replacing R2 resolved the whispering Delco car radio (**Fig. 12-51**).

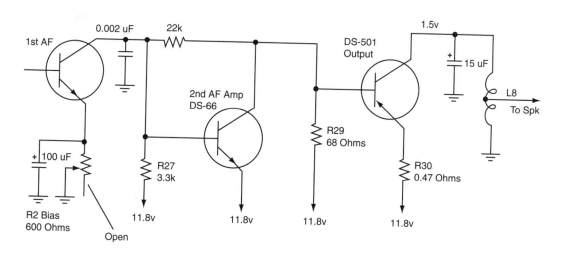

FIGURE 12-51. Only a whisper of sound could be heard in a Delco car radio that had only a rushing noise.

Motorola TM296M Radio That Cuts In and Out

In this case, the volume would cut in and out when the Motorola radio was played really loudly. At first, the audio section was suspected, and an external amp was connected at the volume control to monitor the audio sound. Sometimes the radio would play for days without cutting out. Then one day the radio cut out with the external amp connected, indicating problems within the front-end section of the radio.

After several voltage and transistor tests, the defective component showed up by moving a few parts around with an insulated tool. When C14 (0.047-uF) bypass capacitor within the emitter terminal of the RF Amp (Q1) was prodded, the sound began to act up. Replacing C14 resolved the intermittent car radio (**Fig. 12-52**). It's wise to replace all old paper capacitors, especially those tied to a B+ source, in these antique car radios.

Tough Dog Mercury 6TMM Car Radio

The radio stations at the top end of the dial would disappear, and they seemed to go down the dial as the car radio operated for several hours. Because the stations would quit at the high end of the dial, both the RF and converter transistors were replaced; nevertheless, the same problem still existed.

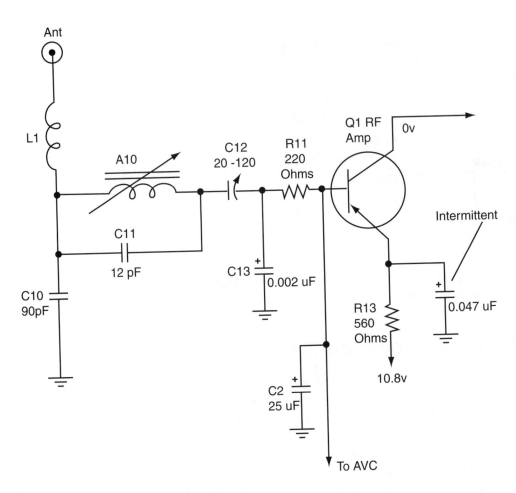

FIGURE 12-52. The intermittent reception in a Motorola car radio was caused by an intermittent bypass capacitor in the emitter circuit of the RF amp transistor (Q1).

The 130-pF trimmer and oscillator coils were checked and prodded with no results. A 0.005-uF bypass capacitor in the leg of the converter transistor (2SA72) was suspected and removed. The capacitor tested as good. A new 0.005-uF capacitor was installed, and the radio operated for weeks without a change in stations. No doubt the bypass capacitor changed value or opened up under load (**Fig. 12-53**).

Hum and Distorted Sound—Automatic Radio

The 7.5-amp fuse was blown in an Automatic model DR4107 car radio, with an overheated output transistor (8P04) and output transformer (T2). The output transistor

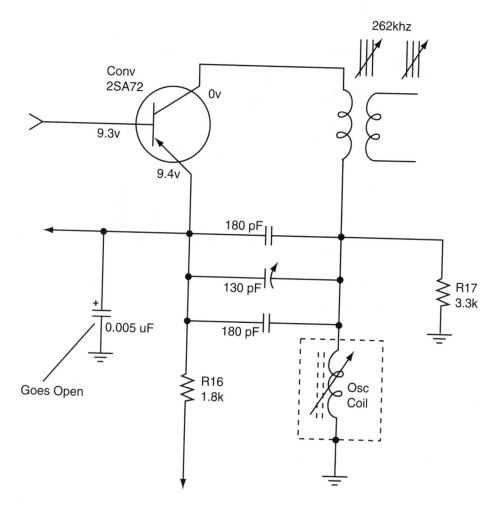

FIGURE 12-53. Replacing a 0.005-uF bypass capacitor in the emitter circuit of a tough dog Mercury car radio stopped the station from fading at the high end of the dial assembly.

was removed, tested, and replaced. Before replacing the output transistor, R28 was found open. The 0.5-ohm fusible resistor was replaced, and all other bias resistors were checked out to be normal. Although the outside winding of T2 showed signs of overheating, the music played loud and clear (**Fig. 12-54**).

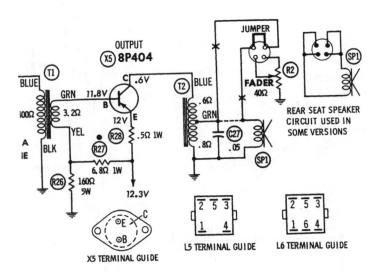

FIGURE 12-54. Replacing a shorted output transistor (8P404) and a fuseable emitter resistor resolved the weak audio.

CHAPTER **13**

Speaker Specifications and Problems

In the early car radios, a round speaker with a 3.2-ohm voice coil provided sound, and a magnetic field winding in the speaker provided filtering action in the power supply. For instance, a Mopar 602 car radio had a $6^7/8$-inch speaker with a 3-ohm voice coil, and a field coil of 4.2 ohms. The field-coil winding served as a filter choke.

The early car radio had separate cables bolted to the front dashboard to operate the radio (**Fig. 13-1**). The car radio was mounted on the firewall with the enclosed speaker in a separate container. Later on, a 4- or 5-inch PM speaker was placed inside the metal radio cabinet that mounted underneath the dashboard with the knobs and dial for easy viewing. A metal bracket that bolted the backside of the car radio in place was attached to the car's firewall.

There are many different size speakers found within the early car radios. Besides the round speakers, square, pin types, and oblong PM speakers were found mounted in the allotted spaces within the automobile. The hybrid car radio had a 7-inch round speaker at 3.2 ohms, a 4x6-inch speaker at 3.2 ohms, and the 6x9-inch speaker at 3.2 ohms impedance. Later, the PM speaker impedance changed

FIGURE 13-1. The early car radio with separate speaker that bolts to the firewall.

Type of Radio	Speaker Size	Impedance
Early Vibrator Car Radio	4, 5, 5x7, 4x6, 7, 8, 6x9 inch	3-4 Ohm
Hybrid Radio	4x6, 4x9, 4x10, 5, 6, 6x9, 7 inch	3-4, 4, 6, 8 Ohm
All Transistor Radio	4x6, 4x10, 5x7, 5, 6, 6x9, 6x10, 7 inch	3-4, 4, 6, 8 Ohm
AM/FM Transistor	4x10, 5x7, 6x9, 6x10 inch	6, 8, 10, 16, 40 Ohm

CHART 13-1. The early car radio speakers and speaker impedance.

to a higher impedance at 4, 6, and 8 ohms (**Chart 13-1**). The actual resistance measured in ohms of a known PM speaker voice coil is close to 1 ohm less than the given impedance, when measured with a good DMM.

TODAY'S SPEAKERS

The speaker is a transducer that converts electrical energy signal into sound waves that we can hear. Today, you might find the PM speaker mounted in the bottom door, under the rear deck, or behind a side panel. Sometimes the factory speaker that appears with the auto receiver might have a thin paper driver for solid bass response. The low-priced car radio speaker might have paper or foam surrounds that can easily deteriorate very quickly. Replacing the old speaker system with high-quality speakers might make a big difference in the quality of sound with today's large and powerful speakers and amplifiers.

The quality name-brand speakers have stronger magnets and heavy-duty voice coils that produce higher quality sound. In addition to a large woofer or subwoofer speaker, a 4-way speaker provides a tweeter, midrange, and super tweeter for better clarity, detail, and higher definition of music. These rubber surrounds are durable, high-performance, delivering great bass every day. Most subwoofer speakers have cones of dense and rigid materials for a more accurate bass response.

Top- or bottom-mount speakers are mounted in several different ways. With the top-mounted speaker, the mounting height must be taken into consideration. How far does the speaker stick out or above the mounting surface? Side- or door-mounted speakers must not stick out enough to be in the road of a passenger. Besides, the top-mounted speaker might not fit in the door panel or in a factory speaker cutout area.

With a top-mounted speaker, the speaker mounts above the mounting surface, and the bottom mount is below the mounting surface. Make sure the speaker to be mounted will fit into the desired space before drilling and cutting out any speaker holes in the door or back panels. A flush-mounted speaker is even with the surface, while a thin mount is slightly above the surface, and a surface-mounted speaker is on the top side of the surface.

Speaker Components

Today's permanent magnetic speakers consist of a framework, surround, cone, voice coil, spider, and a PM magnet. Low-priced speakers have a stamped-out metal frame-work or basket that holds the speaker components together with no special surround material. The framework is bolted or secured to the front dash or back ledge of the automobile.

The cone might also be constructed from paper with the voice coil that operates within a magnetic field of the magnet mounted at the end of the speaker. A voice coil on the replacement speakers might have an impedance of 3.2, 4, or 8 ohms. The spider is a stiffening device that holds the cone into position near the magnet pole piece. A replacement auto radio speaker might have a 2.8- to 9.8-ounce magnet.

The high-powered speaker found in some sound systems from 50 to 1000 watts are constructed of heavy-duty material. The framework or basket found in large subwoofer speakers might be made up of forged aluminum or die-cast aluminum that acts as a heat sink to handle more power. Some high-powered speakers have a standard painted stamped-steel framework or a non-resonance rigid steel framework.

A surround material holds the cone of the speaker to the outside framework. The surround material might consist of a heavy butyl rubber for long-lasting durability and improved linearity. A high-compliant surround for sonic excellence and a thermal-formed polyurethane surround are found in the high-priced subwoofer speakers. The Autotek S1240 speaker has a thermal-formed Santopreme surround, while other speak-ers might have a large black foam surround. You might find a wide surround in a few speakers providing greater bass to the sound system (**Fig. 13-2**).

The speaker cone found in low-priced speakers are made up of paper and light material. A high-powered subwoofer speaker might have high temperature and com-posite cones, poly, graphite, and polypropylene curved linear cones that create sonic realism. The stiff Kelvar paper cone is found in subwoofer speakers. The glass aramide composite cone offers very high stiffness and damping without weight associated with paper or poly cones. The surround speaker might use a silver-injected poly cone with a thermal-paper or fiber poly type of surround. Some speakers have a stiff or thick paper or fiber cone with a plastic-coated surface.

The voice coil is a moving coil found at the end of the speaker cone area that moves through a magnetic field of a large PM magnet or motor. A voice coil found in factory or low-priced replacement speakers might have an impedance of 4, 8, 10, and 16 ohms. The voice coil impedance of a high-powered subwoofer is usually 2 or 4 ohms. Check the continuity of a dead or open voice coil with the low-ohm scale of the DMM.

FIGURE 13-2. The foam surround glued to the speaker paper cone with a coaxial tweeter speaker in the center.

The low-priced and replacement speaker might have Litz wire between the voice coil and the speaker terminals. A voice coil found in large subwoofers might be made up of two-layer dual edge-wound ribbon wire to improve efficiently and transient response. A two-layer voice coil creates smoother highs on a black Kapton former. The dual or 2-layer voice coils have lower harmonic distortion.

The subwoofer might have a dual 4-ohm voice coil to handle extreme amounts of volume or power. A 4-layer black anodized aluminum voice coil can take very high temperatures. Some speakers have a 4- to 6-layer voice coil found on a Kapton former. The aluminum voice coil might be found on a Kapton former for peak power operations. A large subwoofer of 18 inches or more might have a 4- to 6-layer copper wire that can handle 1000 watts of power. Speakers with dual 2-inch voice coils should be wired like any other speaker as two separate speaker connections.

Some voice coils are woven right inside the spider cone to prevent the voice coil from bouncing off of the cone at high volume or power. The voice coil wires might be formed right into the spider cloth. The voice coil wires can be destroyed with too much heat, or too much high power applied and DC voltage across the speaker terminals with one terminal grounded.

The voice coil on some subwoofers are cooled with a three-quarter-inch vent in the pole piece for increased power handling. The direct voice coil venting provides cool air moved over the coil in the subwoofer to maintain lower operating temperatures. A cool subwoofer voice coil can handle more power. Aluminum flat or round voice coils help to keep the coil cool with high wattage amplifiers.

The spider component toward the end of the speaker cone area can provide the apex of the vibrating cone of the PM speaker. Flat spider and two-layer voice coils are designed for low harmonic distortion. A flat and responsive spider proves stiffness that varies across the spider surface. The spider and cone are attached to the Kapton former at the center of the flux core so that the spider can move in both directions symmetrically. A loose or damaged spider can cause a blatting sound when music is played.

The PM magnet found at the center pole piece of a speaker provides a magnetic field for the voice coil to work in. A power neodymium magnet or internal pole piece might also be called the motor component. The subwoofer might have an oversized magnet extended pole piece to increase the magnetic field linearity. A neodymium magnetic motor with a forged aluminum basket provides precision core geometry. Some subwoofers have a double-stacked 20mm-high magnets. Speakers with heavy 120-ounce double-stacked magnets can handle 300-400 RMS watts of power.

Factory Speaker Location and Size

Most speakers that fit in the front side doors are rather small: 5¼ or 6½ inches with a depth of 2 inches. Check the depth before cutting out the hole, or check where the car factory has the speaker location. For instance, in a Buick Regal 89-94, the front door has a 4x6-inch opening and a space of only 2 inches deep. A Chrysler New Yorker 1989-1993 has a 5¼-inch front-door speaker opening with only 2¼ inches of depth. The Oldsmobile 1998-1999 Cutlass has a 4x6-inch speaker front-door mounting area with 2⁵/₈ inches of depth. Always choose the right speaker for the depth and physical size whereever the speaker will be mounted.

The factory car rear-deck speaker openings are much larger with the mounting of a 5x7- or 6x9-inch size speaker. The 1995-2000 Chevrolet Lumina automobile has a rear-deck mounting for a 6x9-inch speaker. A Dodge Dakota (1987-1996) has a 5x7-inch opening for mounting the speaker in the rear deck. The Honda Hatchback (1986-1987) has a rear-side shelf for a 6½-inch speaker, while a Lincoln (1996-2000) Navigator auto has a rear-door mounting of a 6x8-inch PM speaker.

Speakers mounted within the factory front decks or dash might be from 3½, 4½, 5¼, 4x6, and 4x10 inches. In a Dodge Caravan, the car dash might hold a 5¼-inch speaker with only a depth of 2½ inches. The Honda 1985-1998 Ranger can contain a 3½-inch speaker in the front car dash. The Pontiac Bonneville Sedan of 1989 to 1991 can hold a 4x6-inch PM speaker. A 1996-1998 Suzuki Sidekick can mount a 4-inch PM speaker in the front dash (**Fig. 13-3**).

The rear-side panel speaker in the automobile with custom openings can hold a 4, 5¼, 6½, 5x7, 6x8, 6x9 PM speakers. A 1989-1995 Dodge Spirit can handle a 5¼-inch speaker mounted in the side panel, while a Ford Aerorta (1996-1991) can hold a 6x9-inch speaker in the mid-rear side panel. The 1987-1993 Mustang auto has a 6x8-inch mounting hole in the rear-side panel. A GMC Jimmy 1994-1996 2-door truck can mount a 6x9-inch PM speaker in the mid-rear side panel.

The different size PM speakers found in the rear wheel well might be from 4, 5, 5¼, 6½, 6x8, and 6x9-inch custom speaker openings. The Honda LTD Crown Wagon 1990-1991 has a 6x8-inch cutout in the rear wheel well with a 5-inch clearance. A rear

FIGURE 13-3. Replacement speakers may appear rectangular, round, and square, with mounting holes.

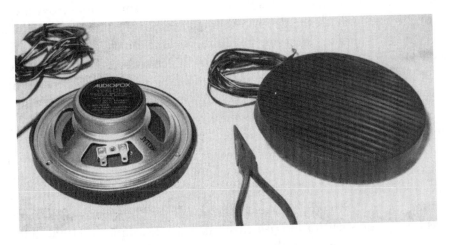

FIGURE 13-4. Check the depth of the speaker to be mounted in doors, front kick panels, and rear-door panels.

panel pillar post speaker of 4 and 5 inches can be mounted in a Jeep Wrangler automobile. The Suzuki 1998-2000 Esteem Wagon has an opening for a 6½-inch PM speaker.

Whether the speaker is mounted in the front dash, rear deck, side panels, door panels, or rear seat panels, check the speaker openings and the depth to mount the PM speaker. Make sure the right custom speaker will fit on top or on the bottom side of the speaker mounting. Compare the depth of the speaker to the allowed mounting space (**Fig. 13-4**). Remember that the various speakers are mounted within the front deck, front kick panels, front-door panels, rear-door or rear-seat panels, rear deck, trunk or hatchback, rear-side panels or rear-heel wells, tailgate, behind the seat, and mid-rear-side panels.

Speaker Frequency Response

The speaker frequency response of a 3½-inch PM speaker might range from 50-28 kHz. A 3½-inch Infinity 3012CF speaker has a frequency response of 100-20 kHz, while a Pioneer TS-A878 has a frequency response of 60-28 kHz. The Polk EX351 speaker has a frequency response of 130-18 kHz in a 3½-inch PM speaker.

A 4x6-inch PM speaker might have a frequency response from 45-32 kHz. The Jensen J462CX speaker has a frequency response of 45-25 kHz, while a Kenwood KFC-4675 speaker has a frequency response of 50-25 kHz. The Pioneer TS-A4615 4x6-inch speaker has a frequency response of 45-32 kHz, and a Rockford FRC1246 has a frequency response of 128-20 kHz for a 4x6-inch speaker.

The 5¼-inch PM speaker might have a frequency response range from 35-30 kHz. A Blaupunkt PCXg542 5¼-inch speaker has a frequency response of 60-20 kHz, while a JBL GT0520 speaker has a 65-21 kHz frequency response. A Panasonic 5¼-inch speaker has a frequency response from 35-25 kHz.

The 6½-inch PM speakers might have a frequency response between 28-32 kHz. An Infinity 612i has a 6½ to 6¾-inch speaker with a frequency response of 60-20 kHz, while a Cerwin-Vega SS-262 6½-inch PM speaker has a frequency response of 37-22 kHz. The larger the cone area of the speaker, the lower the frequency response. The 5x7- or 6x8-inch PM speakers might have a frequency response range from 30-32 kHz. A 4x10-inch speaker might have a frequency response between 38-32 kHz. The 6x9-inch speaker might have a frequency response from 28-30 kHz (**Chart 13-2**).

Speaker Size	Frequency Response
3 1/2 PM Speaker	50-28 kHz
4x6"	45-32 kHz
51/2"	30-35 kHz
61/2"	28-32 kHz
5x7 or 6x8	30-32 kHz
6x9	28-30 kHz
8"	28-50 kHz
10"	30-200 kHz
12"	20-2000 kHz
15"	20-500 kHz
18"	20-150 kHz

CHART 13-2.

The different sizes of speakers and the speaker frequency response.

Power Range and Peak Power

To replace or install a new speaker, make sure the RMS wattage falls within the speaker's recommended power range. The power range indicates the power required for a continuous flow of music at a reasonable volume. The power rating or power range is the specified power required by equipment for normal operation of an amplifier.

A peak power rating indicates how much power the speaker can handle during a brief period of time. Peak power might be called the maximum power handled by the PM speaker. For instance, a Punch 15-inch subwoofer PM speaker has a power range of 50-150 RMS watts and a peak power of 300 watts. The Cerwin-Vega 15-inch subwoofer speaker has a power range of 95-500 RMS watts and a peak power of 1000 watts. Likewise, the Punch HX2DVC 18-inch PM speaker has a power range of 150-600 RMS watts and a peak power of 1200 watts.

A subwoofer 8-inch PM speaker might have a power range of 36-150 RMS watts with a peak power between 200-300 watts, while a 10-inch subwoofer might have a power range of 12-400 RMS watts and a peak power range between 200-1200 watts. The 12-inch subwoofer PM speaker might have a power range of 18-400 RMS watts and a peak power range between 200-2000 watts. A 15-inch subwoofer speaker might have a power range of 42-1000 RMS watts and a peak power from 400-2000 watts. The very large 18-inch PM speaker might have a power range from 150-600 RMS watts and a peak power range from 1200-2000 watts (**Chart 13-3**).

Make	Size	Power Range in Watts	Peak Power in Watts
Blaupunkt PCXG352	31/2"	2-30 Watts	90 Watts
Pioneer 75-A878	4x6	2-15 Watts	60 Watts
Infinity Kappa 42.1.1	4"	2-45 Watts	90 Watts
Kenwood KFC 1387	5-1/4"	2-67 Watts	140 Watts
JBL GTO630	6-1/2"	2-50 Watts	125 Watts
Sony	6x9	2-40 Watts	200 Watts
Punch	8"	50-100 Watts	200 Watts
Thunder 4000	10"	50-200 Watts	400 Watts
Pioneer	12"	35-160 Watts	500 Watts
Punch HE2 DVC15	15"	125-500 Watts	1000 Watts
Punch HE2 DVC18	18"	150-600 Watts	1200 Watts

CHART 13-3. The different speaker sizes with a power range and peak power in watts.

FIGURE 13-5. A six-inch coaxial speaker with tweeter that can be mounted in the dash, back door, or rear-door panel.

Two-Way and Four-Way Speakers

A 2-way speaker might be called a coaxial speaker that contains a single tweeter speaker mounted above the woofer speaker cone area. The coaxial speaker has a very large speaker of low frequency and a smaller speaker mounted within the larger speaker at the high frequencies. This combination of tweeter and subwoofer speakers with a crossover network can provide a good wide-range response for the space it takes to mount it in. The 6x9 2-way flush mount speaker might contain a 2¾-inch tweeter at 60 watts and 180 watts of maximum power.

A three-way speaker system has a tweeter mounted above the subwoofer's cone and adds a midrange driver for greater clarity and power. The 6¼-inch flush-mount 3-way speaker might contain a 2-inch midrange and a 1-inch tweeter with a 40 watts RMS, and 120 watts of peak power at 4-ohms impedance. A 3-way 6x9-inch speaker might have a 2½-inch midrange and a 1½-inch piezo speaker to cover the full range of frequencies.

The 4-way speaker system includes a super tweeter that adds more detail to the higher frequencies. Sometimes a separate subwoofer, tweeter, and crossover network in the speaker system can provide better power handling and performance than the 2- to 4-way combined speaker system (**Fig. 13-5**)

Speaker Enclosures

Full-range truck, vans, wagons, hatchback, and back-ledge enclosures might have up to an 8-inch woofer PM speaker and a piezo horn tweeter that might have a frequency response of 54-20,000 Hz. Each box can handle from 20 to 100 watts of power with an RMS of 200 watts. A three-way box enclosure can handle up to a 10-inch speaker and a 3-inch horn at midrange with a frequency response of 38-20,000 Hz, and each box can handle 20-200 watts of power.

A two-way full-range box enclosure might have two 10-inch speakers that can handle 20-100 watts and 200 peak watts per channel. Another two-way box enclosure can have two 12-inch subwoofers in one container with a 20-100 watts RMS and 200 watts of peak power. There are many different speaker enclosures that you can choose from or you can make your own custom-built containers. Many speaker enclosures are constructed by professional auto installation specialists.

The truck and van speaker commercial enclosures might be mounted in an 8-inch box band pass speaker or in a 10-inch two-way truck speaker with 150 watts RMS and a 450 watts of peak power. Slanted two-way 8-inch speaker boxes that can fit in a hatchback might have 125 watts RMS with 375 watts of peak power. The 8-inch two-way truck/van speaker enclosure contains an 8-inch subwoofer with 100 watts RMS and a peak power of 300 watts. Many of these model boxes can hold speakers with 2- or 4-ohm impedance with carpeted types of enclosures. Commercial 6x9 or 8-inch subwoofer speakers boxes made up of a wooden enclosure, covered with carpet, can add awesome bass to the car radio system.

TWEETERS, MIDRANGE, WOOFERS, AND SUBWOOFERS

The Tweeter Speaker

A tweeter speaker is a speaker that produces high-frequency sound. The tweeter-size speakers with molded corners can be mounted within the doors and side panels ,and can handle up to 50-100 watts of power. These tweeter speakers can be flush, at an angle, or surface mounted, and can be three-fourths of an inch deep. A tweeter pod type of speaker can be mounted in a flush mount for greater angle listening-power volume. The tweeter speaker may very from a 1-inch piezo to a 1½-inch PM speaker or a horn type of speaker. A tweeter speaker can be mounted separately or one found inside a two-, three-, or four-way speaker arrangement. The tweeter speaker might be found with a three- or four-way bass-reflex surface-mounted speakers.

Midrange Speakers

The midrange loudspeaker operates at frequencies in the middle range of the audio spectrum. A midrange speaker is found between a tweeter and a subwoofer loudspeaker. A midrange speaker may be a 4, 5¼, 6, 6¾, and 8-inch diameter speaker that includes rubberized plastic surrounds, polymer/mica composite cones, and an added strength cone with a wider dynamic range and lower distortion. A mica-filed poly cone provides a wider frequency response.

The midrange speakers have a cupped spider to provide excellent suspension and low resonance. The midrange speaker might be mounted separately, located in an enclosure, or in a two-way and three-way speaker system.

Woofers and Subwoofers

The subwoofer is much larger and produces more bass than a midrange or woofer type of speaker. The subwoofer speaker reproduces the bass frequencies and may range from an 8, 10, 12, 15, to an 18-inch PM speaker. The bone-crushing power of the subwoofer might have a voice coil mounted on a black Kapton former for high-power handling. A strontium ferrite magnet yields a very highly efficient speaker for better high-power needs.

The high-powered 15-inch Thunder 6000 speaker has an injection-molded woofer with eight reinforcing ribs to keep the cone stiff so that it can handle loads of power output. The oversized magnet structure four-layer voice coil, Apical voice coil former, and rubber surround combine to produce strong and very deep bass.

The large 10-inch subwoofer can handle up to 400 peak watts, while a 12-inch speaker might handle up to 1200 watts of peak power. A 15-inch PM subwoofer can handle up to 1200 watts of peak power with a frequency range of 20-200 Hz at 4 ohms impedance. The subwoofer speakers can be found in carpeted enclosures, boxes, and surface mounted.

Component Subwoofers

The small-sealed box subwoofers speakers are mounted in a small enclosure for use in sealed, ported, and band pass boxes. A Pioneer subwoofer speaker might have diamond-plate-foamed IMPP cones made out of injection-molded polypropylene. The Polk Audio DX series subwoofers can be built right in sealed boxes with an extra-rigid polymer/graphite cone with durable rubber surround for smooth response and for longer life. The Sony XPLOD subwoofer can be placed in small enclosures, and have high-rigidity compound poly cones for quick and tight response. Double damper suspensions and Santpopreme rubber surround promote linear cone travel and they are very durable.

The Infinity Kappa Perfect is a high-output subwoofer with a lightweight cone that delivers great bass, is extremely rigid, and delivers awesome bass. Die-cast baskets and flared-vertical pole pieces can help crank up great volume of sound. A JLB Power Series subwoofer speaker has a 4-ohm voice coil, Thermalum (TM) woofer cone, a hybrid of polypropylene and nickel alloy. These two natural materials deliver a cone that is extremely light and yet very strong. The power subwoofers have a cast aluminum basket and butyl rubber surround.

The component subwoofer can appear in 8, 10, 12, and 15-inch diameters. The 8-inch subwoofer might have a power range of 35-200 RMS watts with a 300-400 peak watts of power output. A 10-inch subwoofer might have 1200 peak power watts with a frequency range of 20-500 Hz. The 12-inch subwoofer might have a power range between 33-350 RMS watts and a peak power around 450-1300 watts. These really large subwoofer speakers might have a top mount depth of 6½ inches or less. Subwoofer constructed boxes can handle 8- to 15-inch PM speakers.

Component Speaker Systems

The speaker component systems might include all of the required speakers in the standard automobile. Most speakers found in a component system are mounted separately. The speaker component systems might include 4-, 5¼-, or 6½-inch speakers that can handle 8-50 watts with a frequency response of 35 to 30 kHz. The Infinity Component System includes 5¼- and 6½-inch woofers that can handle 2-90 watts of power, while a Polk component system has a 4-, 5¼-, and a 6½-inch woofer that can handle up to 70 watts of power. A Polk/Mono component system might include either a 10- or 12-inch PM speaker that handles 75-400 watts and 75-500 watts respectively with 20-300 Hz frequency response.

A Rainbow CS2/130 speaker system consists of a Cal 25 tweeter, a 5¼-inch woofer and a second-order crossover. The crossover thermal protection enables the speaker to handle 80 watts of power. The 25-mm tweeter has a neodymium magnet and Ferrofluid cooling. The 5¼-inch speaker has a natural-fiber cone, rubber surround, and a 25mm copper voice coil.

Crossover Networks

A crossover network provides a frequency range to the tweeter for highs, woofers for low, and subwoofers for the really low frequencies. Crossovers are designed to allow some frequencies to pass to another speaker while other frequencies are attenuated. The crossover network supplies the correct frequency to each speaker. A tweeter might be installed in the dash or the doors, while the woofers are installed in door panels with matching crossovers. Matching crossover networks provide the correct frequency to the woofers mounted in the trunk, hatch, or anywhere else (**Fig. 13-6**).

A crossover network to a 5¼-inch woofer should not receive frequencies under 100 Hz, while a 6x9-inch speaker should not have a frequency under 80 Hz. A comparative

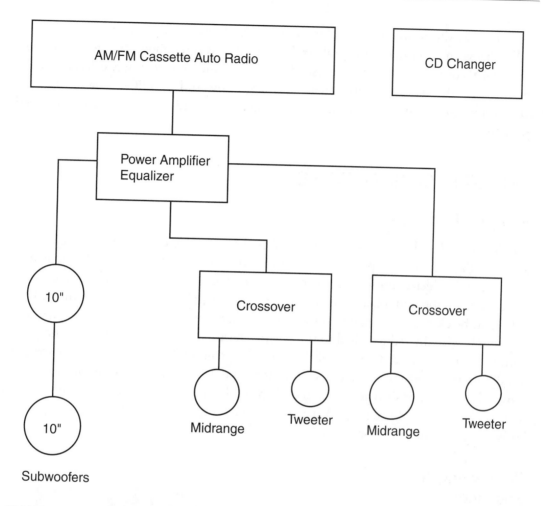

FIGURE 13-6. The crossover networks are found between the midrange and tweeter speakers.

performance of crossovers might be the steepness of the filter slopes, accuracy of summed response, front high-pass range, rear high-pass range, rear low-pass range, subwoofer low-pass range, and accuracy of the frequency setting. A crossover network might be installed into the higher-priced amplifier circuits when it is manufactured.

There are three types of crossovers: digital, active and passive active filters, or crossovers that are made up of active components with an operational amplifier. The active crossover is used in low-level signals and installed between the source component and amplifier. The passive filter might consist of only capacitors, inductors, and resistors. A high passive filter is usually found in midrange and tweeter circuits.

Low-pass filters are connected to subwoofers. Passive filters or crossovers are installed between amplifiers and speakers. The band pass filter allows frequencies to pass for mid-bass or midrange applications. The equalizer and crossover networks are usually mounted in the dash or trunk area. The large powerful amplifier might have the crossover networks built right in the amplifier circuits before connecting to the required speakers.

SPEAKER CONNECTIONS

Speaker Terminals

The low-priced and auto replacement speakers have a soldered connection type of lug terminal (**Fig. 13-7**). The speaker wires are looped into the terminal holes of the soldered terminal and soldered up for a good connection. Some subwoofer speakers have a red button connection and a black one to indicate correct polarity, while others might have a red and black gasket or washer on the speaker terminals to indicate correct polarity.

Simply wire the red terminal to the positive terminal of the amplifier, and the black terminal to the negative terminal of the amplifier. You might find that another speaker has a spring-loaded gold-plated terminal for solid low-resistance connections. Some subwoofers have a push-type terminal and spring-loaded wire sets for easy connections that will hold up to 8-gauge speaker wire. Use larger speaker wire on long-run speaker leads from the amplifier for lower loss of audio power.

Connecting the Speakers

Speakers can be connected in series or parallel to match the required impedance of the amplifier. The speakers are connected in series so that the wire from one side is connected to the speaker and the wire from the other terminal goes to the next speaker. The two outside voice coil terminal connections are now wired to the amplifier. For instance, two 2-ohm speakers are connected in series to equal a 4-ohm impedance of the

FIGURE 13-7. Solder types of speaker terminal connections are found on lower-priced replaceable and mounting speakers.

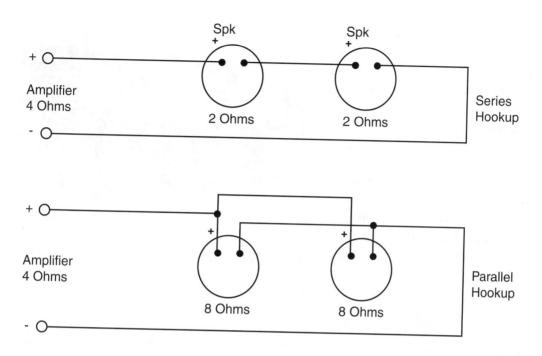

FIGURE 13-8. A pair of 8-ohm PM speakers can be wired in series, and a pair of 4-ohm speakers can be wired in parallel to match a 4-ohm amplifiers output impedance.

output power of the amplifier. The speaker impedance will add when wired in series, and will divide when wired in a parallel hookup.

Several speakers can be wired in parallel to equal the audio amp output impedance. Paralleled speaker wiring requires connecting both speaker leads to the same speaker, to equal the required amplifier output impedance. Two 8-ohm speakers can be wired in parallel to equal 4 ohms of amplifier impedance. The speakers should be polarized when wired in either series or parallel fashion (**Fig. 13-8**).

A series-parallel speaker hookup should equal the amplifier's output impedance for correct transfer of power. The even number of speakers must be added to prevent imbalance of the output impedance of the amplifier. The speaker impedance of 8 ohms at 400 watts, 4 ohms at 780 watts, and another speaker of 2-ohm impedance at 1600 watts should be connected separately.

The length of wire to a given speaker inside the automobile should not cause loss of resistance unless the wire is too small. It's best to use fairly large wire on long speaker runs from the amplifier. A transformer impedance-matching device can be connected between the amplifier and speakers to change the load of an 8-ohm speaker down to 4 ohms or 2 ohms and properly match several different speakers.

A high-impedance speaker load will never harm the amplifier. For instance, a 4-ohm speaker can be connected safely on a 2-ohm output amplifier channel. There can be a loss of wattage but the speaker and amplifier cannot be harmed with this type of hookup. A really low impedance speaker will draw more power from the amplifier, and it can destroy the amplifier if connected to the wrong amplifier output impedance.

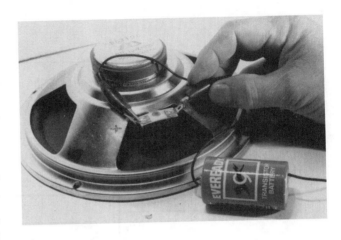

FIGURE 13-9. Check the polarity of the speakers with a 1.5-volt battery and mark the positive terminal on the speaker.

Correct Polarity

All speakers should be wired up for correct polarity; the speaker cones of all speakers should go in and out at the same time. Subwoofer speakers with colored terminals are easy to connect together for the correct polarity. Simply wire the red terminals to the positive and the black terminals to the negative terminal of the amplifier. All speakers should operate in phase for greater listening pleasure.

If the speakers are not polarized with a sign or a colored terminal, you can use a 1.5-volt battery to check the polarity of speakers with solder types of connections. Connect the positive lead of the battery, and quickly touch the other terminal with the other battery clip and notice if the cone is going out or in. Do not leave the battery on too long. If the cone is going outward, mark that terminal as a positive terminal.

Now reverse the polarity by switching the battery terminals and notice if the cone pulls inward. Mark all speakers with positive (+) terminals on the correct speaker terminal (**Fig. 13-9**). The positive terminal of the speaker should be connected to the positive amplifier terminal.

Cable Connectors

Cable connectors, fuse panels, fuse connectors, and power distribution blocks affect all speakers, amplifiers, and other audio components. The cable connector might be found between amplifier and speakers to tie two speaker wires together. Large cable wire and connectors are found between the battery power lead and amplifier. Battery terminators, power rings, fuse holders, power distribution blocks, and power cables may be machined from solid brass for precision fit and finish; gold-plated brass contacts are the best with very little loss.

Some distribution blocks accept 1-4-gauge input wire and supply 4-8-gauge output cable with 24-carat gold-plated connectors. A gold-plated brass power distribution block is very compact which allows it to be mounted in places where space is at a minimum. The plated distribution blocks can accept up to 1-4-gauge and 4-8-gauge power cable.

The battery terminals and cable wire can be 1/0, 2-, 4-, or 8-gauge wire. The car battery voltage should never be under 12 volts to operate the sound system. The alternator and battery should produce 14 volts or more. A poor battery ground and hot cable can produce low voltage and a noisy condition. Battery voltage that falls too low can cause distortion in the amplifier. Also, distortion in the amplifier might be caused by a defective component in the amplifier or too much high voltage applied to the amplifier. High DC voltage (16 volts or more) can sometimes cause the auto CD player to shut off when the headlights are turned on.

Speaker wire cable can be made from 12- to 16-gauge multistrand pure copper wire. All speaker-connecting wire should be flexible for easy mounting and for going under dips and turns in the carpet. You can purchase wire kits and patch types of cords in kit form, according to the power output of the amplifier. Choose double-plated 24-carat gold terminal connections of 100-percent pure solid brass for distribution and fuses blocks, for best results.

A twisted pair of cables should be used to connect auto audio components together in systems with only 2 volts of signal or less. Zero-noise cables are 14 dB quieter than other audio cables. Silver soldered connections ensure maximum signal transfer and sound quality. Use the zero-noise patch cables with twisted pair construction for a noise-free audio system.

Stiffening Capacitors

A stiffening capacitor of 1-farad increases instantaneous power transfer and greatly improves transient power response, better bass impact, and definition. Large electrolytic capacitors are added to stiffen the power supply to the amplifier. The large stiffening capacitors can improve bass response and sound quality. A stiffening capacitor should be installed when the headlights become dim and the power is booming out of the high wattage system.

The 500,000 micro-farad (0.5 Farad) capacitor is recommended for audio systems up to 500 watts of power. Use a 1,000,000 micro-farad (1-Farad) stiffening capacitor for audio systems up to 1000 watts. Some capacitors have an LED readout to show how much energy the amp is pulling, and the handy reverse-polarity warning tells you if the capacitor is hooked up backward. A digital voltage readout on the stiffening capacitor shows the exact voltage and a audible warning if the capacitor is hooked up backward.

SPEAKER PROBLEMS

The defective speaker might have an open voice coil, intermittent voice coil connection, or drag on the center magnet pole piece. Check the suspected open speaker by placing the low-ohm scale of the DMM across the voice coil speaker terminals (**Fig. 13-10**). A digital multimeter resistance reading should be about 1 ohm under the voice coil impedance. No measurement indicates the voice coil is open. Check for a broken Litz wire or poorly soldered connection from the voice coil to the speaker terminals with the DMM. Clip the DMM leads to the speaker terminals and push up and down on the speaker cone to locate an intermittent voice coil or connection. The solid-state amp can be damaged with an open voice coil or poor speaker cable connection and no load on the amplifier.

Speaker damage can be caused by too much volume or power applied to the speaker, bad weather conditions, and a DC voltage applied to the voice coil. Excessive power can cause the speaker to blow out the voice coil and cause the cone to drag on the center pole piece. Too much power can blow the voice coil right out of the cone area. The overheated voice coil from long hours of high power operations can cause the voice coil to overheat and began to drag. Too much power can freeze the voice coil to the motor magnet piece.

Sometimes wet weather or water getting into the speaker cone area can cause the cone to warp and cause a mushy distorted sound. A tinny sound from the speaker with very little cone movement can be caused by the cone dragging or becoming frozen on the metal pole piece. The speaker cone can come loose from the surround framework and produce a blatting sound. A damaged spider or a loose spider from the cone framework can cause a blatting or noisy sound at high volume.

Always check the voltage on the speaker terminals with one

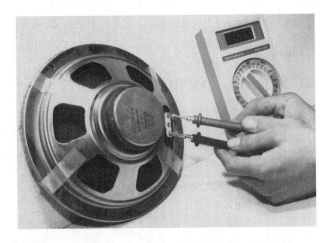

FIGURE 13-10. Check the defective speaker with the DMM attached to the speaker terminals for an open voice coil.

side of the speaker grounded, and
when the voice coil is frozen on
the magnet motor piece. Some
speakers that are not grounded
can have voltage on both speaker
terminals with normal operation.
Check the schematic for the correct speaker connections in the
output circuits. A leaky audio
output transistor or IC can destroy the voice coil by placing a
DC voltage at the speaker terminals. The defective output component can place a DC voltage
on the speaker terminals if an

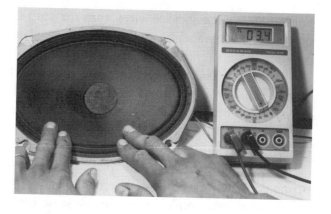

FIGURE 13-11. Push up and down on the speaker
cone and notice if the voice coil is dragging against
the magnet pole piece.

electrolytic capacitor is not connected between the output and speaker. The voice coil
may heat up with voltage applied and may drag or remain frozen against the magnet
pole piece.

A balanced output circuit in the amp should have zero voltage where the speakers
are directly connected to the output transistors or IC component. This only applies to
speakers that have a negative terminal grounded to the auto and amplifier chassis.
Remember to keep a speaker connected when the amplifier is operating to prevent
damage to the amplifier output circuits.

Check the suspected speaker with an ohmmeter test across the voice coil terminals.
Next, inspect the cone for holes or breaks in the speaker cone. Make sure the speaker
cone surround is not loose on the framework. Sometimes you can repair the cone or
spider by gluing it back into position with speaker cement. Place both fingers or
thumbs on the speaker cone and move the cone up and down. If the cone does not
move and it drags, then replace the speaker (**Fig. 13-11**). Small holes punched in the
speaker cone can be repaired with speaker or contact cement.

SPEAKER REPLACEMENT

Locate the bad speaker with the DMM, careful inspection, or subbing another speaker
of any diameter across the speaker terminals. Suspect bad speaker cable wires or a bad
amplifier if the subbed speaker has no sound. Replace the bad speaker when the sub
speaker sounds off. Bad cable wiring, worn cables, or a grounded cable speaker wires
can cause a dead, intermittent, and popping noise in the speaker. Speaker wires
grounding to the auto chassis or a defective audio amp can cause cracking and popping

noise. Remove the speaker wires from the amp that is causing the popping noise and determine if the noise is found in the other speakers. Replace the bad cable wire if the popping noise is only in one speaker cable.

Replace the bad speaker with another one of the same size as the defective speaker, with the exact voice coil impedance, wattage, and magnet weight. You do not want to replace an 8-ohm speaker if the original speaker had 4-ohm impedance, as it will produce less power or volume. Again, do not replace a 100-watt speaker with a 50-watt speaker, as it can damage the new speaker.

Factory auto radios and CD players with factory speakers might have a special size or may only be able to fit in a certain area. These types of speakers can be picked up at your local automobile dealership. Most tweeters, midrange, woofers, and subwoofers can be obtained from local electronic stores, Radio Shack, auto installation, and mail-order firms.

Index